Superoxide Dismutase

Volume II

Editor

Larry W. Oberley, Ph.D.
Associate Professor of Radiobiology
Radiation Research Laboratory
University of Iowa
Iowa City, Iowa

CRC Press, Inc.
Boca Raton, Florida

Library of Congress Cataloging in Publication Data

Main entry under title:

Superoxide dismutase.

Includes bibliographies and indexes.
1. Superoxide dismutase. 2. Active oxygen.
I. Oberley, Larry W.
QP603.S94S97 574.19′258 81-21683
ISBN 0-8493-6240-7 (v. 1) AACR2
ISBN 0-8493-6241-5 (v. 2)

Direct all inquiries to CRC Press, Inc., 2000 Corporate Blvd., N.W., Boca Raton, Florida, 33431.

International Standard Book Number 0-8493-6240-7 (Volume I)
International Standard Book Number 0-8793-6241-5 (Volume II)

Library of Congress Card Number 81-21683
Printed in the United States

PREFACE

This book is about oxygen radicals and the living cell's defense mechanisms against these species. It describes the production of these radicals, their chemistry, the enzyme superoxide dismutase, its structure and function, and the role of these entities in biological systems. The significance of oxygen radicals in various pathological conditions will also be discussed.

The work described herein is mainly due to the vision of one man. As modern physics is based on the work of Bohr and molecular biology is built on the work of Watson and Crick, so the study of oxygen radicals is built on the work of Dr. Irwin Fridovich (realizing, of course, that he had excellent help, especially from Dr. Joe McCord). Others have claimed to have discovered superoxide dismutase before Dr. Fridovich. However, we should remember that Dr. Fridovich's main contribution is not that of being the first to purify the enzyme, but rather that he was the first to explain its function and significance. More than any other person, he has guided subsequent work. This book is a tribute to his genius.

I hope these are the first two of several volumes. The next volumes will deal with superoxide dismutase and disease. The last chapter in these two volumes is a preview of this subject area.

This book is organized as follows. In the first volume, Dr. Fridovich discusses the history of superoxide dismutase in Chapter 1. In Chapter 2, Dr. Steinman elucidates what is known about the structure of this enzyme. Dr. Fridovich describes the assay of this enzyme in Chapter 3. Chapters 4 and 5 are devoted to the function of this enzyme-protection against the destructive effects of superoxide radicals. In Chapter 4, Dr. Fridovich explains the importance of SOD in defense against oxygen toxicity in prokaryotic cells. Dr. Halliwell talks about oxygen toxicity in plant tissues in Chapter 5. It should be remembered when reading these chapters that oxygen toxicity is also seen in higher organisms, like man. The next three chapters are devoted to production of superoxide radical, the substrate for superoxide dismutase, in chemical and biological systems. In Chapter 6, Drs. Allen and Bielski discuss the production of superoxide radicals in aqueous solutions. In Volume II, Drs. Gabig and Babior explain the role of oxygen radicals in killing of microbes in Chapter 1. At the subcellular level, Drs. Boveris and Cadenas discuss the production of superoxide and hydrogen peroxide by mitochondria in Chapter 2. The next two chapters are devoted to the methods for detecting oxygen radicals. In Chapter 3, Dr. Bors outlines the various methods used for measuring superoxide and hydroxyl radicals. In Chapter 4, Dr. Buettner explains one of the newly discovered and most useful techniques for detecting oxygen radicals — spin trapping. Chapter 5 is devoted to the organic chemistry of superoxide anion radical. In this chapter, Dr. Frimer is mainly concerned about the chemistry of superoxide radical in nonaqueous media. The last chapter, written by myself, deals with the role of superoxide dismutase in cancer. This last chapter is built on all the others, because knowledge of the basic chemistry and biology of superoxide dismutase is necessary to understand the role of this enzyme in disease.

It is always dangerous to speculate about the significance of any research, especially when one is involved in that area and is thus somewhat biased. Nevertheless, with all the attention given to other areas of biochemistry, especially recombinant DNA, I feel some speculation is in order. I think the subject of oxygen radicals in disease will be the most important field of research in medical sciences for some time to come, eclipsing some of the present very popular subject areas. This is because any disease must have as its basis some mechanism of damage, and that mechanism will most likely be shown to be the very reactive oxygen radicals. Thus, oxygen radicals have already been implicated in immune diseases, cancer, aging, diabetes, and muscular dystrophy.

Oxygen radicals may not be the initial cause of those diseases, but they may cause the ultimate cell damage. I predict that knowledge of this area will be essential to future medical treatment.

THE EDITOR

Larry W. Oberley, Ph.D., is Associate Professor of Radiology in the College of Medicine of the University of Iowa in Iowa City, Iowa. Dr. Oberley was graduated from Northwestern University, Evanston, Illinois with a B.A. in Physics in 1968. He received an M.S. in physics from the University of Iowa in 1980. During the next five years, Dr. Oberley was employed as a research associate in the Division of Nuclear Medicine, Department of Radiology, at the University of Iowa. It was during this period that Dr. Oberley developed his interest in biochemistry and medicine. In 1974, Dr. Oberley received his Ph.D. in physics from the University of Iowa. Dr. Oberley was appointed Instructor in the Radiation Research Laboratory of the Department of Radiology at the University of Iowa in 1975, Assistant Professor in 1978, and Associate Professor in 1980.

Dr. Oberley is a member of several societies including Sigma Xi, the New York Academy of Sciences, the Society of Nuclear Medicine, the American Society for Photobiology, the Radiation Research Society, and the American Association for the Advancement of Science.

Dr. Oberley has published over 30 articles and has given presentations at more than 20 scientific meetings. His current research interests are the role of free radicals in cancer, radiation, and aging.

CONTRIBUTORS

Augustine O. Allen, Ph.D.
Consultant
Brookhaven National Laboratory
Upton, New York

Bernard M. Babior, M.D.
Professor of Medicine
Tufts University School of Medicine
New England Medical Center Hospital
Boston, Massachusetts

Benon H. J. Bielski, Ph.D.
Chemist
Brookhaven National Laboratory
Upton, New York

Wolf Bors, D. Sc.
Research Scientist
Institut für Biologie
Gesellschaft für Strahlen- und Umweltforschung
Neuherberg
Federal Republic of Germany

Alberto Boveris, Ph.D.
Associate Professor
Department of Biochemistry
University of Rio de Janeiro
Rio de Janeiro, Brazil

Garry R. Buettner, Ph.D.
Assistant Professor of Chemistry
Wabash College
Crawfordsville, Indiana

Enrique Cadenas, M.D., Ph.D.
Research Associate
Johnson Research Foundation
School of Medicine
University of Pennsylvania
Philadelphia, Pennsylvania

Irwin Fridovich, Ph.D.
James B. Duke Professor
Department of Biochemistry
Duke University Medical Center
Durham, North Carolina

Aryeh A. Frimer, Ph.D.
Senior Lecturer
Department of Chemistry
Bar Ilan University
Ramat Gan
Israel

Theodore G. Gabig, M.D.
Chief, Division of Hematology
Veterans Administration Hospital
Ann Arbor, Michigan

Barry Halliwell, D.Phil.
University Lecturer in Biochemistry
Kings College
University of London
London
England

Christa Michel, Ph.D.
Research Assistant
Institut für Biologie
Gesellschaft für Strahlen- und Umweltforschung
Neuherberg
Federal Republic of Germany

Larry W. Oberley, Ph.D.
Associate Professor
Radiation Research Laboratory
Department of Radiology
University of Iowa
Iowa City, Iowa

Manfried Saran, Ph.D.
Research Scientist
Institut für Biologie
Gesellschaft für Strahlen- und Umweltforschung
Neuherberg
Federal Republic of Germany

Howard M. Steinman, Ph.D.
Associate Professor
Department of Biochemistry
Albert Einstein College of Medicine
Bronx, New York

ACKNOWLEDGMENTS

I would like to thank Susan Barnett and Carol Kieffer for their help in manuscript preparation. I would like to thank the contributing authors for their time and effort.

The support of the family of my childhood — Ruby, Jim, and Terry — and my present family — Kathleen, Rebecca, and Christopher — is gratefully acknowledged. Without their encouragement and patience, this volume would never have been completed.

Last, I must thank medical science for allowing me to edit and write this volume. I have been an artificial kidney patient for the last eight years. It is sobering to think that if I had been born 20 years earlier, I would not have been able to work on this book. As my life has been prolonged by the wonders of science, I hope that the lives of others will be prolonged by the fruits of the knowledge contained in this volume.

DEDICATION

To Kathleen, my partner through life,
and to Rebecca and Christopher,
who I hope will benefit
from this work someday

TABLE OF CONTENTS

Volume I

Volume II

Chapter 1

OXYGEN-DEPENDENT MICROBIAL KILLING BY NEUTROPHILS

Theodore G. Gabig and Bernard M. Babior

TABLE OF CONTENTS

I. INTRODUCTION

Studies of the oxygen-utilizing pathways of phagocyte metabolism began with the work of Baldridge and Gerard,[1] who showed in 1933 that a large increase in oxygen consumption accompanied the uptake of particles by neutrophils. This and subsequent studies showed that resting neutrophils consume little oxygen,[2] obtaining most of their energy from glycolysis, even in an aerobic environment. Oxygen consumption, however, was consistently found to increase sharply following exposure to phagocytosable particles or other appropriate stimuli.[1,3-5] In 1959, Sbarra and Karnovsky reported that this increase in oxygen consumption was not blocked by CN^- and that phagocytosis of particles occurs with equal facility under an atmosphere of nitrogen or air.[3] These observations indicated that the extra oxygen consumed by stimulated neutrophils is utilized by a nonmitochondrial pathway and has nothing to do with the production of the energy required for phagocytosis. The fate of this extra oxygen was elucidated by the demonstration in 1961 by Iyer, Islam, and Quastel that stimulated phagocytes release H_2O_2 into the surrounding medium,[6]a fiding that has subsequently been widely confirmed.[7-10] They proposed a potential role for H_2O_2 as an antimicrobial agent of the neutrophil. In other studies, superoxide anion ($O_2^{\cdot -}$ production by the neutrophil[11-16] and an increase in glucose oxidation via the hexosemonophosphate shunt[3,6,17-19] have been shown to accompany the extra oxygen consumption of stimulated neutrophils. These four events — namely, the increases in oxygen consumption, $O_2^{\cdot -}$ and H_2O_2 production, and hexosemonophosphate shunt activity that occur when neutrophils are stimulated — are known as the "respiratory burst". The mechanism by which neutrophils generate oxidant species during the respiratory burst and use them to kill microorganisms is the subject of this review.

II. GENERATION OF OXIDANT SPECIES BY NEUTROPHILS

A. Activation of the Respiratory Burst

At the cellular level, the respiratory burst of neutrophils is initiated by a complex series of events which result in the activation of a membrane-bound oxidase (see below). Many agents are capable of respiratory burst activation, some soluble[15,20-30] and some particulate.[12,31-33] A few of the soluble agents are listed in Table 1. Although there is a remarkable diversity in the types of stimuli which are capable of burst initiation, there seem to be certain common features in their mode of action. All agents studied to date work by interacting with the surface of an intact phagocyte; locomotor activity (i.e., phagocytosis and degranulation) are not prerequisites for burst activation.[25] Although activation occurs anaerobically,[34] the combination of CN^- and 2-deoxyglucose, which blocks energy production by inhibiting both oxidative phosphorylation and glycolysis, prevents burst activation,[35,36] suggesting that energy is required for this process. F^-,[25] A23187,[23] concanavalin A,[37] and f-met-leu-phe[38] require extracellular Ca^{++} to activate the burst, but phorbol does not.[35]

During activation, the following sequence of events has been shown to occur upon exposure of the neutrophil to an appropriate stimulus:[30,37,39,40]

$$\text{Stimulus} \xrightarrow[\text{sec}]{\text{0 to 10}} \text{membrane depolarization} \xrightarrow[\text{sec}]{\text{10 to 60}} \begin{cases} O_2^{\cdot -}\ \text{generation} \\ \text{Biphasic change in cAMP levels} \\ \text{Extracellular lysosomal} \\ \quad \text{enzyme release} \end{cases} \qquad (1)$$

$O_2^{\cdot -}$ production and the changes in intracellular cAMP concentration begin at the same time, suggesting that neither is the cause of the other. Membrane depolarization, how-

Table 1
SOME SOLUBLE AGENTS CAPABLE OF ACTIVATING THE RESPIRATORY BURST

Phorbol myristate acetate
Fluoride
F-met-leu-phe and other chemotactic N-formylated peptides
C5a, a complement-derived peptide
Concanavalin A
A23187, a calcium ionophore

ever, precedes $O_2^{\cdot -}$ generation by several seconds. This would be consistent with a cause-effect relationship between these two, except for the fact that membrane depolarization and $O_2^{\cdot -}$ production in response to A23187 are clearly independent events.[40]

Recently, several synthetic and naturally occurring protease inhibitors have been shown to inhibit extracellular $O_2^{\cdot -}$ generation when the soluble agents concanavalin A and cytochalasin E were used as activating agents.[41] The dose response curves of respiratory burst inhibition by several classes of protease inhibitors led to the conclusion that an endogenous chymotrypsin-like protease may be involved in the activation of the $O_2^{\cdot -}$-generating oxidase of the respiratory burst by these agents. On the other hand, the finding that burst activation by F^- is a reversible process represents evidence against proteolysis as an activating mechanism.[25]

B. The Metabolic Pathway of the Burst and the $O_2^{\cdot -}$-forming Enzyme

The $O_2^{\cdot -}$-forming enzyme responsible for the respiratory burst is thought to be a membrane-bound oxidase, dormant in resting cells, which catalyzes the following reaction:[31,42-49]

$$2\,O_2 + NAD(P)H + H^+ \longrightarrow 2\,O_2^{\cdot -} + NAD(P)^+ \qquad (2)$$

This reaction explains the oxygen consumption and $O_2^{\cdot -}$ generation seen in the burst. The subsequent dismutation of $O_2^{\cdot -}$ and H_2O_2, occurring either spontaneously or through catalysis by superoxide dismutase,[50] accounts for H_2O_2 production during the burst:[16]

$$2\,O_2^{\cdot -} + 2H^+ \xrightarrow[\text{Dismutase}]{\text{Spontaneous}} H_2O_2 + O_2 \qquad (3)$$

Hexosemonophosphate shunt activity reflects the increased availability of $NADP^+$ [51] that results from the NADPH-consuming reactions of the respiratory burst. Some of this $NADP^+$ is generated in the $O_2^{\cdot -}$-forming reaction itself. The rest probably arises from the detoxification of H_2O_2 that leaks into the cytoplasm of the neutrophil during the burst. This detoxification is accomplished largely by the glutathione peroxidase-glutathione reductase couple, an NADPH-consuming system:[52-54]

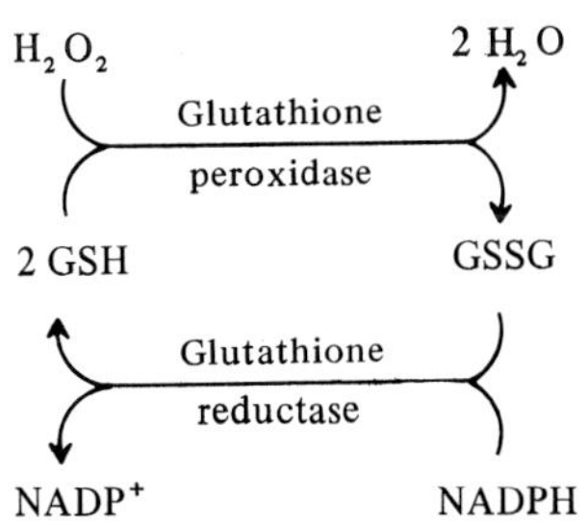

(4)

Catalase plays only a minor role in H_2O_2 detoxification under normal circumstances.[55]

The $O_2^{\cdot-}$-forming enzyme itself has been shown to be associated with the plasma membrane,[56] as well as with phagocytic vacuole walls, which are composed of portions of plasma membrane internalized during the uptake of a particle.[49] This localization assures that $O_2^{\cdot-}$ will be delivered into the region of an ingested microorganism with optimum efficiency. $O_2^{\cdot-}$ formation by this enzyme: requires a reduced pyridine nucleotide substrate, preferring NADPH (K_m 30 μM) over NADH (K_m 0.5 to 1.0 mM);[46,48,57-59] is dependent on flavin adenine dinucleotide (FAD) (K_m 60 nM);[60] and is insensitive to heme enzyme inhibitors.[57] The $O_2^{\cdot-}$-forming enzyme is present in preparations from normal phagocytes which have been exposed to a specific activating stimulus prior to disruption, but is absent from membrane fractions prepared from unstimulated normal neutrophils and from stimulated neutrophils obtained from patients with chronic granulomatous disease (CGD), an inherited condition in which neutrophils are incapable of expressing a respiratory burst.[46,48,57] Its activity can account for all $O_2^{\cdot-}$ production by an equivalent number of intact neutrophils.[57]

Besides the particulate $O_2^{\cdot-}$-forming activity described above, other pyridine nucleotide oxidases, specifically one or more which prefer NADH over NADPH, have been proposed as the enzyme responsible for the respiratory burst.[8,61-65] Because these oxidases are active in neutrophil preparations obtained from patients with chronic granulomatous disease,[44,63,65-67] their relationship to the respiratory burst is less obvious than in the case of the NADPH-preferring oxidase. Useful criteria for evaluating the connection between a candidate oxidase and the respiratory burst have been set forth by Karnovsky and associates.[68]

Recently, the possibility has arisen that a b-type cytochrome is involved in some way in the respiratory burst. This cytochrome was first described in rabbit neutrophils in 1966.[69] At that time its potential role as an electron transferring protein responsible for the pyridine nucleotide diaphorase activity of neutrophils ("NAD_i-oxidase") was discussed. Further work on the function of this cytochrome was not reported until 1977, when Segal described its presence in normal human neutrophils but not in neutrophils from some patients with CGD.[70] Subsequently, other CGD patients were shown to have normal amounts of this cytochrome in their neutrophils.[71] The studies to date have utilized difference spectroscopy of oxidized vs. dithionite-reduced neutrophil preparations. By this technique cytochrome *b* has been demonstrated in both unstimulated and stimulated neutrophils. Objections that these dithionite difference spectra represented myeloperoxidase, eosinophil peroxidase, or contaminating microsomal cytochrome b_5 or P450 have been raised. That the cytochrome *b* difference spectrum can be separated from the myeloperoxidase difference spectrum in subcellular fractionation studies of neutrophil homogenates is evidence against a peroxidase origin.[72] However, evidence that this cytochrome is not a microsomal cytochrome b_5 or altered P450 (i.e., P420) remains to be obtained. Adding to the puzzle are experiments showing that this cytochrome cannot be reduced by either NADH or NADPH.[73] Moreover, carbon monoxide, a compound with no effect on the respiratory burst, forms a well-defined complex with the cytochrome.[69,72,73] These last findings suggest that the cytochrome is not directly involved in oxygen reduction during the respiratory burst. At present the function of this b-type cytochrome of human neutrophils, its relationship to chronic granulomatous disease, and its connection with the respiratory burst remain uncertain.

III. ANTIMICROBIAL OXIDANTS

The $O_2^{\cdot-}$ and H_2O_2 generated during the respiratory burst are used in subsequent reactions to generate more potent oxidizing agents. These oxidant species are employed as antimicrobial agents. They are able to effect rapid killing of most bacteria contained within the phagocytic vacuole of the neutrophil. While the highest concentrations of these oxidant species are presumably achieved within the phagocytic vacuole, extracellular fluxes of $O_2^{\cdot -}$ and the accumulation of H_2O_2 in the extracellular environment can readily be shown in vitro, and oxidant species derived from $O_2^{\cdot -}$ and H_2O_2 have been found to mediate extracellular lysis of tumor cells[74] and endothelial cells[75] by neutrophils. The following sections will focus on the destruction of bacteria by various oxidants produced by the neutrophil.

A. Superoxide

Studies of the microbicidal activity of cell-free enzymatic systems which generate $O_2^{\cdot -}$ suggest that $O_2^{\cdot -}$ itself has negligible antimicrobial activity.[76-78] The possibility that $O_2^{\cdot -}$ alone may have significant antimicrobial activity at the substantially higher concentrations probably achieved within the phagocytic vacuole, however, is difficult to explore experimentally, and remains an unanswered question.

B. Hydrogen Peroxide

The susceptibility of various bacteria to the microbicidal action of H_2O_2 alone in a cell-free system varies somewhat as a function of the levels of catalase and peroxidase within the microbe.[77] At sufficiently high concentrations, however, H_2O_2 shows substantial bactericidal activity.[79,80] Moreover, iodide ion[81] or ascorbic acid plus copper[80] augment the bactericidal power of H_2O_2, so that it is lethal at lower concentrations. The role of these latter systems in the antimicrobial activity of the neutrophil is unknown.

C. Hypochlorite

The microbicidal capacity of HOCl has been known for a long time, and is exploited in the purification of water and in other processes where antiseptic treatment is necessary. Current evidence strongly suggests that HOCl is one of the major antimicrobial oxidants of neutrophils.

In the neutrophil, HOCl is produced by the oxidation of Cl^- by H_2O_2.

$$H^+ + Cl^- + H_2O_2 \longrightarrow HOCl + H_2O \qquad (5)$$

This reaction is catalyzed by myeloperoxidase,[82] a heme enzyme which is confined to the azurophil granules* of the unstimulated cell,[83,84] but can be demonstrated within the phagocytic vacuole following degranulation.[88] Neutrophil myeloperoxidase has a molecular weight of approximately 150,000, has an isoelectric point on the alkaline side of neutrality, and contains two heme prosthetic groups per molecule.[89-91] Myeloperoxidase has been estimated to comprise nearly 5% of the dry weight of human neutrophils.[92] It has an acidic pH optimum whose exact value varies with the concentrations of Cl^- and H_2O_2, but under physiological conditions probably approximates pH 5.[93] Michaelis constants for H_2O_2 and Cl^- each depend in a complex way on the concentration of the other substrate.[94]

Myeloperoxidase can also catalyze the production of microbicidal oxidants from Br^-

* Neutrophils contain two types of lysosome-like structures which are important for bacterial killing. One type, the azurophil granules, contains myeloperoxidase as well as a battery of hydrolytic enzymes of the sort that are found in the lysosomes of many tissues;[83,84] the other type, the specific granules, contains lactoferrin,[85] a cobalamin-binding protein,[86] lysozyme,[84] and a few other constituents. Both types fuse with phagocytic vacuoles, discharging their contents into the space surrounding the microorganism contained within the vacuole. This process of fusion is known as "degranulation". In addition, some of the specific granules secrete their contents into the extracellular medium in response to stimulation.[87]

and I^-,[95] but Cl^- appears to be the physiological substrate because of its concentration in the neutrophil.[96] In an artificial system, Br^- and I^- are better substrates than Cl^-, because the concentrations at which the former two are oxidized to antimicrobial agents are orders of magnitude lower than the concentration required for the generation of HOCl from Cl^-.[95] In the neutrophil, however, Cl^- is present at a concentration more than 100 times that required for maximal microbicidal function in vitro, while Br^- and I^- are virtually absent.

Once formed, HOCl can oxidize many substances of biological significance. Complex reactions with free amino acids lead to decarboxylation with the production of nitriles (in the case of glycine, HCN) or aldehydes.[97-99] Proteins may be degraded by the successive elimination of the N-terminal amino acids.[100] Amino acid side chains are subject to oxidative attack: the phenol of tyrosine is halogenated[101,102] (iodination of this residue by neutrophils is used as a test of the integrity of the oxidative killing systems of neutrophils[81,95,103-105]), and methionine is oxidized to methionine sulfoxide.[106-108] Sulfhydryl groups are also oxidized.[109,110] Lipid peroxidation occurs, possibly mediated at least in part by the myeloperoxidase system.[111,112] The mechanisms of killing by HOCl, however, are incompletely understood, largely because of the uncertainty as to whether a given reaction observed in vitro is or is not relevant to the lethal event(s) which occur within the phagocytic vacuole of the neutrophil. So far, the only evidence connecting a specific oxidative reaction with the lethal event is the finding that sulfhydryl group oxidation correlates with bacterial killing in a model system;[110] whether this finding is applicable to the situation in the neutrophil remains to be determined.

D. Singlet Oxygen

Singlet oxygen is a highly reactive oxidizing species formed by inversion of the spin of one of the two unpaired electrons of the oxygen molecule. Studies of neutrophil chemiluminescence and experiments using singlet oxygen traps and scavengers have led to the proposal that this reactive oxidant is generated and used as a microbicidal oxidant by the neutrophil.[113-116] It has been pointed out, however, that the observations on which the singlet oxygen hypothesis is based can be accounted for by reactions involving $O_2^{\cdot-}$, H_2O_2, and HOCl, all well-characterized products of the respiratory burst.[117,118] For this reason, the question of singlet oxygen participation in oxygen-dependent microbial killing by neutrophils remains open.

E. Oxidizing Radicals

Oxidizing radicals represent the other major category of microbicidal oxidants employed by neutrophils. Under normal circumstances these probably serve mainly as an auxiliary system, backing up the action of HOCl, a more potent antimicrobial agent under the conditions prevailing in neutrophils.[78,119] They are, however, effective enough to provide adequate microbial killing when HOCl cannot be made available, as, for example, in congenital myeloperoxidase deficiency.[119]

These powerful oxidizing radicals are usually detected by virtue of their ability to release hydrocarbon gases from certain sulfur-containing compounds:[120-122]

$$CH_3SCH_2CH_2CHO + Ox\cdot \longrightarrow CH_2 = CH_2 \quad (6)$$

$$CH_3SCH_2CH_2COCOO^- + Ox\cdot \longrightarrow CH_2 = CH_2 \quad (7)$$

$$CH_3SOCH_3 + Ox\cdot \longrightarrow CH_4 \quad (8)$$

Several types of oxidizing radicals are able to carry out this reaction, including ·OH,

$\cdot$OR, RCOO$\cdot$, and R$\cdot$ (but not $O_2^{\cdot-}$),[123,124] so that hydrocarbon release cannot be regarded as evidence for a particular radical species. Spin trapping experiments, however, have shown that OH$\cdot$ is at least one of the oxidizing radicals which is produced by neutrophils during the respiratory burst.[125,126]

The mechanism by which these radicals are produced is not fully understood. The Haber-Weiss reaction, an iron-catalyzed one-electron reduction of H_2O_2 by $O_2^{\cdot-}$, is usually proposed as the mechanism of $\cdot$OH production:[127-129]

$$H_2O_2 + O_2^{\cdot-} \xrightarrow[Fe^{++}(+)]{} \cdot OH + OH^- + O_2 \quad (9)$$

Studies have shown, however, that a substantial amount of oxidizing radical production by whole neutrophils as well as by $O_2^{\cdot-}$-forming subcellular particles does not require H_2O_2, though all of it seems to require $O_2^{\cdot-}$.[120-122,125,126,130] Moreover, oxidizing radical production by neutrophils, at least as measured by hydrocarbon release, is greatly decreased, though not absent, in myeloperoxidase-deficient cells.[131] One explanation for these findings is that much of the oxidizing radical production by neutrophils involves the formation of alkoxy radicals by a Haber-Weiss-like reaction between $O_2^{\cdot-}$ and alkyl hydroperoxides formed, for example, during lipid peroxidation:[132]

$$ROOH + O_2^{\cdot-} \longrightarrow RO\cdot + OH^- + O_2 \quad (10)$$

Evidence that alkoxy radicals are actually generated during the respiratory burst, however, remains to be obtained.

A role for bacterial killing by oxidizing radicals has been demonstrated both with artificial radical-generating systems[76,78,133,134] and with intact neutrophils.[13] In the best recent study, remarkable parallels have been drawn between bacterial killing by a radical-generating system (xanthine oxidase plus acetaldehyde, which produces $O_2^{\cdot-}$, H_2O_2, and $\cdot$OH) in the presence and absence of myeloperoxidase and bacterial killing by normal and myeloperoxidase-deficient neutrophils.[78] Direct examination of bacterial killing by oxidizing radicals in intact cells showed that, under suitable conditions, killing was sharply reduced by the addition of exogenous superoxide dismutase or catalase, implicating $\cdot$OH as a microbicidal agent under these conditions.[13] These studies provide strong evidence supporting the participation of oxidizing radicals in microbial killing by neutrophils. Nothing is known, however, about how these radicals accomplish their lethal effects.

IV. SUMMARY

Neutrophils manufacture powerful oxidizing agents which they use for the destruction of bacteria. The precursors of these oxidizing agents are generated during the respiratory burst, a metabolic event which is initiated by the exposure of neutrophils to appropriate stimuli and is characterized by the production of large quantities of $O_2^{\cdot-}$ and H_2O_2. The enzyme responsible for $O_2^{\cdot-}$ production is a membrane-bound flavoprotein oxidase which is dormant in resting neutrophils but is active in stimulated cells. It catalyzes the reduction of oxygen to $O_2^{\cdot-}$ by NADPH. The H_2O_2 generated during the respiratory burst arises by the dismutation of the $O_2^{\cdot-}$ produced by this enzyme.

The microbicidal oxidants are HOCl and a group of powerful oxidizing radicals, at least one of which is $\cdot$OH. HOCl is produced by the myeloperoxidase-catalyzed oxidation of Cl^- by H_2O_2. The oxidizing radicals may be produced in part by the reduction of H_2O_2 by $O_2^{\cdot-}$ to give $\cdot$OH (the Haber-Weiss reaction), but other as yet unspecified reactions are also involved in oxidizing radical production. HOCl is a more potent microbicidal agent than the oxidizing radicals and appears, therefore, to be the major oxygen-dependent antimicrobial in normal neutrophils, but oxidizing radicals are sufficiently potent to provide adequate antimicrobial activity when HOCl is lacking. Mechanisms of bacterial killing by either of these two oxidizing systems are unknown.

REFERENCES

1. **Baldridge, C. W. and Gerard, R. W.,** The extra respiration of phagocytosis, *Am. J. Physiol.,* 103, 235, 1933.
2. **Karnovsky, M. L.,** The metabolism of leukocytes, *Semin. Hematol.,* 5, 156, 1968.
3. **Sbarra, A. J. and Karnovsky, M. L.,** The biochemical basis of phagocytosis. I. Metabolic changes during the ingestion of particles by polymorphonuclear leukocytes, *J. Biol. Chem.,* 234, 1355, 1959.
4. **Oren, R., Farnham, A. E., Saito, K., Milofsky, E., and Karnovsky, M. L.,** Metabolic patterns in three types of phagocytic cells, *J. Cell Biol.,* 17, 487, 1963.
5. **Weening, R. S., Roos, D., and Loos, J. A.,** Oxygen consumption of phagocytizing cells in human leukocyte and granulocyte preparations: a comparative study, *J. Lab. Clin. Med.,* 83, 570, 1974.
6. **Iyer, G. Y. N., Islam, M. F., and Quastel, J. H.,** Biochemical aspects of phagocytosis, *Nature,* 192, 535, 1961.
7. **Paul, B. and Sbarra, A. J.,** The role of the phagocyte in host parasite interactions. XIII. The direct quantitative estimation of H_2O_2 in phagocytizing cells, *Biochim. Biophys. Acta,* 156, 168, 1968.
8. **Baehner, R. L., Gilman, N., and Karnovsky, M. L.,** Respiration and glucose oxidation in human and guinea pig leukocytes: comparative studies, *J. Clin. Invest.,* 49, 692, 1970.
9. **Homan-Müller, J. W. T., Weening, R. S., and Roos, D.,** Production of hydrogen peroxide by phagocytizing human granulocytes, *J. Lab. Clin. Med.,* 85, 198, 1975.
10. **Root, R. K., Metcalf, J., Oshino, N., and Chance, B.,** H_2O_2 release from human granulocytes during phagocytosis. I. Documentation, quantitation, and some regulating factors, *J. Clin. Invest.,* 55, 945, 1975.
11. **Babior, B. M., Kipnes, R. S., and Curnutte, J. T.,** Biological defense mechanisms: the production by leukocytes of superoxide, a potential bactericidal agent, *J. Clin. Invest.,* 52, 741, 1973.
12. **Curnutte, J. T. and Babior, B. M.,** Biological defense mechanisms. The effect of bacteria and serum on superoxide production by granulocytes, *J. Clin. Invest.,* 53, 1662, 1974.
13. **Johnston, R. B., Keele, B. B., Misra, H. P., Lehmeyer, J. E., Webb, L. S., Baehner, R. L., and Rajagopalan, K. V.,** The role of superoxide anion generation in phagocytic bactericidal activity. Studies with normal and chronic granulomatous disease leukocytes, *J. Clin. Invest.,* 55, 1357, 1975.
14. **Weening, R. S., Wever, R., and Roos, D.,** Quantitative aspects of the production of superoxide radicals by phagocytizing human granulocytes, *J. Lab. Clin. Med.,* 85, 245, 1975.
15. **Goldstein, I. M., Roos, D., Kaplan, H. B., and Weissmann, G.,** Complement and immunoglobulins stimulate superoxide production by human leukocytes independently of phagocytosis, *J. Clin. Invest.,* 56, 1155, 1975.
16. **Root, R. K. and Metcalf, J. A.,** H_2O_2 release from human granulocytes during phagocytosis: relationship to superoxide anion formation and cellular catabolism of H_2O_2: studies with normal and cytochalasin B treated cells, *J. Clin. Invest.,* 60, 1266, 1977.
17. **Stähelin, H., Karnovsky, M. L., Farnham, A. E., and Suter, E.,** Studies on the interaction between phagcytes and tubercle bacilli. III. Some metabolic effects in guinea pigs associated with infection with tubercle bacilli, *J. Exp. Med.,* 105, 265, 1957.
18. **Zatti, M. and Rossi, F.,** Relationship between glycolysis and respiration in surfactant-treated leukocytes, *Biochim. Biophys. Acta,* 148, 553, 1967.
19. **Stjernholm, R. and Manek, R. C.,** Carbohydrate metabolism in leukocytes. XIV. Regulation of pentose cycle activity and glycogen metabolism during phagocytosis, *J. Reticuloendothel. Soc.,* 8, 550, 1970.
20. **Graham, R. C., Jr., Karnovsky, M. J., Shafer, A. W., Glass, E. A., and Karnovsky, M. L.,** Metabolic and morphological observations on the effect of surface-active agents on leukocytes, *J. Cell Biol.,* 32, 629, 1967.
21. **Goetzl, E. J. and Austen, K. F.,** Stimulation of human neutrophil leukocyte aerobic glucose metabolism by purified chemotactic factors, *J. Clin. Invest.,* 53, 591, 1974.

22. **Repine, J. E., White, J. G., Clawson, C. C., and Holmes, B. H.,** The influence of phorbol myristate acetate on oxygen consumption by polymorphonuclear leukocytes, *J. Lab. Clin. Med.,* 83, 911, 1974.
23. **Romeo, D., Zabucchi, G., Miani, N., and Rossi, F.,** Ion movement across leukocyte plasma membrane and excitation of their metabolism, *Nature (London),* 253, 542, 1975.
24. **Cohen, H. J. and Chovaniec, M. E.,** Superoxide generation by digitonin-stimulated guinea pig granulocytes. A basis for a continuous assay for monitoring superoxide production and for the study of the activation of the generating system, *J. Clin. Invest.,* 61, 1081, 1978.
25. **Curnutte, J. T., Babior, B. M., and Karnovsky, M. L.,** Fluoride-mediated activation of the respiratory burst in human neutrophils. A reversible process, *J. Clin. Invest.,* 63, 637, 1979.
26. **Boxer, L. A., Yoder, M., Bonsib, S., Schmidt, M., Ho, P., Jersild, R., and Baehner, R. L.,** Effects of a chemotactic factor, N-formyl-methionyl peptide, on adherence, superoxide anion generation, phagocytosis, and microtubule assembly of human polymorphonuclear leukocytes, *J. Lab. Clin. Med.,* 93, 506, 1979.
27. **Nakagawara, A. and Minakami, S.,** Role of cytoskeletal elements in cytochalasin E-induced superoxide production by human polymorphonuclear leukocytes, *Biochim. Biophys. Acta,* 584, 143, 1979.
28. **Klempner, M. S., Dinarello, C. A., Henderson, W. R., and Gallin, J. I.,** Stimulation of neutrophil oxygen-dependent metabolism by human leukocytic pyrogen, *J. Clin. Invest.,* 64, 996, 1979.
29. **Cohen, M. S., Metcalf, J. A., and Root, R. K.,** Regulation of oxygen metabolism in human granulocytes: relationship between stimulus binding and oxidative response using plant lectins as probes, *Blood,* 55, 1003, 1980.
30. **Smolen, J. E., Korchak, H. M., and Weissmann, G.,** Initial kinetics of lysosomal enzyme secretion and superoxide anion generation by human polymorphonuclear leukocytes, *Inflammation,* 4, 145, 1980.
31. **Patriarca, P., Cramer, R., Moncalvo, S., Rossi, F., and Romeo, D.,** Enzymatic basis of metabolic stimulation in leucocytes during phagocytosis: the role of activated NADPH oxidase, *Arch. Biochem. Biophys.,* 145, 255, 1971.
32. **Gabig, T. G., Bearman, S. I., and Babior, B. M.,** Effects of oxygen tension and pH on the respiratory burst of human neutrophils, *Blood,* 53, 1133, 1979.
33. **Sagone, A. L., Jr., King, G. W., and Metz, E. N.,** A comparison of the metabolic response to phagocytosis in human granulocytes and monocytes, *J. Clin. Invest.,* 57, 1352, 1976.
34. **Jandl, R. C., Andre-Schwartz, J., Borges-DuBois, L., Kipnes, R. S., McMurrich, B. J., and Babior, B. M.,** Termination of the respiratory burst in human neutrophils, *J. Clin. Invest.,* 61, 1176, 1978.
35. **Newburger, P. E., Chovaniec, M. E., and Cohen, H. J.,** Activity and activation of the granulocyte superoxide-generating system, *Blood,* 55, 85, 1980.
36. **Cohen, H. J. and Chovaniec, M. E.,** Superoxide production by digitoxin-stimulated guinea pig granulocytes. The effects of N-ethylmaleimide, divalent cations, and glycolytic and mitochondrial inhibitors on the activation of the superoxide generating system, *J. Clin. Invest.,* 61, 1088, 1978.
37. **Korchak, H. M. and Weissmann, G.,** Changes in membrane potential of human granulocytes antecede the metabolic responses to surface stimulation, *Proc. Natl. Acad. Sci. (U.S.A.),* 75, 3818, 1978.
38. **Simchowitz, L. and Spilberg, I.,** Generation of superoxide radicals by human peripheral neutrophils activated by chemotactic factor. Evidence for the role of calcium, *J. Lab. Clin. Med.,* 93, 583, 1979.
39. **Whitin, J. C., Chapman, C. E., Simons, E. R., Chovaniec, M. E., and Cohen, H. J.,** Correlation between membrane potential changes and superoxide production in human granulocytes stimulated by phorbol myristate acetate, *J. Biol. Chem.,* 255, 1874, 1980.
40. **Seligmann, B. E. and Gallin, J. I.,** Use of lipophilic probes of membrane potential to assess human neutrophil activation. Abnormality in chronic granulomatous disease, *J. Clin. Invest.,* 66, 493, 1980.
41. **Kitagawa, S., Takaku, F., and Sakamoto, S.,** Evidence that proteases are involved in superoxide production by human polymorphonuclear leukocytes and monocytes, *J. Clin. Invest.,* 65, 74, 1980.
42. **Iyer, G. Y. N. and Quastel, J. H.,** NADPH and NADH oxidation by guinea pig polymorphonuclear leucocytes, *Can. J. Biochem. Physiol.,* 41, 427, 1963.
43. **Paul, B. B., Strauss, R. R., Jacobs, A. A., and Sbarra, A. J.,** Direct involvement of NADPH oxidase with the stimulated respiratory and hexosemonophosphate shunt activities in phagocytizing leukocytes, *Exp. Cell Res.,* 73, 456, 1972.
44. **Hohn, D. C. and Lehrer, R. I.,** NADPH oxidase deficiency in X-linked chronic granulomatous disease, *J. Clin. Invest.,* 55, 707, 1975.
45. **Curnutte, J. T., Kipnes, R. S., and Babior, B. M.,** Defect in pyridine nucleotide dependent superoxide production by a particulate fraction from the granulocytes of patients with chronic granulomatous disease, *N. Engl. J. Med.,* 293, 628, 1975.
46. **Babior, B. M., Curnutte, J. T., and McMurrich, B. J.,** The particulate superoxide-forming system from human neutrophils. Properties of the system and further evidence supporting its participation in the respiratory burst, *J. Clin. Invest.,* 58, 989, 1976.

47. **McPhail, L. C., DeChatelet, L. R., and Shirley, P. S.,** Further characterization of NADPH oxidase activity of human polymorphonuclear leukocytes, *J. Clin. Invest.,* 58, 774, 1976.
48. **Suzuki, Y. and Lehrer, R. I.,** NAD(P)H oxidase activity in human neutrophils stimulated by phorbol myristate acetate, *J. Clin. Invest.,* 66, 1409, 1980.
49. **Cohen, H. J., Newburger, P. E., and Chovaniec, M. E.,** NAD(P)H-dependent superoxide production by phagocytic vesicles from guinea pig and human granulocytes, *J. Biol. Chem.,* 255, 6584, 1980.
50. **Salin, M. L. and McCord, J. M.,** Superoxide dismutase in polymorphonuclear leukocytes, *J. Clin. Invest.,* 54, 1005, 1974.
51. **Beck, W. S.,** The control of leukocyte glycolysis, *J. Biol. Chem.,* 232, 251, 1958.
52. **Reed, P. W.,** Glutathione and the hexosemonophosphate shunt in phagocytizing and hydrogen peroxide-treated rat leukocytes, *J. Biol. Chem.,* 244, 2459, 1969.
53. **Baehner, R. L., Murrmann, S. K., Davis, J., and Johnston, R. B., Jr.,** The role of superoxide anion and hydrogen peroxide in phagocytosis-associated oxidative metabolic reactions, *J. Clin. Invest.,* 56, 571, 1975.
54. **Voetman, A. A., Loos, J. A., and Roos, D.,** Changes in the levels of glutathione in phagocytizing human neutrophils, *Blood,* 55, 741, 1980.
55. **Roos, D., Weening, R. S., Wyss, S. R., and Aebi, H. E.,** Protection of human neutrophils by endogenous catalase. Studies with cells from catalase-deficient individuals, *J. Clin. Invest.,* 65, 1515, 1980.
56. **Dewald, B., Baggiolini, M., Curnutte, J. T., and Babior, B. M.,** Subcellular localization of the superoxide-forming enzyme in human neutrophils, *J. Clin. Invest.,* 63, 21, 1979.
57. **Gabig, T. G. and Babior, B. M.,** The O_2^--forming oxidase responsible or the respiratory burst in human neutrophils. Properties of the solubilized enzyme, *J. Biol. Chem.,* 254, 9070, 1979.
58. **Cohen, H. J., Chovaniec, M. E., and Davies, W. A.,** Activation of the guinea pig granulocyte NAD(P)H-dependent superoxide generating enzyme: localization in a plasma membrane enriched particle and kinetics of activation, *Blood,* 55, 355, 1980.
59. **Tauber, A. I. and Goetzl, E. J.,** Structural and catalytic properties of the solubilized superoxide-generating activity of human polymorphonuclear leukocytes. Solubilization, stabilization in solution, and partial characterization, *Biochemistry,* 25, 5576, 1979.
60. **Babior, B. M. and Peters, W. A.,** The O_2^--producing enzyme of human neutrophils: further properties, *J. Biol. Chem.,* 256, 2321, 1981.
61. **Cagan, R. H. and Karnovsky, M. L.,** Enzymatic basis of the respiratory stimulation during phagocytosis, *Nature,* 202, 255, 1964.
62. **Briggs, R. T., Drath, D. B., Karnovsky, M. L., and Karnovsky, M. J.,** Localization of NADH oxidase on the surface of human polymorphonuclear leukocytes by a new cytochemical method, *J. Cell Biol.,* 67, 566, 1975.
63. **Baehner, R. L. and Karnovsky, M. L.,** Deficiency of reduced nicotinamide-adenine dinucleotide oxidase in chronic granulomatous disease, *Science,* 162, 1277, 1968.
64. **Segal, A. W.,** Characterisation of the enzyme defect in chronic granulomatous disease, *Lancet,* 1, 1363, 1976.
65. **Badwey, J. A. and Karnovsky, M. L.,** Production of superoxide and hydrogen peroxide by an NADH-oxidase in guinea pig polymorphonuclear leukocytes, *J. Biol. Chem.,* 254, 11530, 1979.
66. **Holmes, B., Page, A. R., and Good, R. A.,** Studies of the metabolic activity of leukocytes from patients with a genetic abnormality of phagocyte function, *J. Clin. Invest.,* 46, 1422, 1967.
67. **Iverson, D., DeChatelet, L. R., Spitznagel, J. K. and Wang, P.,** Comparison of NADH and NADPH oxidase activities in granules isolated from human polymorphonuclear leukocytes with a fluorimetric assay, *J. Clin. Invest.,* 59, 282, 1977.
68. **Badwey, J. A., Curnutte, J. T., and Karnovsky, M. L.,** The enzyme of granulocytes that produces superoxide and peroxide, *New Engl. J. Med.,* 300, 1157, 1979.
69. **Shinagawa, Y., Tanaka, C., Teraoka, A., and Shinagawa, Y.,** A new cytochrome in neutrophilic granules of rabbit leucocyte, *J. Biochem.,* 59, 622, 1966.
70. **Segal, A. W., Webster, D., Jones, O. T. G., and Allison, A. C.,** Absence of a newly descibed cytochrome b from neutrophils of patients with chronic granulomatous disease, *Lancet,* 2, 446, 1978.
71. **Borregaard, N., Johansen, K. S., Taudorff, E., and Wandall, J. H.,** Cytochrome b is present in neutrophils from patients with chronic granulomatous disease, *Lancet,* 2, 949, 1979.
72. **Segal, A. W. and Jones, O. T. G.,** The subcellular distribution and some properties of the cytochrome b component of the microbicidal oxidase system of human neutrophils, *Biochem. J.,* 182, 181, 1979.
73. **Cross, A. R., Jones, O. T. G., Harper, A. M., and Segal, A. W.,** Oxidation-reduction properties of the cytochrome b found in the plasma membrane fraction of human neutrophils: a possible oxidase in the respiratory burst, *Biochem. J.,* 194, 599, 1981.
74. **Hafeman, D. G. and Lucas, Z. J.,** Polymorphonuclear leukocyte-mediated, antibody-dependent, cellular cytotoxicity against tumor cells: dependence on oxygen and the respiratory burst, *J. Immunol.,* 123, 55, 1979.

75. **Katz, P., Simone, C. B., Henkart, P. A., and Fauci, A. S.,** Mechanisms of antibody-dependent cellular cytotoxicity. The use of effector cells for chronic granulomatous disease patients as investigative probes, *J. Clin. Invest.,* 65, 55, 1980.
76. **Babior, B. M., Curnutte, J. T., and Kipnes, R. S.,** Biological defense mechanisms: evidence for the participation of superoxide in bacterial killing by xanthine oxidase, *J. Lab. Clin. Med.,* 85, 235, 1975.
77. **Mandell, G. L.,** Catalase, superoxide dismutase, and virulence of staphylococcus aureus. In vitro and in vivo studies with emphasis on staphylococcal-leukocyte interaction, *J. Clin. Invest.,* 55, 561, 1975.
78. **Rosen, H. and Klebanoff, S. J.,** Bactericidal activity of a superoxide anion-generating system. A model for the polymorphonuclear leukocyte, *J. Exp. Med.,* 149, 27, 1979.
79. **McRipley, R. J. and Sbarra, A. J.,** Role of the phagocyte in host-parasite interactions. XII. Hydrogen peroxide-myeloperoxidase bactericidal system in the phagocyte, *J. Bacteriol.,* 94, 1425, 1967.
80. **Drath, D. B. and Karnovsky, M. L.,** Bactericidal activity of metal-mediated peroxide-ascorbate systems, *Infect. Immun.,* 10, 1077, 1974.
81. **Klebanoff, S. J.,** Iodination of bacteria: a bactericidal mechanism, *J. Exp. Med.,* 126, 1063, 1967.
82. **Harrison, J. E. and Schultz, J.,** Studies on the chlorinating activity of myeloperoxidase, *J. Biol. Chem.,* 251, 1371, 1976.
83. **Michell, R. H., Karnovsky, M. J., and Karnovsky, M. L.,** The distributions of some granule-associated enzymes in guinea pig polymorphonuclear leukocytes, *Biochem. J.,* 116, 207, 1970.
84. **Bretz, U. and Baggiolini, M.,** Biochemical and morphological characterization of azurophil and specific granules of human neutrophilic polymorphonuclear leukocytes, *J. Cell Biol.,* 63, 251, 1974.
85. **Leffell, M. S. and Spitznagel, J. K.,** Association of lactoferrin with lysozyme in granules of human polymorphonuclear leukocytes, *Infect. Immun.,* 6, 761, 1972.
86. **Kane, S. P. and Peters, T. J.,** Analytical subcellular fractionation of human granulocytes with reference to the localization of vitamin B_{12}-binding proteins, *Clin. Sci. Mol. Med.,* 49, 171, 1975.
87. **Wright, D. G. and Gallin, J. I.,** Secretory responses of human neutrophils: exocytosis of specific (secondary) granules by human neutrophils during adherence in vitro and during exudation in vivo, *J. Immunol.,* 123, 285, 1979.
88. **Stossel, T. P., Root, R. K., and Vaughan, M.,** Phagocytosis in chronic granulomatous disease and the Chediak-Higashi syndrome, *N. Engl. J. Med.,* 286, 170, 1972.
89. **Harrison, J. E. and Schultz, J.,** Myeloperoxidase: confirmation and nature of heme-binding inequivalence. Resolution of a carbonyl-substituted heme, *Biochim. Biophys. Acta,* 536, 341, 1978.
90. **Andrews, P. C. and Krinsky, N. I.,** The reductive cleavage of myeloperoxidase in half, producing enzymically active hemi myeloperoxidase, *J. Biol. Chem.,* 256, 4211, 1981.
91. **Agner, K.,** Verdoperoxidase. A ferment isolated from leucocytes, *Acta Physiol. Scand.,* 2 (Suppl. 8), 1, 1941.
92. **Schultz, J. and Kaminker, K.,** Myeloperoxidase of the leucocyte of normal human blood. I. Content and localization, *Arch. Biochem. Biophys.,* 96, 465, 1962.
93. **Stelmaszynska, T. and Zgliczynski, J. M.,** Myeloperoxidase of human neutrophilic granulocytes as chlorinating enzyme, *Eur. J. Biochem.,* 45, 305, 1974.
94. **Zgliczynski, J. M., Selvaraj, R. J., Paul, B. B., Stelmaszynska, T., Poskitt, P. K. F., and Sbarra, A. J.,** Chlorination by the myeloperoxidase-H_2O_2-Cl^- antimicrobial system at acid and neutral pH, *Proc. Soc. Exp. Biol. Med.,* 154, 418, 1977.
95. **Klebanoff, S. J.,** Myeloperoxidase-halide-hydrogen peroxide antibacterial system, *J. Bacteriol.,* 95, 2131, 1968.
96. **Baron, D. N. and Ahmed, S. A.,** Intracellular concentrations of water and the principal electrolytes determined by the analysis of isolated human leucocytes, *Clin. Sci.,* 37, 205, 1969.
97. **Zgliczynski, J. M., Stelmaszynska, T., Ostrowski, W., Naskalski, J., and Sznajd, J.,** Myeloperoxidase of human leukemic leukocytes: oxidation of amino acids in the presence of hydrogen peroxide, *Eur. J. Biochem.,* 4, 540, 1968.
98. **Strauss, R. R., Paul, B. B., Jacobs, A. A., and Sbarra, A. J.,** Role of the phagocyte in host-parasite interactions. XXII. H_2O_2-dependent decarboxylation and deamination by myeloperoxidase and its relationship to antimicrobial activity, *J. Reticuloendothel. Soc.,* 7, 754, 1970.
99. **Zgliczynski, J. M. and Stelmaszynska, T.,** Hydrogen cyanide and cyanogen chloride formation by the myeloperoxidase-H_2O_2-Cl^- system, *Biochim. Biophys. Acta,* 567, 309, 1979.
100. **Pereira, W. E., Hoyano, Y., Summons, R. E., Bacon, V. A., and Duffield, A. M.,** Chlorination studies. II. The reaction of aqueous hypochlorous acid with α-amino acids and dipeptides *Biochim. Biophys. Acta,* 313, 170, 1973.
101. **Stolc, V.,** Release of monoiodotyrosine and iodoprotein from human leukocytes, *J. Clin. Endocrinol. Metab.,* 37, 397, 1973.
102. **Zgliczynski, J. M. and Stelmaszynska, T.,** Chlorinating ability of human phagocytosing leucocytes, *Eur. J. Biochem.,* 56, 157, 1975.

103. **Klebanoff, S. J.**, A peroxidase-mediated antimicrobial system in leukocytes, *J. Clin. Invest.*, 46, 1078, 1967.
104. **Pincus, S. H. and Klebanoff, S. J.**, Quantitative leukocyte iodination, *N. Engl. J. Med.*, 284, 744, 1971.
105. **Simmons, S. R. and Karnovsky, M. L.**, Iodinating ability of various leukocytes and their bactericidal activity, *J. Exp. Med.*, 138, 44, 1973.
106. **Johnson, D. and Travis, J.**, The oxidative inactivation of human α-1-proteinase inhibitor. Further evidence for methionine at the reactive center, *J. Biol. Chem.*, 254, 4022, 1979.
107. **Carp, H. and Janoff, A.**, In vitro suspression of serum elastase-inhibitory capacity by reactive oxygen species generated by phagocytosing polymorphonuclear leukocytes, *J. Clin. Invest.*, 63, 793, 1979.
108. **Clark, R. A. and Klebanoff, S. J.**, Chemotactic factor inactivation by the myeloperoxidase-hydrogen peroxide-halide system, *J. Clin. Invest.*, 64, 913, 1979.
109. **Thomas, E. L. and Aune, T. M.**, Peroxidase-catalyzed oxidation of protein sulfhydryls mediated by iodine, *Biochemistry*, 16, 3581, 1977.
110. **Thomas, E. L.**, Myeloperoxidase, hydrogen peroxide, chloride antimicrobial system: nitrogen-chlorine derivatives of bacterial components in bactericidal action against *Escherichia coli*, *Infect. Immun.*, 23, 522, 1979.
111. **Stossel, T. P., Mason, R. J., and Smith, A. L.**, Lipid peroxidation by human blood phagocytes, *J. Clin. Invest.*, 54, 638, 1974.
112. **Shohet, S. B., Pitt, J., Baehner, R. L., and Poplack, D. G.**, Lipid peroxidation in the killing of phagocytized pneumococci, *Infect. Immun.*, 10, 1321, 1974.
113. **Allen, R. C., Stjernholm, R. L., and Steele, R. H.**, Evidence for the generation of an electronic excitation state(s) in human polymorphonuclear leukocytes and its participation in bactericidal activity, *Biochem. Biophys. Res. Comm.*, 47, 679, 1972.
114. **Krinsky, N. I.**, Singlet excited oxygen as a mediator of the antibacterial action of leukocytes, *Science*, 186, 363, 1974.
115. **Rosen, H. and Klebanoff, S. J.**, Formation of singlet oxygn by the myeloperoxidase-mediated antimicrobial system, *J. Biol. Chem.*, 252, 4803, 1977.
116. **Piatt, J. and O'Brien, P. J.**, Singlet oxygen formation by a peroxidase, H_2O_2 and halide system, *Eur. J. Biochem.*, 93, 323, 1979.
117. **Cheson, B. D., Christensen, R. L., Sperling, R., Kohler, B. E., and Babior, B. M.**, The origin of the chemiluminescence of phagocytosing granulocytes, *J. Clin. Invest.*, 58, 789, 1976.
118. **Held, A. M. and Hurst, J. K.**, Ambiguity associated with use of singlet oxygen trapping agents in myeloperoxidase-catalyzed oxidations, *Biochem. Biophys. Res. Comm.*, 81, 878, 1978.
119. **Klebanoff, S. J. and Hamon, C. B.**, Role of myeloperoxidase-mediated antimicrobial systems in intact leukocytes, *J. Reticuloendothel. Soc.*, 12, 170, 1972.
120. **Tauber, A. I. and Babior, B. M.**, Evidence for hydroxyl radical production by human neutrophils, *J. Clin. Invest.*, 60, 374, 1977.
121. **Weiss, S. J., Rustagi, P. K., and Lo Buglio, A. F.**, Human granulocyte generation of hydroxyl radical, *J. Exp. Med.*, 147, 316, 1978.
122. **Repine, J. E., Eaton, J. W., Anders, M. W., Hoidal, J. R., and Fox, R. B.**, Generation of hydroxyl radical by enzymes, chemicals, and human phagocytes in vitro. Detection with the anti-inflammatory agent, dimethylsulfoxide, *J. Clin. Invest.*, 64, 1652, 1979.
123. **Pryor, W. A. and Tang, R. H.**, Ethylene formation from methional, *Biochem. Biophys. Res. Comm.*, 81, 498, 1978.
124. **Beauchamp, C. and Fridovich, I.**, A mechanism for the production of ethylene from methional: the generation of hydroxyl radical by xanthine oxidase, *J. Biol. Chem.*, 245, 4641, 1970.
125. **Green, M. R., Hill, H. A. O., Okolow-Zubkowska, M. J., and Segal, A. W.**, The production of hydroxyl and superoxide radicals by stimulated human neutrophils — measurements of EPR spectroscopy, *F.E.B.S. Lett.*, 100, 23, 1979.
126. **Rosen, H. and Klebanoff, S. J.**, Hydroxyl radical generation by polymorphonuclear leukocytes measured by electron spin resonance spectroscopy, *J. Clin. Invest.*, 64, 1725, 1979.
127. **Haber, F. and Weiss, J.**, The catalytic decomposition of hydrogen peroxide by iron salts, *Proc. R. Soc. Edinb. Sect. A. (Math. Phys. Sci.)*, 147, 332, 1934.
128. **Halliwell, B.**, Superoxide-dependent formation of hydroxyl radicals in the presence of iron salts, *F.E.B.S. Lett.*, 96, 238, 1978.
129. **McCord, J. M. and Day, E. D., Jr.**, Superoxide-dependent production of hydroxyl radical catalyzed by iron-EDTA complex, *F.E.B.S. Lett.*, 86, 139, 1978.
130. **Tauber, A. I., Gabig, T. G., and Babior, B. M.**, Evidence for production of oxidizing radicals by the particulate O_2^--forming system from human neutrophils, *Blood*, 53, 666, 1979.
131. **Klebanoff, S. J. and Rosen, H.**, Ethylene formation by polymorphonuclear leukocytes, *J. Exp. Med.*, 148, 490, 1978.

132. **Peters, J. W. and Foote, C. S.,** Chemistry of superoxide ion. II. Reaction with hydroperoxides, *J. Am. Chem. Soc.,* 98, 873, 1976.
133. **Gregory, E. M., Yost, F. J., Jr., and Fridovich, I.,** Superoxide dismutase of *Escherichia coli:* intracellular localization and functions, *J. Bacteriol.,* 115, 987, 1973.
134. **Van Hemmen, J. J. and Meuling, W. J. A.,** Inactivation of *Escherichia coli* by superoxide radicals and their dismutation products, *Arch. Biochem. Biophys.,* 182, 743, 1977.

Chapter 2

PRODUCTION OF SUPEROXIDE RADICALS AND HYDROGEN PEROXIDE IN MITOCHONDRIA

Alberto Boveris and Enrique Cadenas

TABLE OF CONTENTS

I. INTRODUCTION

The demonstration by Warburg and Negelein[1] that the "Atmungsferment" was the carbon monoxide- and cyanide-sensitive catalyst responsible for the main part of oxygen uptake in cell and tissue, eclipsed for decades the idea of biological univalent and bivalent oxygen reduction.[2] Wieland had favored a theory of substrate activation in which hydrogen atoms from the substrate combine with oxygen to form H_2O_2;[3] a test of this hypothesis depended on the demonstration of H_2O_2 formation during substrate oxidation in model and biological systems. Despite numerous attempts, Wieland was never able to identify H_2O_2 in animal tissues, probably due to the failure to comprehend how rapidly H_2O_2 disappears in tissue extracts.[2] However, the idea that cytochrome oxidase was the unique mitochondrial activity consuming oxygen under physiological conditions collided with the demonstration of the cyanide- and antimycin-insensitive respiration which is observed to a limited extent in mitochondria isolated from eukaryotic cells[4] and as a major pathway in bacterial respiratory systems and in mitochondria from higher plants.[5,6]

During 1966 to 1967, Jensen[7] and Hinkle et al.[8] reported that the mitochondrial respiratory chain was capable of producing H_2O_2. However, these results were taken with caution, in the sense that they might reflect an artificial activity induced by the ultrasonic or alkaline treatment used in the preparation of the submitochondrial particles. In 1971 Chance and Oshino[9] demonstrated variations in the level of the catalase intermediate of the peroxisomal-mitochondrial fraction of rat liver following the addition of mitochondrial substrates and uncouplers. In the same year, Loschen et al.[10] showed H_2O_2 formation in pigeon heart mitochondria and its relationship to the mitochondrial metabolic state by using the peroxidase-scopoletin method; since this assay can be easily interfered with by endogenous hydrogen donor of the horseradish peroxidase and by exogenous hydrogen donors in the mitochondrial preparations, Boveris et al.[11] applied cytochrome *c* peroxidase assay to determine the general properties of the mitochondrial generation of H_2O_2 in pigeon heart mitochondria.[12] The initial observations have been now extended to mitochondria isolated from a multitude of sources, including mammalian and avian organs, invertebrates, eukaryotic unicellular organisms, and plant tissues. Moreover, changes in H_2O_2 production rates observed upon supplementation with inhibitors and substrates in isolated rat liver mitochondria correlate with the changes in H_2O_2 production measured in the intact liver.[2]

Production of $O_2^{\cdot-}$ by mitochondrial membranes was first reported by Loschen et al.,[13] and thereafter Boveris and Cadenas[14] and Dionisi et al.[15] showed that $O_2^{\cdot-}$ can be considered as a stoichiometric precursor for mitochondrial H_2O_2. The discovery of the role of mitochondria in $O_2^{\cdot-}$ and H_2O_2 generation was most surprising and afforded a conceptual basis for a marked change in the consideration of the problem of hydrogen transfer reactions in mitochondria and the mechanism of intracellular $O_2^{\cdot-}$ and H_2O_2 generation.[2]

In the present chapter we will describe the properties of $O_2^{\cdot-}$ and H_2O_2 production by mitochondrial membranes and discuss the identity of the generators; we will also consider the role of mitochondrial $O_2^{\cdot-}$ and H_2O_2 production in the generation of other free radicals and excited states in the cell, as well as the nature of the protective mechanisms. A short section considers critically the assays which are currently used to determine mitochondrial $O_2^{\cdot-}$ and H_2O_2 formation.

II. METHODS OF DETECTION OF HYDROGEN PEROXIDE AND SUPEROXIDE RADICAL

A. Methods for Hydrogen Peroxide

1. Cytochrome c Peroxidase

Cytochrome *c* peroxidase (CCP) reacts with H_2O_2 to form a stable enzyme-substrate compound that has an absorption maximum at 419 nm, whereas that of the free enzyme is at 407 nm ($\varepsilon_{419-407}$ = 50 mM^{-1} cm^{-1})[16] (Reaction 1). The formation

$$CCP + H_2O_2 \rightarrow CCP \cdot H_2O_2 \qquad (1)$$

of the intermediate is measured by dual-wavelength spectrophotometry at 419—407 nm or, for more sensitivity, at 424—400 nm.[12] Thus, the measurement of the rate of formation of the enzyme-substrate (ES) complex in an aborted reaction in which the second step is avoided by not providing the hydrogen donor, reduced cytochrome *c* (see Reaction 2), allows the determination of H_2O_2 formation rates as low as 0.1 μM/min with accuracy because of the essentially irreversible reaction of enzyme and substrate.[11]

$$CCP \cdot H_2O_2 + 2\ \text{cyt.}\ c^{2+} + 2\ H^+ \rightarrow CCP + 2\ H_2O + 2\ \text{cyt.}\ c^{3+} \qquad (2)$$

Catalase present in the sample can be inhibited with azide without interference with the assay. In cases where the wavelengths 419—407 nm are inconvenient due to interfering pigments, the assay can also be used to measure H_2O_2 production in the presence of catalytic amounts of cytochrome *c* peroxidase and substrate amounts of reduced cytochrome *c* (Reactions 1 and 2). Such samples must be free of cytochrome oxidase or cytochrome *c* reductase and indeed of $O_2^{\cdot-}$ which also reduces cytochrome *c*. Acetylated ferrocytochrome *c*, which can act as a substrate for cytochrome *c* peroxidase, can be used; unlike native cytochrome *c*, acetylated cytochrome *c* is slowly reduced by mitochondrial cytochrome *c* reductase and slowly oxidized by cytochrome oxidase. Yonetani's method[16] of isolating yeast cytochrome (c) peroxidase yields a useful preparation. It has been noted that the enzyme isolated by inexperienced hands sometimes does not form the H_2O_2-enzyme intermediate; however, the preparation does have a high catalytic capability and can be used to measure ferrocytochrome *c* oxidation.

2. Horseradish Peroxidase and the Coupled Oxidation of Hydrogen Donors

Horseradish peroxidase oxidizes various hydrogen donors in the presence of H_2O_2 as indicated in Reactions 3 and 4, where AH_2 and A denote the reduced and oxidized forms of the hydrogen donor. Fluorescent hydrogen donors or oxidation products are particularly suitable for such measurements in samples with low H_2O_2 formation rates.

$$HRP + H_2O_2 \rightarrow HRP \cdot H_2O_2 \qquad (3)$$

$$HRP \cdot H_2O_2 + AH_2 \rightarrow HRP + 2\ H_2O + A \qquad (4)$$

Scopoletin loses its fluorescence when oxidized in the peroxidase reaction[10,17] and diacetyldichlorofluorescein is oxidized to a fluorescent product.[8,18] Formation of the horseradish peroxidase-H_2O_2 intermediate ($HRP \cdot H_2O_2$) can also be measured to determine H_2O_2 generation[10,19] with peroxidase preparations that are free of endogenous hydrogen donors. However, the broader specificity of the horseradish enzyme for various hydrogen donors, compared with cytochrome *c* peroxidase, may cause underestimation, and controls with cytochrome *c* peroxidase are advisable. Catalase interference may be evaluated by determining the peroxidase-to-catalase heme ratio.[19] The fluorometric method can be calibrated with appropriate systems such as glucose and glucose oxidase or urate and uricase, and rates of H_2O_2 formation in the range of 1 μM/min can be detected easily.

3. Monitoring of the Catalase Intermediate

The direct optical measurement ($\varepsilon_{660-640}$ = 4 mM^{-1} cm^{-1}) of the steady-state concentration of catalase compound I has the widest application for the measurement of the rate of H_2O_2 formation.[2] The method is based on the fact that the formation of catalase intermediate depends on the rate of H_2O_2 formation and on the amount of hydrogen donor present. The assay is applicable to isolated rat liver mitochondria and indeed to any mitochondria supplemented with exogenous catalase. The advantages of the method are its specificity, the compound measured being in equilibrium with H_2O_2, and the measurement may be extended for long periods of time since the indicator is never saturated.[2]

B. Methods for Superoxide Radicals

Chemical methods integratively measure the rates of $O_2^{\cdot-}$ formation by trapping it with suitable spectrophotometric indicators in reactions that are sensitive to superoxide dismutase addition. The concentration of the trapping agent is adjusted to compete effectively with the spontaneous dismutation reaction so that nearly all the superoxide anion can be detected. Since there are no inhibitors of mitochondrial superoxide dismutase, it is not possible to ascertain directly $O_2^{\cdot-}$ production by intact mitochondria. To detect maximal mitochondrial $O_2^{\cdot-}$ formation, endogenous superoxide dismutase must be removed by repetitive washing of submitochondrial particles obtained by sonication or by other means.

1. Cytochrome c and Acetylated Cytochrome c

$O_2^{\cdot-}$ reduces cytochrome *c*[20,21] (Reaction 5) ($\varepsilon_{550-540}$ for reduced-oxidized cytochrome $c = 19\ \text{m}M^{-1}\ \text{cm}^{-1}$).

$$O_2^{-} + \text{cyt. } c^{3+} \rightarrow O_2 + \text{cyt. } c^{2+} \qquad (5)$$

Superoxide dismutase-sensitive initial rate of exogenous cytochrome *c* reduction provide an estimation of the rate of $O_2^{\cdot-}$ formation in submitochondrial particles.[14] The major problem with this method results from the great capacity of mitochondria to reduce added cytochrome *c* by nonsuperoxide-dependent processes. As acetylated cytochrome *c* is more slowly reduced by mitochondrial cytochrome *c* reductase, it can be used to overcome this obstacle.[22]

2. The Adrenochrome Method

The $O_2^{\cdot-}$-dependent oxidation of epinephrine to adrenochrome[23] is conveniently followed spectrophotometrically at 480 nm or 480-575 nm by double beam spectrophotometry ($\varepsilon_{480-575} = 2.96\ \text{m}M^{-1}\ \text{cm}^{-1}$ and $\varepsilon_{480} = 4.0\ \text{m}M^{-1}\ \text{cm}^{-1}$). However, specificity of the reaction is rather poor,[24] and autoxidation of epinephrine must be controlled by lowering the pH. Yet, superoxide dismutase-sensitive adrenochrome formation provides a useful assay for $O_2^{\cdot-}$ generation rates of about 1 to 5 μM/min. Plant mitochondria contain phenol oxidase that strongly interferes with this assay.[25,26] In the absence of side reactions, the assay detects one adrenochrome formed per one $O_2^{\cdot-}$ formed.[27]

III. THE MITOCHONDRIAL PRODUCTION OF SUPEROXIDE RADICALS AND HYDROGEN PEROXIDE

A. General Properties

The mitochondrial respiratory chain consists of a series of redox catalysts (pyridine nucleotide, flavoproteins, iron-sulfur proteins, ubiquinone, and cytochromes) which, when assembled, constitute the mitochondrial inner membrane. These respiratory carriers are arranged spatially according to their redox potentials which extend from −320 to +380 mV.[28] Hence, on thermodynamic grounds, the reduced forms of all these

Table 1
MITOCHONDRIAL METABOLIC STATES

State	Metabolite requirements: ADP	Metabolite requirements: Substrate	Metabolite requirements: O_2	Relative rates: O_2 uptake	Relative rates: ATP production	Relative steady-state of respiratory chain components
4	Low	High	Normal	Slow	Slow	Reduced
3	High	High	Normal	Fast	Fast	Oxidized
3u (uncoupler)	Irrelevant	High	Normal	Fast	Zero	Oxidized

Taken from Chance, B. B. and Williams, G. R., *Adv. Enzymol.*, 17, 65, 1956. With permission.

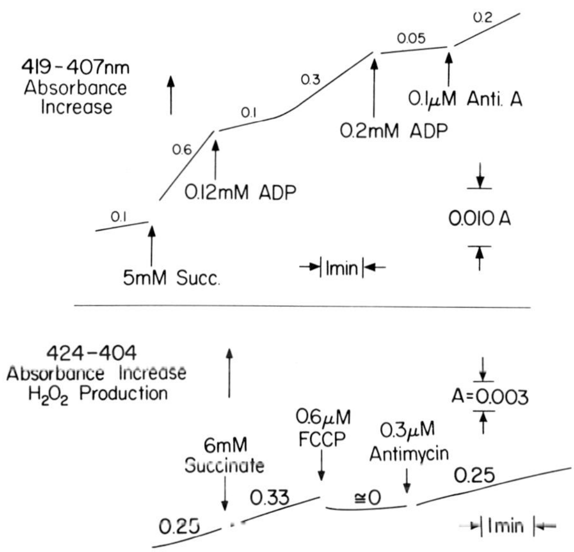

FIGURE 1. Production of H_2O_2 by rat liver (upper trace) and pigeon lung (bottom trace) mitochondria. The cytochrome *c* peroxidase assay detects H_2O_2 production as an upward deflection. The numbers near the trace indicate nmol H_2O_2/min/mg protein. (From Boveris, A., *Adv. Exp. Med. Biol.*, 78, 67, 1977. With permission.)

respiratory carriers are likely to be oxidized by molecular oxygen ($E'_o = +820$ mV). However, oxygen is only taken up effectively by cytochrome oxidase which accounts for 96 to 99% of the oxygen consumption by mitochondrial preparations, and production of $O_2^{\overline{\cdot}}$ and H_2O_2 accounts for the remaining 1 to 4%.[3]

The rate of H_2O_2 production by isolated mitochondria depends on the mitochondrial metabolic state. The regulation of respiration and oxidative phosphorylation is usually described in terms of metabolic states, following the original classification by Chance and Williams[29] (Table 1).

The H_2O_2 generation rate is higher in State 4, where the absence of the phosphate acceptor adenosine 5′-diphosphate (ADP) sets a slow respiratory rate with a high reduction of the components of the respiratory chain, and is lower in State 3, where there is a fast oxygen uptake and where the respiratory carriers are largely oxidized.[10,12] Figure 1 shows H_2O_2 production by rat liver and pigeon lung mitochondria. The low rate of H_2O_2 formation supported by endogenous substrates is increased upon addition of succinate; subsequent supplementation with ADP or uncoupler decreases H_2O_2 for-

Table 2
PRODUCTION OF HYDROGEN PEROXIDE IN ISOLATED MITOCHONDRIA

		nmol H_2O_2/min/mg protein			
Mitochondria from	Substrate	State 4	State 3 or 3u	State 3 or 3u and antimycin	Ref.
Rat liver	Succinate	0.50	0.10	0.20	11
	Malate-glutamate	0.40	0.08	0.20	11
Rat heart	Succinate	0.60	0	3.7	34—35
Rat kidney	Succinate	0.40	0.10	1.6	36
Rat lung macrophage	Succinate	0.60	—	1.4	37
	Malate-glutamate	0.40	—	0.8	37
Pigeon heart	Succinate	0.80	0	16.0	10—12
	Malate-glutamate	0.80	0	15.0	12
Pigeon breast muscle	Succinate	—	—	0.2	38
Pigeon lung	Succinate	0.33	0	0.25	30
Ascaris lumbricoides muscle	Succinate	15.0	15.0	18.0	36
Crithidia fasciculata	Succinate	4.9	4.1	8.5	39
	α-Ketoglutarate	8.6	9.6	13.6	39
Saccharomyces cerevisiae	Succinate	1.5	1.3	2.2	40
	Ethanol	1.4	1.1	1.5	40
Trypanosoma cruzi	NADH	2.3	2.3	2.3	41
Potato tuber	Succinate	0.71	0.34	0.81	42
	NADH	1.80	1.80	1.75	42
Mung bean hypocotyls	Succinate	1.12	0.02	0.74	43
	NADH	1.75	1.75	1.70	43
Skunk cabbage spadix	Succinate	3.6	—	—	43

mation. The transient effect of ADP observed in rat liver mitochondria in Figure 1 is due to ADP exhaustion after its phosphorylation to ATP. Addition of antimycin exerts a moderate stimulatory effect on both rat liver and pigeon lung mitochondria (Figure 1).

Succinate and NADH-dependent substrates (such as malate-glutamate) are equally effective in supporting maximal H_2O_2 production.[12,30] It is noteworthy that when malate-glutamate is used as the source of reducing equivalents, H_2O_2 production is inhibited significantly by rotenone.[11,12,30] Antimycin is especially effective in heart mitochondria where it promotes high rates of H_2O_2 production.[10,12] Concerning other mitochondrial substrates, dihydroorotate is able to support $O_2^{\cdot-}$ and H_2O_2 production by brain and liver mitochondria;[28,29] fatty acids and fatty acid derivatives, such as steroyl-CoA or palmitylcarnitine, do not support significant H_2O_2 production in heart mitochondria;[12] however, they are effective in liver mitochondria, probably due to the peroxisomal NADH-dependent H_2O_2-forming system.[33] It is then clear that any mitochondrial substrate entering the respiratory chain either at NADH or at ubiquinone level is able to support H_2O_2 production. Consequently, it is apparent that H_2O_2 generation constitutes a physiological event in aerobic conditions.[2,11,12] Table 2 lists State 4 and State 3 rates of H_2O_2 formation by mitochondria of several sources with different substrates.

It is apparent that the mitochondrial generator of H_2O_2 is a member of the respiratory chain, since rates are maximal in the highly reduced states and minimal in the oxidized states. From the effect of mitochondrial inhibitors such as antimycin and rotenone, it can be concluded that the mitochondrial generation of H_2O_2 originates mainly from the oxidation of a component located on the substrate site of the antimycin block and on the oxygen side of the rotenone block.[12,32]

B. Superoxide Radical as a Precursor of Mitochondrial Hydrogen Peroxide

Loschen et al.[13] first detected $O_2^{\cdot-}$ production in washed mitochondrial membranes,

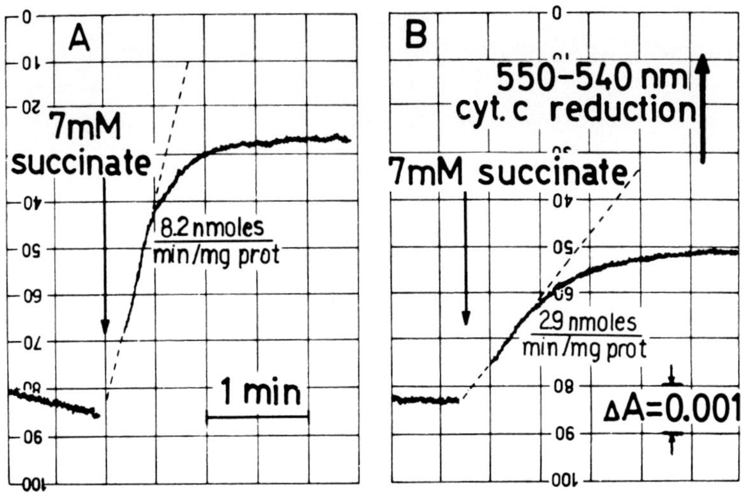

FIGURE 2. Production of $O_2^{\cdot -}$ by beef heart submitochondrial particles in the presence of succinate and antimycin. In B, 0.3 μM superoxide dismutase was added. Cytochrome *c* assay: the rate of $O_2^{\cdot -}$ production is calculated from the superoxide dismutase-sensitive rate of cytochrome *c* reduction (5.3 nmol $O_2^{\cdot -}$/min/mg protein.) (From Boveris, A. and Cadenas, E., *F.E.B.S. Lett.*, 54, 311, 1975. With permission.)

Table 3
PRODUCTION OF SUPEROXIDE RADICALS AND HYDROGEN PEROXIDE AND ANTIMYCIN-INSENSITIVE RESPIRATION IN MITOCHONDRIAL PREPARATIONS

	nmol/min/mg protein		
	Oxygen uptake	$O_2^{\cdot -}$ production	H_2O_2 production
Sonicated submitochondrial particles	5.0	0	1.8
Idem + four washings	10.8	3.8	3.7
Ubiquinone-reconstituted membranes	2.1	0.48	0.17

[a] Data from Boveris et al.[44]

a significant finding since mitochondria are likely to be the most important physiological source of $O_2^{\cdot -}$.[2] Submitochondrial particles in the presence of succinate and antimycin show values of 1.1,[44] 1.6 to 2.1,[14] and 2.0[15] for the ratio of the rates of $O_2^{\cdot -}$ and H_2O_2 generation, varying somewhat with the different experimental conditions. Nevertheless, these values are close enough to 2, especially taking into account that some superoxide dismutase may be trapped inside the submitochondrial vesicles, considering $O_2^{\cdot -}$, if not a stoichiometric precursor, at least the main intermediate of mitochondrial H_2O_2. Figure 2 shows $O_2^{\cdot -}$ generation in submitochondrial particles as detected by the cytochrome *c* assay. The production of $O_2^{\cdot -}$ and H_2O_2 accounts for about 40% of the antimycin-insensitive oxygen uptake in submitochondrial particles (Table 3) and for about 90% of the antimycin-resistant respiration in intact mitochondria.[12] Production of $O_2^{\cdot -}$ has been documented in mitochondrial membranes from a variety of sources (Table 4).

Table 4
PRODUCTION OF SUPEROXIDE RADICALS BY SUBMITOCHONDRIAL PARTICLES

Source	Substrate	Respiratory inhibitor	pH	$O_2^{\overline{\cdot}}$ production (nmol/min/mg protein)	Ref.
Beef heart	Succinate	Antimycin	7.4	1.7	13
			8.1	3.4	13
			7.5	0.8	14
			8.2	4.7	14
Rat heart	Succinate	Antimycin	7.4	1.3	15
			8.2	5.2	15
Morris hepatoma	Succinate	Antimycin	7.5	3.8	15
Rat lung macrophage	Succinate	Antimycin	7.8	0.7	37
Ehrlich ascites	Succinate	Antimycin	7.5	0.9	15
Rat liver	Dihydroorotate	Antimycin	7.8	2.8	32
			8.5	0.6	31
Rat brain	Dihydroorotate	Antimycin	7.8	2.5	32
Beef heart	NADH	Rotenone	7.4	0.9	52
			7.4	4.0	50
Potato tuber	NADH	Rotenone	7.4	1.0	25

C. Sites of Superoxide Anion Production

Identification of the $O_2^{\overline{\cdot}}$ generator in mitochondrial membrane, acting between the rotenone- and the antimycin-sensitive sites, was made arduous because the rate of $O_2^{\overline{\cdot}}$ generation is about 10^3 times slower than electron equilibration between members of the isopotential pool constituted by succinate-dehydrogenase, ubiquinone, cytochromes *b*, and iron-sulfur centers.[45,46] However, extraction and fractionation methods as well as model reactions overcame this problem.

Mitochondrial membranes depleted of endogenous ubiquinone and reconstituted with variable amounts of ubiquinones show a linear relationship between quinone content and H_2O_2 production.[12,44] Similarly, NADH-ubiquinone reductase supplemented with ubiquinone analogs shows H_2O_2 production linearly dependent on the amount of exogenous ubiquinone, with an analog specificity similar to the one required for biochemical activity[27] (Figure 3). Parallel measurements of succinate dehydrogenase and succinate-cytochrome *c* reductase activities with H_2O_2 production in ubiquinone-reconstituted membranes show that the two former activities reach a plateau at a low level of reducible ubiquinone, whereas H_2O_2 generation is linearly related to the ubiquinone content over a wide range of ubiquinone levels.[44] Such evidence points to the reduced forms of ubiquinone as a source of mitochondrial $O_2^{\overline{\cdot}}$ and H_2O_2.

Further evidence for the same conclusion was obtained after fractionation of the mitochondrial respiratory chain. Cadenas et al.[27] found the NADH-ubiquinone reductase (Complex I) and ubiquinol-cytochrome *c* reductase (Complex III), that contain ubiquinone as the sole common component, are effective generators of $O_2^{\overline{\cdot}}$ and H_2O_2. Isolated succinate dehydrogenase (Complex II) is much less effective in producing H_2O_2 (Table 5). In agreement with this low activity of the solubilized flavoprotein, it was reported that submitochondrial particles depleted of succinate dehydrogenase are still able to produce H_2O_2.[44]

Moreover, production of $O_2^{\overline{\cdot}}$ and H_2O_2 by model systems of quinol oxidation[27] and of ubiquinone analogs supports the reduced forms of ubiquinone as responsible for $O_2^{\overline{\cdot}}$ and H_2O_2 generation.[27] Figure 4 illustrates $O_2^{\overline{\cdot}}$ and H_2O_2 formation during the autoxidation of Q_O, the water-soluble ubiquinone analog without isoprene units in the lateral chain.

Cyanide inhibits $O_2^{\overline{\cdot}}$ production in submitochondrial particles in the presence of

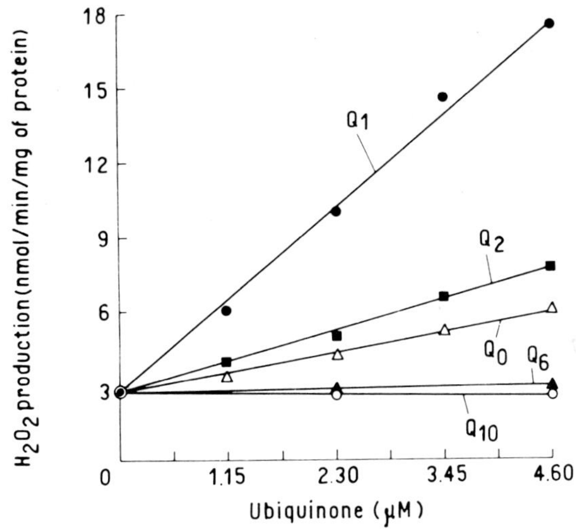

FIGURE 3. Effect of ubiquinone analogs in the H_2O_2 production by isolated NADH-ubiquinone reductase (Complex I). (From Cadenas, E., Boveris, A., Ragan, C. I., and Stoppani, A. O. M., *Arch. Biochem. Biophys.*, 180, 248, 1977. With permission.)

Table 5
HYDROGEN PEROXIDE PRODUCTION BY ISOLATED COMPLEXES OF THE MITOCHONDRIAL RESPIRATORY CHAIN[a]

	H_2O_2 production (nmol/min/mg protein)
NADH-ubiquinone reductase (complex I)	4.62
Idem + rotenone	1.10
Ubiquinol-cytochrome *c* reductase (complex III)	4.24
Succinate dehydrogenase (complex II)	0.03

[a] Data from Cadenas et al.[27] and Boveris et al.[44]

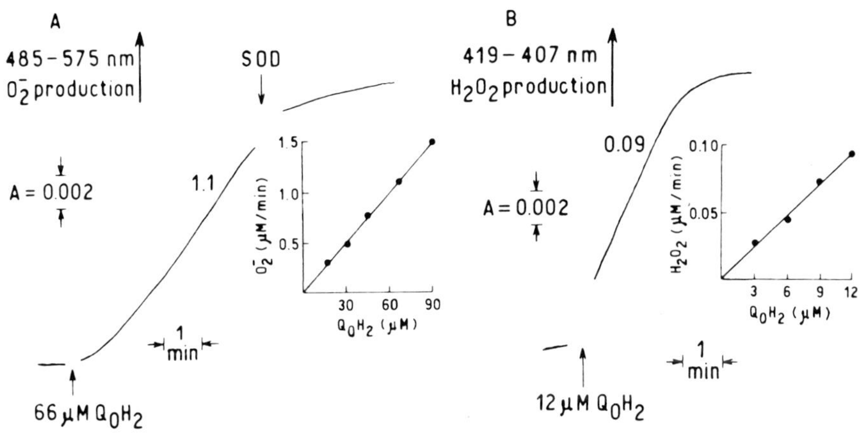

FIGURE 4. The autoxidation of the reduced form of ubiquinone$_0$ (Q_0H_2) produces $O_2^{\cdot -}$ and H_2O_2. From Cadenas, E., Boveris, A., Ragan, C. E., and Stoppani, A. O. M., *Arch. Biochem. Biophys.*, 180, 248, 1977. With permission.)

succinate and antimycin paralleling the level of oxidized cytochrome *c*.[30,34] In these experiments cyanide directly inhibits cytochrome oxidase, increasing the steady-state concentration of reduced cytochrome *c* which is produced by normal electron flow through leaks in the antimycin block. On this experimental basis, a role of ferricytochrome *c* in the generation of ubisemiquinone from ubiquinol was suggested (Reaction 6).[30,34] Ubisemiquinone would be the autooxidizable component yielding $O_2^{\cdot-}$ (Reaction 7).

$$UQH_2 + \text{cyt. } c^{3+} \rightarrow UQH^{\cdot} + \text{cyt. } c^{2+} + H^+ \quad (6)$$

$$UQH^{\cdot} + O_2 \rightarrow UQ + H^+ + O_2^{\cdot-} \quad (7)$$

An alternative explanation is possible; cytochrome b_T may interact with cytochrome *c* through heme-heme interaction; reduction of cytochrome *c* would cause oxidation of cytochrome b_T that would inhibit ubisemiquinone generation.[45]

Moreover, Trumpower and Simmonds[47] and Forman[48] pointed to ubisemiquinone as a source of $O_2^{\cdot-}$ in isolated succinate-cytochrome *c* reductase[47] and dihydroorotate dehydrogenase.[48] It is worth noting that ubisemiquinone radicals in mitochondrial membranes seem to be highly stabilized by electronic delocalization and binding to specific proteins.[34,49] At the present time, Ohnishi et al.[49] have been able to describe three different types of ubisemiquinone in the mitochondrial respiratory chain; these three species are associated with NADH-dehydrogenase (SQn), succinate-ubiquinone reductase (SQs) and ubiquinol-cytochrome *c* reductase (SQc). It is possible to identify which of these ubisemiquinones is able to react effectively with oxygen yielding $O_2^{\cdot-}$. However, the effects of thenoyltrifluoroacetone and antimycin on $O_2^{\cdot-}$ and H_2O_2 generation[12,34] are similar to the effects of the same inhibitors on the size of the esr signal corresponding to the ubisemiquinone associated with succinate-ubiquinone reductase (SQs; 100 mW).[49] Recent evidence from three laboratories[25,50-52] has added the NADH-dehydrogenase of the mitochondrial respiratory chain as another source of mitochondrial $O_2^{\cdot-}$. Rich and Bonner[25] showed that the NADH-dehydrogenase of potato mitochondria generates $O_2^{\cdot-}$ and noted that cyanide inhibited $O_2^{\cdot-}$ production at the ubiquinone-cytochrome *b* site, but did not affect $O_2^{\cdot-}$ formation at the NADH-dehydrogenase. Takeshige and Minakami[50] observed $O_2^{\cdot-}$ production by the NADH dehydrogenase of bovine heart mitochondria and of isolated Complex I. Turrens and Boveris[51,52] measured the $O_2^{\cdot-}$ production by NADH dehydrogenase in submitochondrial particles in which the NADH dehydrogenase flavin was reduced by reversed electron transfer, and they concluded that $O_2^{\cdot-}$ production at the NADH dehydrogenase is about 1 nmol/min/mg protein at pH 7.4. The two sites, ubiquinone-cytochrome *b* and NADH dehydrogenase, show additive rates. A scheme showing the sites of $O_2^{\cdot-}$ production in the mitochondrial respiratory chain, as well as the relative rates at near neutral pH, is presented in Figure 5.

D. Oxygen Dependence of Superoxide Radical Generation

Production of $O_2^{\cdot-}$ in succinate- and antimycin-supplemented submitochondrial particles is proportional to oxygen concentration; the oxygen profile seems to indicate two components, one that saturates at low oxygen levels (below 0.1 m*M*) and the other that functions linearly with oxygen concentration (Figure 6). Apparently, ubisemiquinone reacts nonenzymatically with molecular oxygen yielding $O_2^{\cdot-}$. Disregarding for the moment the rate of the back reaction (quinones can act as oxidants for $O_2^{\cdot-}$ [27]), Equation 7 gives the following differential equation for the rate of $O_2^{\cdot-}$ generation:

$$dO_2^{\cdot-}/dt = (O_2)(UQH^{\cdot})k \quad (8)$$

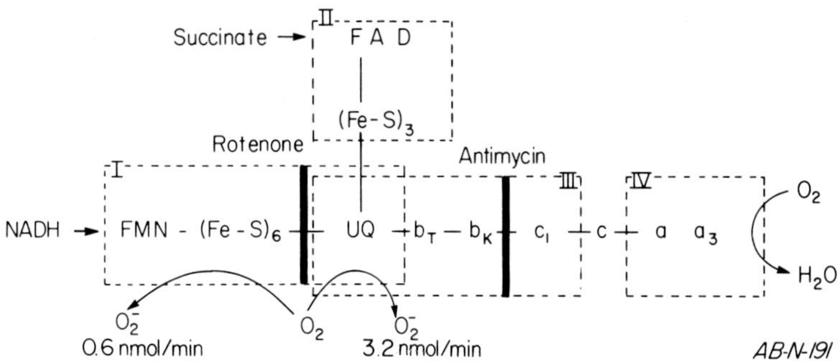

FIGURE 5. Scheme showing the sites of $O_2^{\bar{\cdot}}$ production in the mitochondrial respiratory chain.

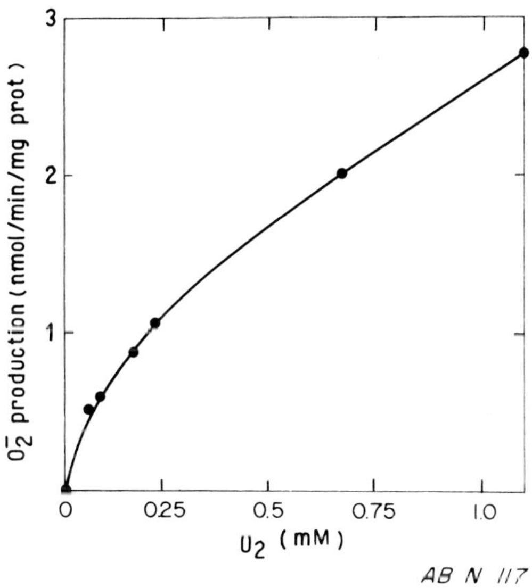

FIGURE 6. Oxygen dependence of $O_2^{\bar{\cdot}}$ production by beef heart submitochondrial particles. (From Boveris, A., *Adv. Exp. Med. Biol.*, 78, 67, 1977. With permission.)

A similar dependence of mitochondrial H_2O_2 production that extends to the hyperbaric region has been reported.[12] This oxygen dependence supports the idea that increased rates of $O_2^{\bar{\cdot}}$ and H_2O_2 production by mitochondria contribute to oxygen toxicity.[2]

IV. ENERGY-LINKED GENERATION OF HYDROGEN PEROXIDE

The energy state of mitochondrial membranes regulates the redox levels of the respiratory carriers and, consequently, it regulates H_2O_2 production by isolated mitochondria. An energized condition (State 4) corresponds to a highly reduced steady-state level of the respiratory carriers and to a high H_2O_2 generation, and similarly, a deenergized condition (State 3 or 3u) corresponds to a highly oxidized level of the respiratory chain component and to a low H_2O_2 production (Tables 1 and 2).[10-12] However, under conditions of low electron flow, as in antimycin-supplemented mitochondria, it has been observed that addition of uncouplers (protophores, ionophores, and Ca^{++}) stimulates H_2O_2 formation up to 13-fold.[12,34] Although the electron flow in antimycin-supplemented mitochondria is relatively low, there is experimental evidence that antimycin insensitive oxygen uptake produces proton extrusion and generates a protonmotive force.[53,54] This effect of uncouplers on H_2O_2 production by antimycin-supplemented mitochondria has been explained in terms of a collapse of membrane potential.[34] It might be inferred that membrane potential would inhibit the rate of ubisemiquinone formation as shown by the decreased rate of H_2O_2 formation. Interestingly enough, Autor and co-workers[55] have observed that the phenomenon may occur also in vivo; rats treated with 2,4-dinitrophenol show a 50-fold increase in H_2O_2 production by isolated liver mitochondria which is accompanied by a 5-fold increase in mitochondrial superoxide dismutase activity.

V. PROTECTION MECHANISMS

The generation of $O_2^{\cdot-}$ and H_2O_2 within the mitochondria may pose a threat to the mitochondria themselves. In addition to the potential for oxidizing essential lipids, through the radical reaction chain of lipid peroxidation, the components of the respiratory chain may also be affected. Tyler[56] has shown that NADH dehydrogenase is inhibited by $O_2^{\cdot-}$ generated by the addition of NADH to submitochondrial particles. The location of the protective enzymes, superoxide dismutase, catalase, and peroxidases in relation to mitochondria is therefore important. Superoxide dismutase exists in two forms in eukaryotic cells: a cytosolic Cu- and Zn-containing enzyme and an Mn-containing enzyme in the mitochondrial matrix.[57] Catalase is located in the peroxisomes or other subcellular microbodies but not in the mitochondria.[2,58] On the other hand, glutathione peroxidase, which also is important in mammalian antioxidant defense since it can utilize both H_2O_2 and lipid and organic hydroperoxides, is found in both the cytosol and mitochondrial matrix.[2,59] The major form of glutathione peroxidase in mitochondria contains selenium and will reduce effectively both H_2O_2 and organic hydroperoxides, while the non-Se-containing enzyme has little affinity for H_2O_2.[60] Cytochrome *c* peroxidase, found in yeast mitochondria,[16] catalyzes physiologically Reactions 1 and 2 that lead to H_2O_2 utilization and cytochrome *c* oxidation.[40]

Thus the enzymatic defenses necessary to remove $O_2^{\cdot-}$ and H_2O_2 exist both in and outside of mitochondria.[2,37] The steady-state level of $O_2^{\cdot-}$ the more reactive species, is maintained at 10^{-11} to 10^{-12} *M* by superoxide dismutase,[2,61] whereas the level of H_2O_2, which is less reactive, is regulated basically by catalase at concentrations up to three orders of magnitude higher, 10^{-7} to 10^{-9}, depending on H_2O_2 production.[2] Biological membranes seem highly permeable to H_2O_2, with values for the permeability constant close to that for water. At the present there are no descriptions of mechanisms that could account for $O_2^{\cdot-}$ permeation through the inner mitochondial membrane. However, Nohl and Hegner[35] have proposed that 20% of the $O_2^{\cdot-}$ produced by mitochondria can escape mitochondrial superoxide dismutase. Several studies have shown that H_2O_2 escapes from intact mitochondria.[10-12,35,37] Therefore, the protective enzymes located outside of mitochondria may function to prevent damage due to mitochondrial $O_2^{\cdot-}$ and H_2O_2 generation. Figure 7 shows the spatial relations of the various "antioxidant" enzymes.

VI. THE MITOCHONDRIAL GENERATION OF PHOTOEMISSIVE SPECIES LINKED TO SUPEROXIDE RADICAL AND HYDROGEN PEROXIDE GENERATION

In addition to the $O_2^{\cdot-}$ and H_2O_2 generation by mitochondria, there is evidence that the formation of other molecules in the excited state, such as singlet oxygen, could

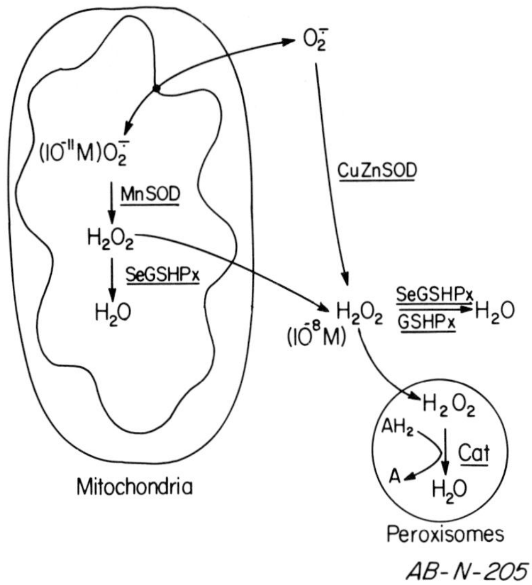

FIGURE 7. The enzymatic antioxidant defense for mitochondrial production of $O_2^{\cdot -}$ and H_2O_2.

depend on the primary production of the former oxygen intermediates. The association between the mitochondrial formation of photoemissive species under oxidative stress (i.e., by supplementation with organic hydroperoxides) and the metabolism of oxygen by-products, such as $O_2^{\cdot -}$ and H_2O_2, is likely considering: (1) the enhancing effect on light emission of respiratory inhibitors such as rotenone and antimycin; and (2) the inhibitory effect of superoxide dismutase on light emission.[62,63]

However, low level chemiluminescence does not have the same requirements as that for $O_2^{\cdot -}$ and H_2O_2 production.[62,63] In addition to hydroperoxides and molecular oxygen, ferricytochrome *c* seems to be of primary importance to the mitochondrial photoemission,[63] and can even replace the mitochondrial membranes.[64-66] A sequence of radical reactions originating in the homolytic rupture of hydroperoxides by cytochrome *c* can account partially for the mechanism of the hydroperoxide-induced light emission in mitochondria. In this case, the reactions do not indicate a participation of $O_2^{\cdot -}$ and H_2O_2 in the reactions leading to photoemission. The involvement of $O_2^{\cdot -}$ and H_2O_2 in mitochondrial chemiluminescence could be understood in terms of a free radical chain reaction involving $O_2^{\cdot -}$ and polyunsaturated lipids,[67] or a metal-catalyzed ·OH radical formation[68,69] able to initiate a free radical process, leading to the formation of more effective oxidant species.

In view of the rates of $O_2^{\cdot -}$ and H_2O_2 production in the subcellular organelles,[2] it is likely that the mitochondrial production of $O_2^{\cdot -}$ and H_2O_2 contributes significantly to the generation of other free radicals, like ·OH and excited molecular states similar to singlet molecular oxygen or excited carbonyl groups in the in vivo situation. The physiological production of these photoemissive excited species affords a satisfactory explanation for the spontaneous light emission of isolated cells and *in situ* organs.[70-73]

VII. CONCLUSIONS

The production of $O_2^{\cdot -}$ in mitochondria appears to be due to the one-electron autoxidation of components of the respiratory chain between NADH and the site of the antimycin block. Ubisemiquinone seems to serve as the primary direct electron donor, affording about 80% of $O_2^{\cdot -}$ formation with the remaining 20% from the NADH dehydrogenase flavoprotein. The production of H_2O_2 by mitochondria appears to result mostly, if not completely, from the dismutation of $O_2^{\cdot -}$ catalyzed by mitochondrial superoxide dismutase. Possible hazards resulting from mitochondrial H_2O_2 and $O_2^{\cdot -}$ production are prevented by "antioxidant" enzymes, such as superoxide dismutase and glutathione peroxidase in the matrix space, and catalase in the peroxisomes. Steady-state levels of $O_2^{\cdot -}$ and H_2O_2 seem regulated at about 10^{-11} and 10^{-8} *M*, respectively.

REFERENCES

1. **Warburg, O. and Negelein, E.**, Uber das Absorptionspektrum des Atmungsferments, *Biochem. Z.*, 214, 64, 1929.
2. **Chance, B., Sies, H., and Boveris, A.**, Hydroperoxide metabolism in mammalian organs, *Physiol. Rev.*, 59, 527, 1979.
3. **Wieland, H.**, Uber den Mechanismus der Oxidationsvorgange, *Liebig's Ann.*, 445, 181, 1925.
4. **Slater, E. C.**, Applications of inhibitors and uncouplers for a study of oxidative phosphorylation, *Meth. Enzymol.*, 10, 48, 1966.
5. **Henry, M. F. and Nyns, E. J.**, Cyanide-insensitive respiration. An alternative mitochondrial pathway, *Sub-Cell. Biochem.*, 4, 1, 1975.
6. **Degn, H., Lloyd, D., and Hill, G. C., Eds.**, *Functions of Alternative Terminal Oxidases*, Pergamon Press, Oxford, 1978.
7. **Jensen, P. K.**, Antimycin-insensitive oxidation of succinate and reduced nicotinamide adenine dinucleotide in electron transport particles, *Biochim. Biophys. Acta*, 122, 157, 1966.
8. **Hinkle, P. C., Butow, R. A., Racker, E., and Chance, B.**, Partial resolution of the enzymes catalyzing oxidative phosphorylation, *J. Biol. Chem.*, 242, 5169, 1967.
9. **Chance, B. and Oshino, N.**, Kinetics and mechanism of catalase in peroxisomes of the mitochondrial fraction, *Biochem. J.*, 122, 225, 1971.
10. **Loschen, G., Flohe, L., and Chance, B.**, Respiratory chain-linked H_2O_2 production in pigeon heart mitochondria, *F.E.B.S. Lett.*, 18, 261, 1971.
11. **Boveris, A., Oshino, N., and Chance, B.**, The cellular production of hydrogen peroxide, *Biochem. J.*, 128, 617, 1972.
12. **Boveris, A. and Chance, B.**, The mitochondrial production of hydrogen peroxide, *Biochem. J.*, 134, 707, 1973.
13. **Loschen, G., Azzi, A., Richter, C., and Flohe, L.**, Superoxide radicals as precursors of mitochondrial hydrogen peroxide, *F.E.B.S. Lett.*, 42, 68, 1974.
14. **Boveris, A. and Cadenas, E.**, Mitochondrial production of superoxide anions and its relationship to antimycin insensitive respiration, *F.E.B.S. Lett.*, 54, 311, 1975.
15. **Dionisi, O., Galeotti, T., Terranova, T., and Azzi, A.**, Superoxide radicals and hydrogen peroxide formation in mitochondria from normal and neoplastic tissues, *Biochim. Biophys. Acta*, 403, 292, 1975.
16. **Yonetani, T.**, Cytochrome c peroxidase (Bakers' yeast), *Meth. Enzymol.*, 10, 336, 1966.
17. **Andreae, W. A.**, A sensitive method for the estimation of hydrogen peroxide in biological materials, *Nature*, 175, 859, 1955.
18. **Keston, A. S. and Brandt, R.**, The fluorometric analysis of ultramicro quantities of hydrogen peroxide, *Anal. Biochem.*, 11, 1, 1965.
19. **Boveris, A., Martino, E., and Stoppani, A. O. M.**, Evaluation of the horseradish peroxidase-scopoletin method for the measurement of hydrogen peroxide formation in biological systems, *Anal. Biochem.*, 80, 145, 1977.
20. **McCord, J. M. and Fridovich, I.**, Superoxide dismutase. An enzymic function for erythrocuprein (hemocuprein), *J. Biol. Chem.*, 244, 6049, 1969.
21. **Forman, H. J. and Fridovich, I.**, Superoxide dismutase: a comparison of rate constants, *Arch. Biochem. Biophys.*, 158, 396, 1973.
22. **Azzi, A., Montecucco, C., and Richter, C.**, The use of acetylated cytochrome c for the detection of superoxide radicals produced in biological membranes, *Biochem. Biophys. Res. Coun.*, 65, 597, 1975.

23. **Misra, H. P. and Fridovich, I.,** The role of superoxide anion in the autooxidation of epinephrine and a simple assay for superoxide dismutase, *J. Biol. Chem.*, 247, 3170, 1972.
24. **Bors, W., Michel, C., Saran, M., and Lengfelder, E.,** The involvement of oxygen radicals during the autoxidation of adrenalin, *Biochim. Biophys. Acta*, 540, 162, 1978.
25. **Rich, P. R. and Bonner, W. D., Jr.,** The sites of superoxide anion generation in higher plant mitochondria, *Arch. Biochem. Biophys.*, 188, 206, 1978.
26. **Boveris, A., Sanchez, R. A., and Beconi, M. T.,** Antimycin and cyanide resistant respiration and superoxide anion production in fresh and aged potato tuber mitochondria, *F.E.B.S. Lett.*, 92, 333, 1978.
27. **Cadenas, E., Boveris, A., Ragan, C. I., and Stoppani, A. O. M.,** Production of superoxide radicals and hydrogen peroxide by NADH-ubiquinone reductase and ubiquinol-cytochrome c reductase from beef heart mitochondria, *Arch. Biochem. Biophys.*, 180, 248, 1977.
28. **Wilson, D. F., Dutton, P. L., Erecinska, M., Lindsay, J. G., and Sato, N.,** Mitochondrial electron transport and energy conservation, *Acc. Chem. Res.*, 5, 234, 1972.
29. **Chance, B. and Williams, G. R.,** The respiratory chain and oxidative phosphorylation, *Adv. Enzymol.*, 17, 65, 1956.
30. **Boveris, A.,** Mitochondrial production of superoxide radical and hydrogen peroxide, *Adv. Exp. Med. Biol.*, 78, 67, 1977.
31. **Forman, H. J. and Kennedy, J.,** Superoxide production and electron transport in mitochondrial oxidation of dihydroorotic acid, *J. Biol. Chem.*, 250, 4322, 1975.
32. **Forman, H. J. and Kennedy, J.,** Dihydroorotate-dependent superoxide production in rat brain and liver. A function of the primary dehydrogenase, *Arch. Biochem. Biophys.*, 173, 219, 1976.
33. **Lazarow, P. B. and De Duve, C.,** A fatty-acyl CoA oxidizing system in rat liver peroxisomes: enhancement by clofibrate, a hypolipidemic drug, *Proc. Natl. Acad. Sci. U.S.A.*, 73, 2043, 1976.
34. **Cadenas, E. and Boveris, A.,** Enhancement of hydrogen peroxide formation by protophores and ionophores in antimycin-supplemented mitochondria, *Biochem. J.*, 188, 31, 1980.
35. **Nohl, H. and Hegner, D.,** Do mitochondria produce oxygen radicals in vivo?, *Eur. J. Biochem.*, 82, 563, 1978.
36. **Boveris, A.,** unpublished results.
37. **Autor, A. P. and Stevens, J. B.,** Regulation of mitochondrial superoxide dismutase by mitochondrially produced superoxide anions and hydrogen peroxide in a higher eukaryote, in *Chemical and Biochemical Aspects of Superoxide and Superoxide Dismutase*, Bannister, J. V. and Hill, H. A. O., Eds., Elsevier, New York, 1980, 104.
38. **Erecinska, M. and Wilson, D. F.,** The effect of antimycin A on cytochromes b_{561} and b_{566} and their relationship to ubiquinone and the iron-sulfur centers S 1 (+ N 2) and S 3, *Arch. Biochem. Biophys.*, 174, 143, 1976.
39. **Kusel, J. P., Boveris, A., and Storey, B. T.,** H_2O_2 production and cytochrome c peroxidase activity in mitochondria isolated from the trypanosomatid hemoflagellate *Crithidia fasciculata*, *Arch. Biochem. Biophys.*, 158, 799, 1973.
40. **Boveris, A.,** Mitochondrial production of hydrogen peroxide in *Saccharomyces cerevisiae*, *Acta Physiol. Lat. Am.*, 26, 303, 1976.
41. **Boveris, A., Docampo, R., Turrens, J. F., and Stoppani, A. O. M.,** Effect of lapachone on superoxide anion and hydrogen peroxide production in *Trypanosoma cruzi*, *Biochem. J.*, 175, 431, 1978.
42. **Rich, P. R., Boveris, A., Bonner, W. D., Jr., and Moore, A. L.,** Hydrogen peroxide generation by the alternate oxidase of higher plants, *Biochem. Biophys. Res. Commun.*, 71, 695, 1976.
43. **Boveris, A., Rich, P. R., and Moore, A. L.,** unpublished results.
44. **Boveris, A., Cadenas, E., and Stoppani, A. O. M.,** Role of ubiquinone in the mitochondrial generation of hydrogen peroxide, *Biochem. J.*, 156, 435, 1976.
45. Erecinska, M., **Chance, B., Wilson, D. F., and Dutton, P. L.,** Aerobic reduction of cytochrome b_{566} in pigeon heart mitochondria, *Proc. Natl. Acad. Sci. U.S.A.*, 69, 50, 1972.
46. **Boveris, A., Erecinska, M., and Wagner, M.,** Reduction kinetics of cytochromes b, *Biochim. Biophys. Acta*, 256, 223, 1972.
47. **Trumpower, B. L. and Simmonds, Z.,** Diminished inhibition of mitochondrial electron transfer from succinate to cytochrome c by thenoyltrifluoroacetone induced by antimycin, *J. Biol. Chem.*, 254, 4608, 1979.
48. **Forman, H. J.,** The role of ubisemiquinone in superoxide production by dihydroorotate dehydrogenase: a proposed mechansim, in *Chemical and Biochemical Aspects of Superoxide and Superoxide Dismutase*, Bannister, J. V. and Hill, H. A. O., Eds., Elsevier, New York, 1980, 76.
49. **Ohnishi, T., Blum, H., and Salerno, J. C.,** Ubisemiquinones in electron transfer systems of mitochondria, in *Function of Quinones in Energy-Conserving Systems*, Trumpower, B. L., Ed., Academic Press, New York, in press, 1980.

50. **Takeshige, K. and Minakami, S.,** NADH- and NADPH-dependent formation of superoxide anions by bovine heart submitochondrial particles and NADH-ubiquinone reductase preparation, *Biochem. J.,* 180, 129, 1979.
51. **Boveris, A. and Turrens, J. F.,** Production of superoxide anion by the NADH-dehydrogenase of mammalian mitochondria, in *Chemical and Biochemical Aspects of Superoxide and Superoxide Dismutase,* Bannister, J. V. and Hill, H. A. O., Eds., Elsevier, New York, 1980, 84.
52. **Turrens, J. F. and Boveris, A.,** Generation of superoxide anion by the NADH dehydrogenase of bovine heart mitochondria, *Biochem. J.,* 191, 421, 1980.
53. **Mitchell, P.,** The protonmotive Q cycle: a general formulation, *F.E.B.S. Lett.,* 59, 137, 1975.
54. **Klingenberg, M. and Rottenberg, H.,** Relation between the gradient of the ATP/ADP ratio and the membrane potential across the mitochondrial membrane, *Eur. J. Biochem.,* 73, 125, 1977.
55. **Dryer, S. E., Dryer, R. L., and Autor, A. P.,** Enhancement of mitochondrial, cyanide-resistant superoxide dismutase in the livers of rats treated with 2,4-dinitrophenol, *J. Biol. Chem.,* 255, 1054, 1980.
56. **Tyler, D. D.,** A protective function of superoxide dismutase during respiratory chain activity, *Biochim. Biophys. Acta,* 396, 335, 1975.
57. **Weisiger, R. A. and Fridovich, I.,** Superoxide dismutase: organelle specificity, *J. Biol. Chem.,* 248, 3582, 1973.
58. **De Duve, C. and Baudhuin, P.,** Peroxisomes (microbodies and related particles), *Physiol. Rev.,* 46, 323, 1966.
59. **Lawrence, R. A. and Burk, R. F.,** Glutathione peroxidase activity in selenium-deficient rat liver, *Biochem. Biophys. Res. Commun.,* 71, 952, 1976.
60. **Burk, R. F., Nishiki, K., Lawrence, R. A., and Chance, B.,** Peroxide removal by selenium-dependent and selenium-independent glutathione peroxidases in hemoglogin-free perfused rat liver, *J. Biol. Chem.,* 253, 43, 1978.
61. **Tyler, D. D.,** Polarographic assay and intracellular distribution of superoxide dismutase in rat liver, *Biochem. J.,* 147, 493, 1975.
62. **Cadenas, E., Boveris, A., and Chance, B.,** Low-level chemiluminescence of bovine heart submitochondrial particles, *Biochem. J.,* 186, 659, 1980.
63. **Cadenas, E., Boveris, A., and Chance, B.,** Hydroperoxide-dependent chemiluminescence of submitochondrial particles and its relationship to superoxide anion and other oxygen radicals, in *Chemical and Biochemical Aspects of Superoxide and Superoxide Dismutase,* Bannister, J. V. and Hill, H. A. O., Eds., Elsevier, New York, 1980, 92.
64. **Cadenas, E., Boveris, A., and Chance, B.,** Low-level chemiluminescence of hydroperoxide-supplemented cytochrome c, *Biochem. J.,* 187, 131, 1980.
65. **Cadenas, E., Boveris, A., and Chance, B.,** Chemiluminescence of lipid vesicles supplemented with cytochrome c and hydroperoxide, *Biochem. J.,* 188, 577, 1980.
66. **Cadenas, E., Boveris, A., and Chance, B.,** Spectral analysis of the low level chemiluminescence of H_2O_2-supplemented ferricytochrome c, *F.E.B.S. Lett.,* 112, 285, 1980.
67. **Lynch, R. E. and Fridovich, I.,** Effects of superoxide on the erythrocyte membrane, *J. Biol. Chem.,* 253, 1838, 1978.
68. **Halliwell, B.,** Superoxide-dependent formation of hydroxyl radicals in the presence of iron chelates. Is it a mechanism for hydroxyl radical production in biochemical systems?, *F.E.B.S. Lett.,* 92, 321, 1978.
69. **McCord, J. M. and Day, E. D., Jr.,** Superoxide-dependent production of hydroxyl radical catalyzed by iron-EDTA complex, *F.E.B.S. Lett.,* 86, 139, 1978.
70. **Boveris, A., Cadenas, E., Reiter, R., Filipkowsky, M., Nakase, Y., and Chance, B.,** Organ chemiluminescence: noninvasive assay for oxidative radical reactions, *Proc. Natl. Acad. Sci. U.S.A.,* 77, 347, 1980.
71. **Lloyd, D., Boveris, A., Reiter, R., Filipkowsky, M., and Chance, B.,** Chemiluminescence of *Acanthamoeba castellanii, Biochem. J.,* 184, 149, 1979.
72. **Boveris, A., Cadenas, E., and Chance, B.,** Ultraweak chemiluminescence: a sensitive assay for oxidative radical reactions, *Fed. Proc.,* 40, 23, 1981.
73. **Cadenas, E., Danielle, R. P., and Chance, B.,** Spectral analysis of the chemiluminescence of macrophages during phagocytosis, *F.E.B.S. Lett.,* in press.

Chapter 3

ASSAYS OF OXYGEN RADICALS: METHODS AND MECHANISMS

Wolf Bors, Manfred Saran, and Christa Michel

TABLE OF CONTENTS

I. INTRODUCTION

One important aspect in studying oxygen activation is the detection of the actual intermediates involved. While a number of reviews have already covered this subject,[1-4] there is continuous impetus both towards finding more suitable assay methods and towards the elucidation of the reaction mechanisms — which in most cases turn out to be complicated and nonspecific.

The main problem in assaying for either $O_2^{\cdot-}$ or $\cdot OH$ is their radical nature, with the quite divergent reactivities of these species in biological systems[2,5,6] rather than in organic solvents.[4,7-9] The short lifetime is somewhat compensated by the wide occurrence (at least of $O_2^{\cdot-}$), yet quite sensitive assay procedures are required due to the competing effect of endogenous substrates acting as scavengers.

Recently another complication arose from the question of whether the "free" oxygen radicals, known from radiolytic studies, are actually identical with the intermediates during biochemical or biological oxygen activation.[10-12] This raises a serious injunction against the specificity of most assays and should be kept in mind during all studies on activated oxygen species.

The literature was compiled up to August 1980, using a computer file and a software program developed in our laboratory.[13]

II. GENERAL ASPECTS

For the discussion of the various methods, we shall leave aside the previously mentioned doubts and shall concentrate on a description of individual procedures. The methods listed are limited to those applied in aqueous solutions, since the chemistry of $O_2^{\cdot-}$ in organic solvents is the subject of a separate chapter.[14]

As a start, some general comments are warranted. First of all, the problem with most compounds is their limited specificity. Detection of $O_2^{\cdot-}$ by *reducing* reactions is very often obscured by direct *an*aerobic reduction of the assay compounds. *Oxidation* of assay compounds by either $O_2^{\cdot-}$ or $\cdot OH$ radicals may result in the same reaction products.

A further handicap, on top of the limited specificity, is the sometimes equivocal radical source, which is used either for control measurements or complementary SOD assays. A case in point is the controversy on the hydroxylating agent of the NADH/phenazine methosulfate system. Here the originally proposed $O_2^{\cdot-}$[15] has now been replaced by $\cdot OH$.[16] The problem of sometimes identical products with $O_2^{\cdot-}$ or $\cdot OH$ is exacerbated by the wide diversity of the respective rate constants with the assay compounds (Table 1). Therefore, even low amounts of $\cdot OH$, being formed subsequent to or as a consequence of $O_2^{\cdot-}$-formation, might supersede an $O_2^{\cdot-}$-reaction.

To overcome the unspecificity of practically all known assay procedures, it became common practice to use the blocking of a certain reaction by supposedly specific inhibitors as additional control. While in the case of superoxide dismutase (SOD) this is a valid and *absolutely* necessary method to test for $O_2^{\cdot-}$, a similar certitude cannot be presumed for scavenging $\cdot OH$ radicals, particularly in light of the doubtful intermediacy of this species in biochemical systems.[11,12]

With the specificity of SOD for $O_2^{\cdot-}$ taken for granted, the mere inhibition of diverse reactions by the enzyme has even been considered sufficient evidence for the involvement of $O_2^{\cdot-}$, e.g.,

1. The breakdown of red blood cells during the autoxidation of dihydroxyfumaric acid[45]
2. The peroxidatic function of catalase during the oxidation of NADH[46]

Table 1
RATE CONSTANTS OF ASSAY COMPOUNDS FOR $O_2^{\cdot-}$ WITH BOTH $O_2^{\cdot-}$ AND $\cdot OH$

Assay compound	$k_{O_2^{\cdot-}}$ $M^{-1}s^{-1}$	pH	Ref., comments	$k_{\cdot OH}$ $M^{-1}s^{-1}$	pH	Ref., comments
Reduction by Superoxide Anions						
Ferri-cytochrome *c*	1.4×10^6	4.7—7.2	—[a]	$1.4—6.1 \times 10^{10}$	5.4—7	—[b]
	1.9×10^5	7.5—9.2				
Nitroblue tetrazolium	5.0×10^4	7.8	25	—		No reaction
chloride	1.0×10^4	10.0	25			
	5.9×10^4	9.8	22			
Tetranitromethane	1.9×10^9	>4.8	26	—		No reaction
	5.0×10^5	<4.8	26			
Dichlorophenol	2.1×10^8	7.0	27	4.0×10^{10}	7.0	27
Indophenol	$1.5—1.7 \times 10^8$		28			
Benzoquinone	9.6×10^8		—[c]	1.2×10^9		29
Oxidation by Superoxide Anions (and Hydroxyl Radicals)						
Epinephrine	5.6×10^4	7.8	—[d]	1.65×10^{10}	1—3	33
	7.0×10^3	9.5	—[d]	2.2×10^{10}	9.2	34
Tiron	1.0×10^7	7.0	35[e]	1.0×10^9	7.0	35
Hydroxylamine	5.9×10^4	7.8	30	9.5×10^9		37
	3.4×10^4	9.5	30			
Sulfite	—		Not known	9.5×10^9	3—6 (HSO_3^-)	38
				5.5×10^9	8.5 (SO_3^{2-})	38
Enzymatic Reactions of Superoxide Anions						
Peroxidase (HRP)	1.6×10^6	7.2, 8.1	39			
Compound I	2.2×10^8	3.8	40			
NADH	27	>4.8	41	"Fast"		41
	1.8×10^5	4.8	42			
NADH..LDH[f]	3.6×10^4	7—9	43			
	1.2×10^6	<4.8	43			
NADH..GDH[f]	2.7×10^7	<4.8	44			

[a] Average of literature data[17-21] obtained by pulse radiolysis, reflecting the effect of pH due to structurally diverse forms of cytochrome *c* (no reaction of the form predominating above pH 9.2;[19] neither does $HO_2\cdot$ react with ferri-cytochrome *c*;[19,20] however, steady-state radiolytic experiments at pH 9.6 arrive at 2.6×10^5 $M^{-1}s^{-1}$;[22] additional effect of ionic strength (at pH 7.9) results in 8.2×10^5 $M^{-1}s^{-1}$ at low vs. 4.4×10^6 $M^{-1}s^{-1}$ at high ionic strength (average of several pulse-radiolytic experiments).[17-19] Steady-state experiments using xanthine/xanthine oxidase as source of $O_2^{\cdot-}$ and SOD as competitor[23] result in too low values of 2.5, 2.1×10^4 $M^{-1}s^{-1}$ at pH 7.0 and 8.5.

[b] Range of values taken from a literature compilation, determined by various methods.[24]

[c] Average of closely agreeing literature values from entry 5.47.[24,28]

[d] Steady-state (autoxidation) experiments with SOD as competitor,[30] taking into account the revised rate constant of SOD + $O_2^{\cdot-}$ at high ionic strength of 3.7×10^9 $M^{-1}s^{-1}$.[31] otherwise the same value of 4.0×10^4 $M^{-1}s^{-1}$ [32] is obtained.

[e] The previously reported values of either 1.5 or 5.0×10^8 $M^{-1}s^{-1}$ [36] are too high.

[f] LDH = lactate dehydrogenase; GDH = glyceraldehyde-3-phosphate dehydrogenase.

3. The vitamin K-dependent carboxylation of prothrombin[47]
4. The strand breakage of covalently closed circular DNA[48]
5. The formation of covalent bonds of catechol metabolites with proteins[49] as explanation of the cytotoxicity of such compounds[50,51]
6. The photodestruction of pheomelanin[52]
7. The autoinactivation of xanthine oxidase[53]

While this list is certainly far from complete, for most of these reactions the sole dependence on inhibition by SOD leaves open the question of whether $O_2^{\cdot-}$ is the ultimate reactant, merely a precursor, or whether $O_2^{\cdot-}$ formation is only an adventitious side reaction.[54-56] A particularly valid example is the first system of the list, the autoxidation of dihydroxyfumarate. In model reactions the partial inhibition by SOD, yet full inhibition by SOD + catalase, respectively, and the abolishment of a lag-phase by H_2O_2, were considered evidence for $\cdot OH$ radicals formed from the precursor $O_2^{\cdot-}$,[57] whereas kinetic studies using the bleaching of *p*-nitrosodimethylaniline made the participation of ·OH radicals unlikely and only supported the existence of an "·OH-analogous oxidant".[58]

In addition, other functions of SOD, aside from $O_2^{\cdot -}$-dismutation, have been argued for[59] and have already been reported,[60] thus leaving some doubts on even this presumably specific reaction. This is even more true for inhibition studies with small Cu-complexes which are suggested to mimic SOD,[61] since quite different reactions than imitation of SOD have been reported.[62-64] Therefore, the argument that the inability of SOD as opposed to small Cu-complexes in inhibiting certain reactions is due to the larger size of the protein or compartimentization of the $O_2^{\cdot -}$-generating site,[65,66] should be reconsidered.

Generally the inhibition of $O_2^{\cdot -}$-dependent reactions is used vice versa as assay for the presence of the enzyme SOD. These methods have been reviewed before[67,68] and are covered in a separate chapter.[69] Since reciprocal application is often merely an adaptation of an existing method, both alternatives may be listed. In some cases, however, only one method should be applied. This is especially the case for both 6-hydroxydopamine[70] and pyrogallol,[71] where rapid autoxidation results in generation of $O_2^{\cdot -}$ in the reaction itself.[71,72]

Inhibition studies with SOD, however, were not limited to specific assay reactions, but have also been applied to evaluate nonspecific oxygen consumption,[73,74] even though this process is more likely to be accelerated by SOD.[75,76] Other "positive" effects of SOD (i.e., acceleration or enhancement of product yields) pertain to H_2O_2 formation,[77,78] the reaction of phenothiazine derivatives with oxygen,[79] and the stimulation of dianisidine oxidation, the latter method being applied in novel SOD assays.[80-82]

A list similar to the quoted one for SOD inhibition with inferred assay specificity can equally be compiled — albeit on a smaller scale — for the identification of ·OH radicals by scavenging methods:

1. Inhibition of alloxan-induced diabetes[83,84]
2. Diminishing of microsomal ethanol oxidation rate[85,86]
3. Blocking of the activation of guanyl cyclase by SOD (aside from catalase or Mn^{2+})[87]
4. Inhibitory effects of scavengers, in addition to SOD + catalase and stimulation by H_2O_2, in inflammatory processes[88]

To evaluate such experiments by merely correlating inhibition efficiency vs. rate constants of the scavengers, sometimes even without knowing some values,[83,84,89,90] is not to be recommended. First, these rate constants may only be valid in aqueous solutions, with changes in a more complex biological system due to micellar effects.[91,92] Second, a homogeneous distribution of such scavengers in in vivo systems, allowing calculations using known concentrations of the reactants, must not be the case.[12,93] It should be emphasized, however, that such doubts are limited to in vivo systems and do not generally extend to in vitro experiments.

Basically, and to summarize these general comments, the use of a battery of inhibitors for several types of reactive oxygen species is always to be preferred. Furthermore, any atttempts to quantitate the data, be they correlation of the stoichiometry of H_2O_2 with $O_2^{\cdot -}$-formation,[94,95] kinetic studies of $O_2^{\cdot -}$-generation[30,96] to obtain reaction rates,[23,97] or even an action spectrum for photodynamic $O_2^{\cdot -}$-production,[98] should be considered with care.

III. INDIVIDUAL ASSAY PROCEDURES — SUPEROXIDE ANIONS

The discussion of the individual assay procedures and the mechanisms pertaining to them shall be done separately for $O_2^{\cdot-}$ and $\cdot OH$ radicals, the occasional overlap notwithstanding. Methods of detecting $O_2^{\cdot-}$ can be characterized as physical, chemical, or enzymatic processes, with sometimes fleeting borderlines.

A. Physical Methods

It has to be emphasized that with the exception of chemiluminescence as an almost chemical method, most of these processes are limited to pure solutions — aqueous or nonaqueous — and are thus not applicable to complex biological systems.

1. Optical Detection

The above limitations are especially true for the determination of $O_2^{\cdot-}$ by its optical absorption at 250 nm in aqueous solution, since most biochemical substances also absorb at this wavelength region. Yet with the accurate molar absorbtivity (ε_{250} = 2350 $M^{-1}cm^{-1}$) known from radiolytic experiments,[99] monitoring of the absorption change of $O_2^{\cdot-}$ can still be used in kinetic and especially pulse-radiolytic studies to determine the reactivity of nonabsorbing substrates with $O_2^{\cdot-}$.[22,100] Such kinetic measurements have also been done with KO_2 as source of $O_2^{\cdot-}$.[101,102] In organic solvents, the molar absorbtivity of 2686 ± 29 $M^{-1}cm^{-1}$ [103] is considerably higher than in water. This fact, in combination with the high yield of $O_2^{\cdot-}$, obtained by solubilization of KO_2 in nonaqueous solution by the addition of crown ether,[104] makes kinetic and product studies of $O_2^{\cdot-}$ far easier than in aqueous solution.

IR spectroscopy as a detection method has only been used at low temperatures, to determine the state of oxygen or superoxide bound to cobalt complexes.[105] Also the IR frequencies of the free $HO_2\cdot$ are known only in frozen solutions.[106]

2. Chemiluminescence Measurements

Determination of superoxide-dependent chemiluminescence (CL) is a rather popular assay method. However, to keep to the systematic arrangement, only the weak direct CL measured by single photon or liquid scintillation counting shall be discussed here, whereas the sensitizer-stimulated CL is part of the assay systems based on oxidation reactions (see below).

The first suggestions that $O_2^{\cdot-}$ might be involved in CL emission came from studies of model[107] or enzymatic systems,[108-110] the latter using the inhibitory effect of SOD. These investigations took already into account that $O_2^{\cdot-}$ has to be the precursor of the light-emitting species.[111] Singlet oxygen, in particular during phagocytosis, was first considered to be this agent,[112] later joined by $O_2^{\cdot-}$-derived $\cdot OH$ radicals.[113-115] Yet the regularly applied assay method to study the mechanism of phagocytosis, i.e., SOD inhibition of CL, may have to be reconsidered, since it has been shown that $O_2^{\cdot-}$ production is *not* correlated to the actual process of phagocytosis.[116,117]

Only a few other CL systems are at least partially inhibited by SOD: the lipoxygenase-dependent reaction[109,119] and the low-level CL of bovine heart submitochondrial particles,[120] whereas the enzyme is without effect in cytochrome *c*-stimulated model systems.[121] In conclusion, SOD-inhibitable CL is a prime example of a rather indirect $O_2^{\cdot-}$-assay alluded to before, i.e., $O_2^{\cdot-}$ in such systems can never be more than a precursor or a fortuitous side-product.

3. EPR Spectroscopy

The direct observation of $O_2^{\cdot-}$ by this method has been achieved only at low temperatures; however, it was one of the early evidences of $O_2^{\cdot-}$-formation by xanthine oxidase,[122,123] in particular after labeling experiments with ^{17}O.[124] The latter approach (i.e., combination of low-temperature EPR and ^{17}O-labeling) has also been used, aside from the previously mentioned IR studies,[105] to detect the presence of $O_2^{\cdot-}$ in cobalt complexes[125,126] (a subject which has biological implications and was reviewed by Nishinaga[127]) or to propose a copper-bound O^- (the dissociated form of ·OH) in laccase.[128] EPR-parameters of $O_2^{\cdot-}$ are also known for organic solvents,[129,130] with the latter report giving important new data on the actual structure of solvated $O_2^{\cdot-}$.

Indirect EPR evidence for the oxygen radicals can be obtained by complex formation with heavy metal ions such as Th^{4+}, V^{5+} for $O_2^{\cdot-}$,[131,132] or Ti^{4+} for ·OH radicals.[133] In the case of $O_2^{\cdot-}$, such complex formation can even be applied for kinetic spectroscopy in pulse radiolysis studies.[134] Far more popular, however, is the indirect EPR method of spin trapping, which is dealt with in a separate chapter.[135] Several limitations of this method, in particular concerning the detection of $O_2^{\cdot-}$, became known only recently: very low reactivities of some spin traps with $O_2^{\cdot-}$ (e.g., 7 $M^{-1}s^{-1}$ for TMPO and 10 M^{-1} s^{-1} for DMPO), aggravated by the occasionally low stability of the $O_2^{\cdot-}$-adduct, and poor resolution of the EPR-resonances of some adduct radicals with $O_2^{\cdot-}$.[136]

4. *Other Physical Methods*

Processes which might also be or have been used to detect $O_2^{\cdot-}$ are electrochemical processes, either conductivity[137] which can be effectively applied in pulse radiolysis[138,139] or cyclic voltammetry.[140,141] Finally, mass spectrometry after reaction of ^{18}O-labeled superoxide has recently been introduced to demonstrate $O_2^{\cdot-}$-specific products. After synthesis of $K^{18}O_2$,[142] it has been applied both in enzymatic studies[143] and in organic chemistry of $O_2^{\cdot-}$.[144] The basic kinetic equations for such isotope studies of oxygen radicals have been delineated quite early.[145]

B. Chemical Methods — Reduction

Superoxide radicals can be detected both by reducing and oxidizing reactions, of which reductions are more often used than oxidation reactions. However, as mentioned earlier, reducing reactions of the assay compounds can similarly be achieved anaerobically without involvement of $O_2^{\cdot-}$.

1. *Cytochrome c*

The most commonly used assay method for $O_2^{\cdot-}$ comprises the reduction of ferricytochrome *c*. The method was introduced together with the identification of SOD[146] and was described in great detail.[68] Quantitation of the data is based on the difference of the molar absorbtivity of reduced minus oxidized cytochrome *c* of 21,000 $M^{-1}cm^{-1}$ at 550 nm.[147,148]

Not only the method, but also the kinetics[17-23] and the mechanism[17,149] of the reaction with $O_2^{\cdot-}$ have been studied extensively. This includes both a quantitation of the reaction of the xanthine/xanthine oxidase system — the most commonly used biochemical $O_2^{\cdot-}$-source — with cytochrome *c* [150,151] and extends to the reoxidation of ferro-cytochrome *c* by $O_2^{\cdot-}$.[152] Basically the reduction process, as studied mainly by pulse radiolysis, consists of an internal electron transfer to the heme group from an unknown primary electron acceptor.[153,154] An additional advantage of the method lies in the possibility of continuously monitoring the reduction,[3,76,155] which at the same time enhances the sensitivity and allows the study of rapid initial bursts of $O_2^{\cdot-}$-formation.[150,156]

The unspecificity of the reaction, i.e., the direct enzymatic reduction without involvement of $O_2^{\cdot-}$,[157-159] can either be overcome by the obligatory control experiments using the inhibition by SOD or, alternatively, by the use of derivatized cytochrome *c*. Both acetylated[160] and succinylated[161] cytochrome *c* have been introduced, showing a dramatic decrease of the enzymatic reduction in favor of reduction by $O_2^{\cdot-}$. In the case of acetylated cytochrome *c*, the pH- and temperature-dependence of the SOD inhibition of the reaction have been investigated.[162]

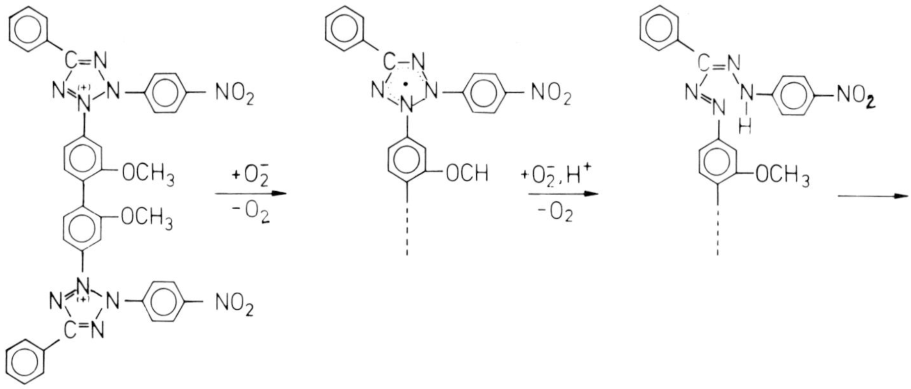

FIGURE 1. Formation of half-reduced formazan by $O_2^{\overline{\cdot}}$.

Another drawback might lie in the reported *reduction* of the ferric iron of oxidized cytochrome *c* by ·OH radicals.[163,164] While only 40 to 50% reduction occurs when based on the yield of ·OH radicals,[164] the much higher reactivity of ·OH compared to $O_2^{\overline{\cdot}}$ (see Table 1) could conceivably result in competing reactions of both radicals. Assuming ·OH radicals to be formed as a consequence of $O_2^{\overline{\cdot}}$, inhibition by SOD alone would therefore not allow differentiation between these two species.

2. Nitroblue Tetrazolium Chloride (NBT)

Tetrazolium salts have been applied in histology for years and are well defined in their reduction behavior.[165] The title compound is most often used in phagocytosis assays[166,167] by determining the production of the water-insoluble diformazan at 530 to 580 nm. The broad absorption peak has a molar absorbtivity at 560 nm of 28,600 $M^{-1}cm^{-1}$.[168]

That $O_2^{\overline{\cdot}}$ might be responsible for the reduction of NBT was first suggested by Rajagopalan and Handler in 1964[169] and substantiated by Miller in 1970.[170] It is therefore conceivable that the previously reported reaction of catecholamines and other hydroxylated aromatic compounds with NBT[171] is equally due to $O_2^{\overline{\cdot}}$-formation by the catechol derivatives. Only very recently the detailed mechanism of NBT reduction was elucidated by pulse radiolysis, and it was shown that the four-electron reduction sequence to yield di-formazan involves a rather complex, strongly pH-dependent mechanism (Figure 1).[172]

In the scheme, only two of the four univalent reduction steps with intermediate tetrazolinyl radical formation and pH-dependent ring opening to the formazan structure are depicted (adapted from Bors et al.[2] and Bielski et al.[172]). Indeed, under limiting reducing conditions, the yellow color of the half-reduced formazan can be observed.

While some improvement of the method has been obtained from working in micellar systems due to an enhanced reactivity with $O_2^{\overline{\cdot}}$ [173,174] — still requiring the control of SOD inhibition — on the whole, application of NBT reduction as assay for $O_2^{\overline{\cdot}}$ is considered obsolete: (1) several reports exist on unspecific reduction of NBT;[157,175-178] (2) the putative $O_2^{\overline{\cdot}}$-source NADH/phenazine methosulfate[179] has been shown to generate ·OH radicals[16,180] and to directly reduce NBT;[181] and (3) NBT reacts with organic oxygen radicals RO· or ROO·,[182,183] in particular in systems in which alternatively $O_2^{\overline{\cdot}}$-formation has been invoked.[184]

3. Tetranitromethane (TNM)

As Table 1 demonstrates, this compound has the fastest recorded rate constant with $O_2^{\overline{\cdot}}$ [26] and was early on applied to assay for $O_2^{\overline{\cdot}}$.[146] The basis of the procedure is the formation of the stable, strongly absorbing nitroform anion (ε_{350} = 15,000 $M^{-1}cm^{-1}$):[26]

$$e^-_{aq}\ (O_2^-) + C(NO_2)_4 \rightarrow C(NO_2)_3^- + NO_2\ (+\ O_2) \qquad (1)$$

However, since the substance is quite noxious and is reduced unspecifically by reaction with organic radicals,[185] it never became a popular assay compound. The best example of its use was the demonstration of the reverse reaction of SOD,[186] aided by the high sensitivity of this substance for $O_2^{\cdot-}$.

4. Electron Acceptors

Dichlorophenol-indophenol (DCIP) represents a number of rather unspecific electron acceptors, of which it is, together with methyl viologen (paraquat), the most commonly used. Introduced as an electron acceptor for the dihydroorotate dehydrogenase reaction,[187] the subsequent suggestion of $O_2^{\cdot-}$ formation by this enzyme[188] led to an isolated application as $O_2^{\cdot-}$ assay in combination with SOD inhibition.[189] The reaction is based on the bleaching of the compound at 600 nm ($\varepsilon_{600} = 21,000\ M^{-1}cm^{-1}$).[190]

Despite its exclusive use as $O_2^{\cdot-}$-assay with this particular enzyme reaction and its unspecific role as electron acceptor,[157] we still propose it as an effective and sensitive competitor for $O_2^{\cdot-}$ reactions due to its high molar absorbtivity and the high rate constant.[27,28] However, its use has to be limited to pure $O_2^{\cdot-}$ systems such as radiolysis, since the unresolved bleaching process can also be initiated by RO· and/or ROO· radicals.[191] A similar qualification pertains to other electron acceptors such as quinones,[28,192] i.e., their use in $O_2^{\cdot-}$ studies should be limited to competition kinetics with known sources of the radicals.

C. Chemical Methods — Oxidation

Except for the oxidation of epinephrine, these reactions are less popular than the previously discussed reducing reactions. Basically all oxidations, excluding hydroxylamine and sulfite, involve aromatic compounds and mainly catechol derivatives. Part of the reason may be the limited reactivity of $O_2^{\cdot-}$ due to the low oxidation potential of −330 mV[6] in aqueous solutions, with catechols among the few likely target molecules. As shall become clear from the discussion of the mechanisms, the earlier mentioned identical products after oxidation by either $O_2^{\cdot-}$ or ·OH radicals is a severe handicap of these reactions.

The general oxidation mechanism of catechols by oxygen radicals is depicted in Figure 2. The scheme demonstrates that up to the semiquinone formation, the mechanism is considered to be identical for all catechols. Further reactions are subject to the respective substituents, e.g., formation of either *o*- or *p*-quinones, rapid vs. slow cyclization,[193,194] and subsequent oxidation to melanins.[195]

1. Epinephrine

This is the best-known catechol compound whose oxidation has been shown to involve $O_2^{\cdot-}$. The original proposal called for the xanthine/xanthine oxidase-induced cooxidation of epinephrine in combination with inhibition studies as an assay for SOD.[196] By omitting the $O_2^{\cdot-}$ source and inhibiting epinephrine oxidation with known SOD preparations, the method could be employed vice versa as an $O_2^{\cdot-}$ assay.[146]

Monitoring of the weak absorption of adrenochrome at 485 nm ($\varepsilon = 4.470\ M^{-1}cm^{-1}$ at pH 7.8)[197] was always a major nuisance. Higher sensitivities were obtained by observing the absorption of adrenochrome at 310 nm ($\varepsilon = 22,000\ M^{-1}cm^{-1}$)[197] rather than at 485 nm,[198,199] in which case care should be taken to limit the observation period to the initial reaction, since oxidation products of adrenochrome absorb at the same wavelength.[198] Also, conference reports on fluorescence detection of the adrenochrome oxidation,[200,201] which have yet to find the way into the literature, offer a higher sensitivity of the method.

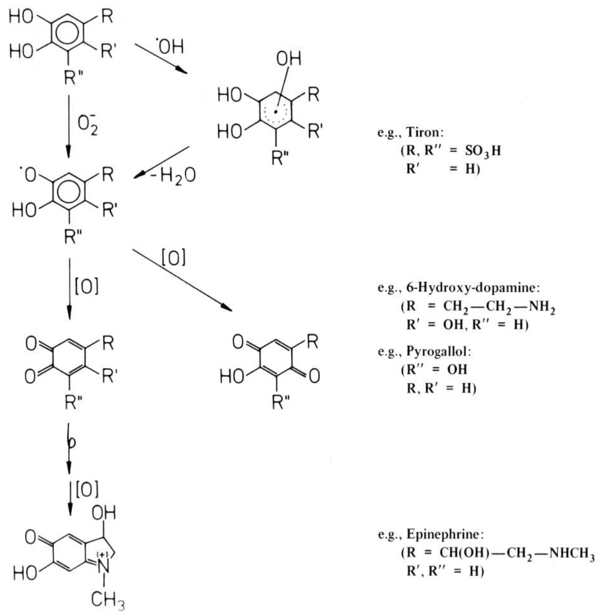

FIGURE 2. General oxidation mechanism of catechols by oxygen radicals.

With numerous reports in the literature on the oxidation mechanism of epinephrine, both radiolytically,[202] chemically,[30,198,203,204] or enzymatically,[204-207] serious deficiencies of the method become evident:

1. Both major intermediates and adrenochrome as end product of the four-electron oxidation sequence are formed both by $O_2^{\cdot-}$ and ·OH radicals.
2. During the autoxidation of epinephrine, $O_2^{\cdot-}$ is both generated and consumed.

Furthermore, even the inhibition by SOD is not a clear-cut scavenging of $O_2^{\cdot-}$, since evidence exists for complex formation of the prosthetic copper with either the catecholamine itself[208] or the semiquinone.[30,209] Alternatively, Winterbourn, et al.[210] suggest that SOD shifts the equilibrium of the reaction of semiquinones with oxygen — one of the endogenous sources of $O_2^{\cdot-}$ in the case of epinephrine — to an *accelerated* quinone formation and thus actually stimulates the reaction!

Taken together, these arguments strongly curtail the unqualified use of the, admittedly, second-most common assay for $O_2^{\cdot-}$, in particular if it is applied as the only method.

2. 1,2-Dihydroxybenzene-3,5-Disulfonate (Tiron)

Compared with the four-electron sequence of epinephrine oxidation to adrenochrome, the one-electron oxidation to the semiquinone of Tiron[211] represents the simplest mechanism for any catechol assay compound for $O_2^{\cdot-}$. Since this semiquinone is quite stable, it is amenable to EPR spectroscopy and has thus been used as a kinetic test for $O_2^{\cdot-}$-formation.[65,209] Originally the catechol compound was only considered to function as a metal chelator.[169,212] After postulating its role in scavenging $O_2^{\cdot-}$,[213] its main application was the use as scavenger of the radical instead of SOD.[65,188,214]

However, even though the rate constant with $O_2^{\cdot-}$ is relatively high (1.0×10^7 $M^{-1}s^{-1}$), it is still only one hundredth of the reaction rate with ·OH radicals and, as has been pointed out before for epinephrine and was shown in Figure 2, both $O_2^{\cdot-}$ and ·OH form the semiquinone.[35] Thus, both competition kinetics and EPR spectroscopy could be liable to misinterpretations.

3. Hydroxylation of Aromatic Compounds

These reactions as assay for $O_2^{\overline{\cdot}}$ should be avoided under all circumstances. While the early studies leave open the nature of the oxidizing species[215,216] and an extensive review favors chemically generated ·OH radicals,[217] the best that was said for $O_2^{\overline{\cdot}}$ is participation in a complex-activated form.[218] In addition, the favorite "$O_2^{\overline{\cdot}}$" source for such hydroxylation reactions, NADH/phenazine methosulfate,[15,219] has been disputed,[16] and the suggested formation of ·OH radicals was substantiated by recent pulse-radiolytic studies.[180]

Some flavoproteins may actually involve $O_2^{\overline{\cdot}}$ during their hydroxylation reaction,[220,221] as originally proposed for flavoprotein dehydrogenases (flavoprotein oxidases or hydroxylases were considered to form only complex-bound $O_2^{\overline{\cdot}}$).[222] Yet even the inhibition by SOD, in particular in complex systems such as isolated chloroplasts,[223] does not allow to differentiate whether $O_2^{\overline{\cdot}}$ is (1) the ultimate hydroxylating agent; (2) participating as a complex-bound species;[218] or (3) merely precursor or cofactor of such reactions.[224-226]

Since it has repeatedly been proposed that ·OH radicals rather than $O_2^{\overline{\cdot}}$ are involved in such hydroxylation reactions,[16,180,224,225] this subject will again be taken up under the aspect of ·OH-assays (see below).

4. Emission of Chemiluminescence by Oxidized Sensitizers

We already discussed the SOD-inhibitable weak direct CL as one of the physical methods of detecting $O_2^{\overline{\cdot}}$-involvement. To enhance such weak CL, e.g., in phagocytosis, the use of fluorescing sensitizers such as luminol or luciferin[227] was considered an advantage. The first application of such compounds in combination with detection of $O_2^{\overline{\cdot}}$ was reported in the early 1960s,[228,229] and eventually led to very sensitive SOD assays,[230,231] as well as a detailed kinetic study of the radical-generating reactions of xanthine/xanthine oxidase.[150]

The mechanism of the luminol reaction was investigated as early as 1928,[232] with chemical,[233,234] photolytic,[235] and radiolytic studies[236,237] even very recently. As reviewed repeatedly,[238-240] the mechanism features the steps in Figure 3.

The basic chemical oxidation mechanism involves alkaline dissociation and oxidative denitrogenation.[233,234] The involvement of oxygen radicals as studied by pulse radiolysis[236] or inhibition by SOD[240,241] is depicted in the lower part. Both mechanisms demonstrate that cyclic peroxide intermediates are *not* obligatory,[233] only the di-cation of aminophthalate is considered to emit light,[234,235] and the reaction has to be induced by hydroxyl anions. The strong dependence of the CL yield on the structure of the luminol derivatives has been noted[242,243] as well as the wide variety of oxidants capable of inducing CL of luminol.[242-245] Since $O_2^{\overline{\cdot}}$ is only indirectly produced in these systems,[240,241] other radicals such as $CO_2^{\overline{\cdot}}$ or $CO_3^{\overline{\cdot}}$ are presently favored to initiate the reaction.[246]

The luminescence mechanism of luciferin from *Pholad dactylus,* a marine mollusc,[247] as a potential $O_2^{\overline{\cdot}}$ assay, has been intensively investigated by Michelson and co-workers.[248] However, Michelson himself[240] critically reviewed the aspect of $O_2^{\overline{\cdot}}$-involvement in the sensitizer-stimulated CL and concluded that even the sensitivity of a CL reaction combined with SOD inhibition might still not be good enough to offer a specific assay for $O_2^{\overline{\cdot}}$.

5. Hydroxylamine

This substance belongs, together with sulfite and the chemiluminescence emitters

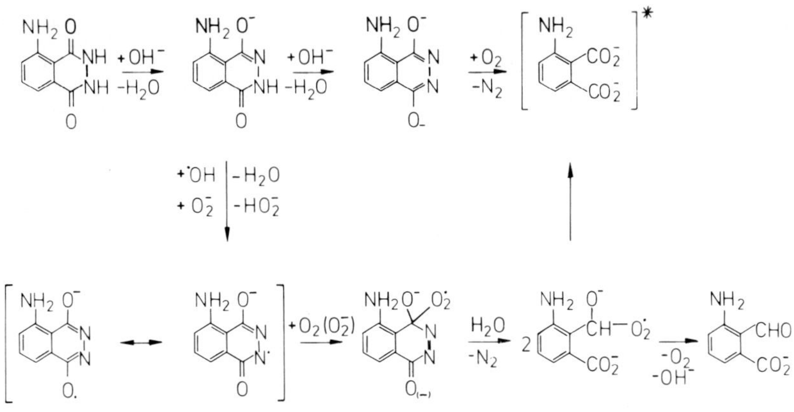

FIGURE 3. Mechanism of luminol oxidation.

luminol and luciferin,[227] to the small group of noncatechol substances which are oxidized by $O_2^{\cdot-}$ in aqueous solution. The assay method is based on the spectrophotometric determination of the oxidation product nitrite as diazo compound after reaction with α-naphthylamine and salicylic acid (ε_{530} = 35,800 $M^{-1}cm^{-1}$).[249]

A study of hydroxylamine oxidation by illuminated isolated chloroplasts[250] subsequently resulted in the identification of an $O_2^{\cdot-}$-dependent reaction,[251] corroborated by parallel studies with another method.[252] Mechanism and kinetics of the oxidation of hydroxylamine were examined both by pulse radiolysis[37,253] and during autoxidation reactions.[30,254,255] The difference in the rate constants of attack of $O_2^{\cdot-}$ and $\cdot OH$ radicals (see Table 1) make a combined attack of both radicals likely,[253] especially in complex systems where $O_2^{\cdot-}$ might function as the precursor of $\cdot OH$ radicals (SOD would therefore completely inhibit the reaction).

$$NH_2OH + \cdot OH \rightarrow \cdot NHOH + H_2O \tag{2}$$

$$NH_2OH + O_2^{\cdot -} \rightarrow \cdot NHOH + HO_2^- \tag{3}$$

$$\cdot NHOH + O_2^{\cdot -} \rightarrow [\cdot NHOH \ldots O_2^{\cdot -}] \tag{4}$$

$$[\cdot NHOH \ldots O_2^{\cdot -}] \rightarrow NO_2^- + H_2O \tag{5}$$

The overall stoichiometry of the reaction is

1. When only superoxide is present:

$$NH_2OH + 2\,O_2^{\cdot -} + H^+ \rightarrow NO_2^- + H_2O_2 + H_2O \tag{6}$$

2. When superoxide and hydroxyl radicals are present:

$$NH_2OH + \cdot OH + O_2^{\cdot -} \rightarrow NO_2^- + 2\,H_2O \tag{7}$$

Since $O_2^{\cdot-}$ seems also to be generated during the autoxidation of hydroxylamine by reaction of oxygen with either contaminating trace metals or the $\cdot NHOH$ radical,[254] SOD inhibition could also be due to indirect interference in an autoxidation process rather than scavenging of external $O_2^{\cdot-}$ to start the reaction.

Despite the disadvantage of the color reaction, (no *continuous* monitoring of $O_2^{\cdot-}$-formation is possible) and some unresolved questions on the mechanism of the reaction with $O_2^{\cdot-}$, the method is simple and sufficiently specific to allow kinetic studies of $O_2^{\cdot-}$-generation in autoxidation processes.[30,254]

6. Sulfite Oxidation

This reaction, monitored by the spectrophotometric detection of sulfite disappearance at 235 nm, was the earliest one to be proposed to involve biochemically generated $O_2^{\cdot-}$.[256] The subsequent corroboration by inhibition with SOD[257] led to the application as assay for $O_2^{\cdot-}$,[188] subsequently modified as SOD assay by polarographic determination of sulfite.[258]

The mechanism of sulfite oxidation is very complicated, since it involves chain processes, initiated by externally generated $O_2^{\cdot-}$, e.g., by metal ions,[259] organic radicals,[260] or xanthine/xanthine oxidase.[261] Chain propagators are in particular $HSO_3\cdot/SO_3^-$ and $\cdot OH$ radicals, interrelated by the following reaction:[259-261]

$$SO_3^{2-} + O_2^{\cdot-} + 3\,H^+ \rightarrow HSO_3^{\cdot} + 2\,\cdot OH \qquad (8)$$

Recently, the probable involvement of organic radicals has been taken into account,[262,263] again limiting the specificity of this assay method.

D. Enzymatic Detection of Superoxide

By implication, enzymatic processes ought to be more specific than mere chemical reactions. Yet the methods involved, observing either the formation of the rather inactive compound III of peroxidases[264,265] or the enzyme-catalyzed oxidation of NADH by $O_2^{\cdot-}$,[266] lack sensitivity. Unfortunately only few enzymes are known whose mechanism might involve $O_2^{\cdot-}$ as cosubstrate as determined by inhibition studies with SOD, e.g., indoleamine dioxygenase,[267] nitropropane dioxygenase,[268] and possibly hydroxybenzoate hydroxylase,[221] and could thus constitute the backbone of highly specific and sensitive enzymatic assays for $O_2^{\cdot-}$. Yet their potential has never been tested and would depend strongly on the ease of isolation and stability of the enzymes in question.

Turning back to the known enzymatic assay procedures, formation of compound III was first used with lactoperoxidase (λ_{max} = 598 nm)[269] and later with myeloperoxidase (λ_{max} = 630 nm).[270] An interesting application was the determination of $O_2^{\cdot-}$ formation in suspensions of rat liver microsomes.[271] In the case of horseradish peroxidase, the mechanism of the reaction with $O_2^{\cdot-}$ was studied by pulse radiolysis,[39,40] and the relative inactivity of the compound III formed with $O_2^{\cdot-}$ was suggested as a "positive staining" assay method for SOD.[80]

The catalysis of NADH oxidation by $O_2^{\cdot-}$ in the presence of lactate dehydrogenase has been extensively investigated and summarized in the paper by Bielski and Chan.[266] The background of the reaction is the very slow uncatalyzed reaction of $O_2^{\cdot-}$ with NADH[41] (ε_{340} = 6,200 $M^{-1}cm^{-1}$)[272] as compared to the strong stimulation after binding NADH to the enzyme, and it should therefore be highly specific. Unfortunately, in a system where several types of oxygen radicals are formed, the fast oxidation of NADH by $\cdot OH$ radicals,[41] — both in the absence and presence of lactate dehydrogenase,[249] — would not allow discrimination between $O_2^{\cdot-}$ and $\cdot OH$ radicals.

The chemistry of free and enzyme-bound NAD· radicals — the obligatory intermediate in one-electron oxidation — was studied by pulse radiolysis,[273] and it was shown that the stability of NAD· is independent of the catalytic activity of several oxidases to oxidize NADH.[44] Thus, only lactate dehydrogenase[266] and glyceraldehyde-3-phosphate dehydrogenase[44] are potential candidates for this assay method, with the former enzyme already used repeatedly.[274,275] Whether a membrane-associated malate dehydrogenase would, in contrast to the soluble enzyme,[44] indeed be capable of oxidizing NADH with $O_2^{\cdot-}$,[276] or whether this implication by analogy is invalid, requires further studies.

Hopefully the list of enzymatic assay methods can be extended in the future, since only such reactions have the potential of high specificity combined with good sensitivity.

IV. INDIVIDUAL ASSAY PROCEDURES — HYDROXYL RADICALS

As pointed out repeatedly, the detection of ·OH radicals in complex biological systems is difficult, since the high reactivity of ·OH radicals entails that (1) only few escape from the site of generation to be detectable; and (2) indirect scavenger experiments, even correlation of rate constants with ·OH with inhibition efficiency, are unreliable. Assay procedures which are based on determination of oxidation products are unspecific since other oxidizing species are similarly capable of yielding such products. In addition, specific hydroxylated products are not exclusively formed and may even result from minor side reactions. Hydroxylation of aromatic compounds is also prone to differential product pattern (isomer distribution) for radiolytically, chemically, or biochemically generated ·OH radicals or analogous species.[217]

Basically, assay methods for ·OH radicals can be arranged in three groups: (1) gas chromatograph (GC) detection of gaseous hydrocarbons after oxidation of sulfur-containing substances; (2) detection of ·OH-specific aromatic or heterocyclic hydroxylation products; (3) indirect bleaching experiments in combination with ·OH-specific scavengers. As a background to the following discussions, a compilation of all known rate constants of ·OH radicals with its assay compounds is given (Table 2). The values were mainly taken from literature compilations,[24,277] with an attempt to select the most accurate ones.

A. GC-Determination of Gaseous Hydrocarbons

1. Methyl-Mercapto-Propanal (Methional)

The formation of ethylene from this derivative of methionine was the first assay to be used in conjunction with ·OH radicals.[282] The radicals were assumed to be formed in the so-called Haber-Weiss reaction, since ethylene formation in the xanthine/xanthine oxidase system could be inhibited synergistically by SOD + catalase, by ·OH-scavengers, and a lag-phase could be abolished by prior addition of H_2O_2. This has now been retracted (see discussion in Koppenol et al.[283]), as it was shown that the uncatalyzed reaction of H_2O_2 with $O_2^{\cdot-}$ is too slow.[284] In addition the oxidizing species generated by xanthine/xanthine oxidase behaves differently than genuine. OH radicals during kinetic studies of the bleaching of *p*-nitroso dimethylaniline (see below).[55]

Since the original report,[282] ethylene formation from methional has been used repeatedly to assay for ·OH radicals. Pulse-radiolytic studies did reveal a high rate constant with ·OH radicals of 8.2×10^9 $M^{-1}s^{-1}$, and no reactivity of $O_2^{\cdot-}$ could be ascertained.[278] Yet, the mechanism of the reaction with ·OH was far more complicated than anticipated,[278,285] including an unknown factor of sulfoxide formation.[286,287] Subsequent analysis of ethylene formation from methionine and derivatives by radiolytically generated oxygen radicals[288] demonstrated that $O_2^{\cdot-}$ does generate ethylene from methional by an unknown mechanism, albeit at a far lower yield than ·OH radicals. This is not too surprising in light of the findings by Pryor and Tang,[289] that methional fragmentation to ethylene can be achieved by practically every conceivable radical. Thus, we advise strongly that the use of methional as an assay compound for ·OH radicals should be discontinued.

2. 4-Keto-Mercapto-Butyric Acid (KMBA)

In view of the unspecificity of methional,[289] the title substance is now increasingly used as an assay compound for ·OH radicals,[89-290] sometimes in combination with

Table 2
RATE CONSTANTS OF ASSAY COMPOUNDS AND SCAVENGERS WITH HYDROXYL RADICALS

Substance	k $M^{-1}s^{-1}$	Ref.[a]
Formation of Gaseous Hydrocarbons		
Methional	8.2×10^9	278
Keto-mercapto butyric acid	—	Not known
Methionine	8.1×10^9	24
Dimethylsulfoxide	5.8×10^9	24
Formation of Specific Products		
Benzene	7.8×10^9	24
p-Cresol	1.2×10^{10}	24
p-Nitrophenol	3.8×10^9	24
Benzoate	5.7×10^9	24
Salicylate	1.2×10^{10}	24
o-Hydroxybenzoic acid	8.7×10^9	24
p-Coumaric acid	—	Not known
Phenylalanine	6.6×10^9	24
Thymine	4.7×10^9	24
Thymidine	5.3×10^9	24
Observation of Bleaching Kinetics		
p-Nitrosodimethylaniline	1.25×10^{10}	24
Crocin	3.3×10^{10}	279
Tryptophan	1.4×10^{10}	24
Scavengers of Hydroxyl Radicals		
Ethanol[b]	1.8×10^9	24
Glycerol	1.6×10^9	24
Mannitol	1.0×10^9	27
Pentaerythritol	3.2×10^9	24
Dimethylurea	2.7×10^9	280
Thiourea[c]	4.7×10^9	277
Formate	2.8×10^9	24

[a] Values, if possible, taken from literature compilations[24,277] with an attempt to select the most accurate ones.
[b] Average of numerous values.
[c] Reacts with hydrogen peroxide as potential precursor of $\cdot$OH radicals.[281]

methional and other assay methods amenable to GC analysis,[291,292] as well as with control inhibition by $\cdot$OH-scavengers.[291] A lack of kinetic and mechanistic data — it probably reacts similarly to other thioethers with potential sulfoxide formation[285-287] — argues for caution in the use of this substance.

3. Methionine

A biologically more relevant assay for $\cdot$OH radicals was proposed by Konze and Elstner,[293] who investigated the pyridoxal phosphate-stimulated ethylene formation from methionine after illumination of isolated chloroplasts. First doubts on the involvement of genuine $\cdot$OH radicals[294] were later corroborated by radiolytic studies:[288]

in the presence of biochemical oxidants, pyridoxal phosphate strongly stimulates ethylene production,[293] whereas ·OH radicals themselves enhance the fragmentation of pyridoxal phosphate-supplemented methionine only by a factor of 1.4.

As was the case for methional, the mechanism of the oxidation of methionine by radiolytically produced ·OH radicals is quite complicated and *different* from simple thioethers.[285-287] After the initial oxidation of the sulfur atom, the presence of the amino group causes a pH-dependent intramolecular oxidation of this moiety, and at higher concentrations of methionine formation of a bimolecular radical cation at the sulfur bridgenly inferred for methional,[286,287] has actually been demonstrated for methionine after oxidation in a biochemical reaction.[259]

To conclude, none of the assay systems based on ethylene formation may actually detect ·OH radicals. Either other radical species are similarly capable of initiating the reaction[289] or, as seems to be the case with methionine + pyridoxal phosphate,[288,293] may react far more specifically than ·OH radicals. Logical control experiments by combining these assays with scavenging studies necessarily assume that these scavengers are specific for ·OH radicals, a point which will be taken up later.

4. *Dimethylsulfoxide (DMSO)*

The GC-analysis of methane after oxidation of DMSO has been introduced as still another method of detecting ·OH radicals, either combined with ethylene-evolving reactions[291,292] or as a single assay.[297] In the latter case, detailed inhibition studies were combined with labeling experiments, which demonstrated that CD_4 is indeed derived from deuterated DMSO.

However, similar to the case of the ·OH-induced oxidation of other sulfur-containing substances, the reaction mechanism is quite complicated. As studied by EPR with Fenton-derived ·OH radicals[298] or by pulse radiolysis,[299,300] methane is probably only a minor product, already evident from the low yield of methane as compared to ethylene-generating systems.[301] Indeed, formaldehyde has recently been identified as product of DMSO after reaction with the putative ·OH-source xanthine/xanthine oxidase.[301] While mannitol partially inhibited the reaction, the fact that the oxidation is initiated by hydrogen abstraction makes it likely that other oxidizing species with similar capacity as ·OH might be responsible.

B. Detection of Specific Products

Hydroxylations of aromatic compounds as demonstrably most specific assay methods for ·OH radicals have previously been discussed under the aspect of potential involvement of $O_2^{\cdot -}$; yet if participating, it was considered that further activation by metal complexes was required.[218] Early mechanistic studies generally favored ·OH radicals,[215-217] however, with a strong dependence of the hydroxylated isomers on the radical-generating systems. Hydroxylation of basic aromatic substances such as benzene and related hydrocarbons by ·OH radicals were studied in great detail both by radiolysis[302-305] and Fenton chemistry.[306,307]

Benzene itself, however, is an unlucky choice as assay compound. Phenol, while easily detectable by high pressure liquid chromatography (HPLC)[308] or gas liquid chromatography (GLC)[309], is not the only or even the major product. During Fenton reactions it is formed with various ratios of biphenyl,[307] whereas radiation in the presence of oxygen results in ring opening to mucon- or hydroxymuconaldehyde.[310] The complexity of the hydroxylation mechanism of naphthalene and the multitude of products after reaction with dioxygen activated by ferrous iron (synonymous to a ferryl/perferryl reaction) was demonstrated by Lindsay Smith et al.[311] While they compared the yield of mono- and dihydroxylated products for various chemical hydroxylation systems, unfortunately experiments with radiolytically generated ·OH radicals have only been performed by pulse radiolysis and not by product analysis.[312]

The hydroxylation of phenol derivatives, *p*-cresol and *p*-nitrophenol, to catechol compounds was applied by Goscin and Fridovich[57] in the dihydroxyfumarate/ferrous sulfate/oxygen system. All nitrophenol isomers have been studied in great detail by pulse and steady-state radiolysis, in combination with conductivity, EPR-measurements, and product analysis.[313] An electrophilic attack of ·OH does indeed preferentially occur at the *o*- and *p*-position to the hydroxyl group, resulting eventually in 4-nitrocatechol from attack at the *o*-position. Addition of ·OH in *p*-position, on the other hand, yields nitrite and *p*-quinone and hydroquinone from intermediary ·NHOH elimination and *p*-semiquinone formation.[313]

A better assay substance is benzoate, for which three analytic techniques have thus far been applied: (1) spectrophotometric determination of salicylate after paper-chromatographic separation;[314] (2) the highly sensitive fluorimetric detection of salicylate;[315] and (3) CO_2 detection after decarboxylation of benzoate,[316,317] again demonstrating that hydroxylation is not the only reaction. Due to its high rate constant with ·OH radicals (5.7×10^9 $M^{-1}s^{-1}$),[303] however, benzoate was mainly applied as scavenger of these radicals.

Hydroxylated aromatic acids — *p*-hydroxybenzoic acid,[16,318] *o*-hydroxybenzoic acid (= salicylic acid),[16,224,318] or 4′-hydroxy-*β*-phenylpropenoic acid (= *p*-coumaric acid)[318,319] — have exclusively been used by Halliwell and co-workers. A comparison showed that *p*-coumaric acid was the optimal substrate for hydroxylation by biochemical systems.[318] Unfortunately no rate constant of this substance with ·OH radicals is known, which would probably allow correlation of this hydroxylation efficiency with its reactivity with the radicals.

A very recent product study of cumene demonstrated the full potential of such methods.[320] While ·OH radicals from photolyzed H_2O_2 preferentially form ring hydroxylation products (hydroxyisopropyl benzenes), organic alkoxy (RO·) and peroxy (ROO·) radicals almost exclusively oxidize the side chain with the formation of formaldehyde, cumene hydroperoxide, cumyl alcohol, and acetophenone via a chain reaction. This allowed the authors, by comparing the ratio of side-to-ring oxidation, to quantitate the formation of either ·OH or ROO· radicals in aquatic systems.

The fact that different hydroxylating agents result in different isomer distribution has been noted before. To actually use this product pattern as an assay method for ·OH radicals was already suggested in 1962,[321] with a recent addition by the use of phenylalanine hydroxylation to tyrosine isomers.[322] However, the view that isomer distribution would allow conclusions as to the nature of the hydroxylating species has been contested,[185] with the ensuing controversy as to whether anisol (methoxybenzene) is hydroxylated by radicals[185] or a catechol-iron-oxygen complex (the Hamilton system).[323,324]

The only nonbenzene-derived substance whose hydroxylation was suggested as an assay compound for ·OH radicals is thymine or thymidine.[325] An investigation of the oxidation products with ·OH or singlet oxygen — $O_2^{\cdot-}$ does not react — demonstrated that the 5,6-dihydroxy-5,6-dihydro derivatives are specific for ·OH radicals. The method has already been applied in another laboratory.[326]

C. Bleaching Experiments

The chemistry of such reactions does not require detailed elucidation, as the interest is limited to the kinetics of the disappearance of a certain absorption band. Both oxidizing and reducing processes can be responsible. However, in combination with *specific* scavengers this assay method may become a powerful tool and need not be limited to kinetic studies and/or determination of rate constants.

1. *p*-Nitrosodimethylaniline (*p*-NDA, RNO)

Extensive use of the bleaching of this compound at 440 nm (ε = 34,400 $M^{-1}cm^{-1}$)[327]

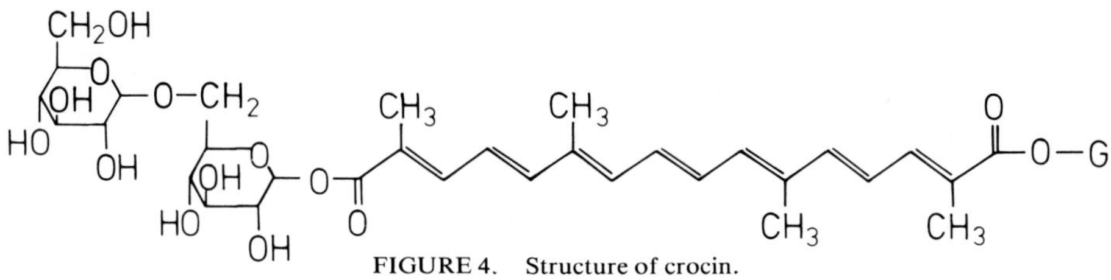

FIGURE 4. Structure of crocin.

in radiation and photochemistry to determine rate constants with ·OH radicals (see Farhataziz and Ross, entry 3.582[24]) constitutes the prime example of this approach. After the initial suggestion,[328] the compound was studied extensively both from kinetic and mechanistic viewpoints.[327,329-331]

The bleaching of *p*-NDA, however, cannot be recommended as assay method in complex biochemical systems: (1) as a reactive nitroso compound, *p*-NDA adds to polyunsaturated fatty acids to form stable N-oxyl radicals;[332] (2) it can equally be bleached by reducing processes;[333] and (3) in addition to dihydroxyfumarate- or xanthine/xanthine oxidase-derived oxidizing species,[55] reacts with more clearly defined oxidants such as singlet oxygen-derived peroxides[334] or organic oxygen radicals.[100]

2. *Crocin*

This compound, which is presently the main focus of our own investigations, was first isolated from saffron.[335] It is one of the rare water-soluble carotenoids due to the two ester-bound gentobiose molecules at the dicarboxylic acid crocetin (Figure 4).

The bleaching at 440 nm, where the substance has an extremely high molar absorptivity of 133,500 $M^{-1}cm^{-1}$,[332] was repeatedly used during cooxidation studies of the reaction of lipoxygenase with its substrates,[336,337] and, during the reaction of other oxidizing enzymes.[338-340]

An application in $O_2^{\cdot-}$-generating systems[341] led to the impression that it might be used as a specific $O_2^{\cdot-}$-detecting substance.[4] Yet subsequent pulse-radiolytic investigations revealed that crocin is not attacked by $O_2^{\cdot-}$ at all,[279] and demonstrated a superior bleaching efficiency by alkoxy radicals as compared to ·OH radicals.[342,343] The results with crocin demonstrate amply that bleaching experiments are by nature quite unspecific and require obligatory control experiments with specific inhibitors. As compared to *p*-NDA, crocin offers the advantage that the substance itself is chemically unreactive and is bleached only by radical reactions. Also the formation of rather stable products from either ·OH or RO· attack can be observed by HPLC.[342]

3. *Tryptophan*

The bleaching of this substance at 278 nm was used in another attempt to discover ·OH radicals formed in the Haber-Weiss reaction.[344] Unfortunately, pulse radiolysis experiments combined with kinetic spectroscopy, despite detailed mechanistic studies, have been limited to wavelengths above 300 nm[345,346] and, therefore, absorption of potential products or reformation of tryptophan[346] would not be accounted for by mere observation of the bleaching at the absorption maximum. The same qualification, incidently, holds true for the application of the bleaching of *p*-NDA in steady-state as compared to pulse radiolytic competition experiments.[329]

V. CONCLUSIONS

A comparison of all known and listed assay procedures for $O_2^{\cdot-}$ demonstrated the preferential use of cytochrome *c* reduction, with epinephrine oxidation, nitroblue tetrazolium reduction, and chemiluminescence methods following. All other methods have been used less frequently, in some cases only once.

The doubts pertaining to the individual procedures have been voiced before,[4,160] aside from the present discussion. All methods lack specificity if used without control inhibition experiments with SOD, yet such studies still would only demonstrate the *formation* of $O_2^{\cdot-}$ but not its actual involvement as the active reactant. Only reduction of derivatized cytochrome *c* — since reduction by $O_2^{\cdot-}$ is far less diminished than enzymatic reduction[161] — and some enzymatic assays, hopefully to be expanded in the near future, offer a relative specificity for $O_2^{\cdot-}$.

For ·OH-specific reactions, inhibition studies may only become pertinent, if one can unequivocally demonstrate the specificity of scavengers for ·OH radicals. Polyhydroxylated compounds such as mannitol or pentaerythritol exhibit high rate constants with ·OH radicals (see Table 2), while their reactivity with organic radicals (RO·) is very much lower.[342,347] The desired specificity can, however, only be established when reactions of the potential biological "·OH-mimicking agents",[12] ferryl or perferryl species, with these substances can be neglected for the given system. Then we shall be able to prove or disprove once and for all the biochemical or biological generation of genuine ·OH radicals. Therefore, mere comparison of inhibition efficiencies with, sometimes assumed, rate constants of ·OH radicals[84,85,90,91] should never be the basis for identification of ·OH radicals.

Momentarily, approaches which compare several assay procedures are most reliable,[56,348,349] even though they mainly demonstrate the quoted lack of specificity[157] or even the inapplicability of some assays under certain conditions.[74,350]

VI. REACTIVITY OF $O_2^{\cdot-}$/·OH-MIMICKING SPECIES

It was stated in the introduction that the role of free $O_2^{\cdot-}$ or ·OH radicals, as they are known from radiolytic studies, has been questioned recently for biochemical or biological systems,[10-12] as many alternative species — alkoxy, peroxy radicals, ferryl/perferryl complexes, "Fenton-type oxidants", etc. — presumably show similar chemical behavior.[12]

To cover this extended concept we thought it advisable to add this section to the review of presently known assay procedures for $O_2^{\cdot-}$ and/or ·OH radicals. While the most specific of these methods may still be used to verify the occurrence of the "free" oxygen radicals, the term "specificity", however, has to be emphasized, since a major problem is the proof that oxygen species other than the inorganic oxygen radicals $O_2^{\cdot-}$ and ·OH do not participate in such assay reactions. A case in point is the identity of the "oxygen-centered radical" responsible for the self-deactivation of prostaglandin synthetase.[351] Doubts on the involvement of free ·OH radicals[352] have now led to the postulation of alkoxy radicals derived from fatty acids (arachidonic acid).[353]

At the present, nothing is known whether $O_2^{\cdot-}$ bound in a metal complex can equally be scavenged by SOD, or whether an equilibrium between free and bound $O_2^{\cdot-}$ might facilitate such a scavenging. This is even more difficult to predict for ·OH-analogs, occurring during Fenton-type or metal-catalyzed reactions.[10,354,355] The only evidence thus far for different behavior of these species was obtained by kinetic studies,[55,356] and by different isomer distribution, extending from model hydroxylating systems all the way to cytochrome P450.[216,354]

Such differentiations may be somewhat easier for alkoxy radicals (RO·), in which case only hydrogen abstraction but not addition to double bonds would result in identical products as with ·OH. This has been demonstrated excellently by the product analysis of cumene oxidation by either ·OH radicals or RO·/ROO· species.[320]

Since the rate constants for H abstraction by either ·OH or RO· are quite different, a selection could also be made by kinetic methods.[100,342,343] Crocin, which was used in these experiments, showed a higher bleaching yield with RO· than with ·OH radicals, both generated by pulse radiolysis,[100] even though the attack of ·OH radicals is certainly faster. We attribute this surprising behavior to the fact that the gentobiose side chains of crocin (see Figure 4), similar to the polyhydroxylated compounds mannitol or pentaerythritol, may act as transparent (i.e., not observable) "radical sinks" for ·OH but far less so for RO· radicals.

There are no known reports on specific RO·- or ROO·-assays; rather, isolated investigations merely demonstrate the reactivity of supposedly specific $O_2^{\cdot -}$-assays with these species, a fact which was mentioned often during the course of this article. Methods, specific for organic oxygen radicals, may have to be deduced eventually from reviews on these organic oxygen radicals themselves.[357-359] Altogether we hope that a clear picture will emerge when more experimental data and mechanistic details on the reactions of the various types of activated oxygen species become available in the future.

REFERENCES

1. **McCord, J. M., Beauchamp, C. O., Goscin, S., Misra, H. P., and Fridovich, I.,** Superoxide and superoxide dismutase, in *Oxidases and Related Redox Systems,* Vol. 1, King, T. E., Mason, H. S., and Morrison, M., Eds., University Park Press, Baltimore, 1973, 51.
2. **Bors, W., Saran, M., Lengfelder, E., Spöttl, R., and Michel, C.,** The relevance of the superoxide anion radical in biological systems, *Curr. Top. Radiat. Res. Q.,* 9, 247, 1974.
3. **Michelson, A. M.,** Chemical production of superoxide anions by reaction between riboflavin, oxygen, and reduced nicotinamide adenine dinucleotide. Specificity of tests for O_2^-, in *Superoxide and Superoxide and Superoxide Dismutases,* Michelson, A. M., McCord, J. M., and Fridovich, I., Eds., Academic Press, London, 1977, 257.
4. **Bors, W., Saran, M., Lengfelder, E., Michel, C., Fuchs, C., and Frenzel, C.,** Detection of oxygen radicals in biological reactions, *Photochem. Photobiol.,* 28, 629, 1978.
5. **Fridovich, I.,** Oxygen radicals, hydrogen peroxide, and oxygen toxicity, in *Free Radicals in Biology,* Vol. 1, Pryor, W. A., Ed., Academic Press, New York, 1976, 239.
6. **Fee, J. A. and Valentine, J. S.,** Chemical and physical properties of superoxide, in *Superoxide and Superoxide Dismutases,* Michelson, A. M., McCord, J. M., and Fridovich, I., Eds., Academic Press, London, 1977, 19.
7. **Lee-Ruff, E.,** The organic chemistry of superoxide, *Chem. Soc. Rev.,* 6, 195, 1977.
8. **Wilshire, J. and Sawyer, D. T.,** Redox chemistry of dioxygen species, *Acc. Chem. Res.,* 12, 105, 1979.
9. **Sawyer, D. T. and Gibian, M. J.,** The chemistry of superoxide ion, *Tetrahedron,* 35, 1471, 1979.
10. **Hill, H. A. O.,** The chemistry of dioxygen and its reduction products, in *Oxygen Free Radicals and Tissue Damage, Ciba Found. Symp.,* 65 (new series), 1979, 5.
11. **Willson, R. L.,** Hydroxyl radicals and biological damage in vitro: what relevance in vivo?, in *Oxygen Free Radicals and Tissue Damage, Ciba Found. Symp. 65* (new series), 1979, 19.
12. **Bors, W., Saran, M., and Czapski, G.,** Nature of intermediates during biological oxygen activation, in *Biological and Clinical Aspects of Superoxide and Superoxide Dismutase,* Bannister, W. H. and Bannister, J. V., Eds., Elsevier, Amsterdam, 1980, 1.
13. **Saran, M., Erben-Russ, M., and Bors, W.,** Literature storage and retrieval system using a table-top computer, *Inf. Proc. Manage.,* 17, 317, 1981.
14. **Frimer, A. A.,** The organic chemistry of superoxide anion radical, in *Superoxide Dismutase,* Vol. II, Oberley, L. W., Ed., CRC Press, Boca Raton, Fla., 1982, chap. 5.
15. **Kumar, A. A., Rao, B. S. S. R., Vaidyanathan, C. S., and Rao, N. A.,** Mechanism of aromatic hydroxylation: involvement of perhydroxyl radicals and semiquinones in hydroxylation of aromatic amino acids by a model system, *Indian J. Biochem. Biophys.,* 12, 163, 1975.
16. **Halliwell, B.,** Hydroxylation of aromatic compounds by NADH and phenazine methosulfate requires hydrogen peroxide and hydroxyl radicals, but not superoxide, *Biochem. J.,* 167, 317, 1977.
17. **Koppenol, W. H., van Buuren, K. J. H., Butler, J., and Braams, R.,** The kinetics of the reduction of cytochrome *c* by the superoxide anion radical, *Biochim. Biophys. Acta,* 449, 157, 1976.

18. **Simic, M. G., Taub, I. A., Tocci, J., and Hurwitz, P. A.**, Free radical reduction of ferri-cytochrome c, *Biochem. Biophys. Res. Comm.*, 62, 161, 1975.
19. **Butler, J., Jayson, G. G., and Swallow, A. J.**, The reaction between the superoxide anion radical and cytochrome c, *Biochim. Biophys. Acta.*, 408, 215, 1975.
20. **Land, E. J. and Swallow, A. J.**, One-electron reactions in biochemical systems as studied by pulse radiolysis. V. Cytochrome c, *Arch. Biochem. Biophys.*, 145, 365, 1971.
21. **Seki, H., Ilan, Y. A., Ilan, Y., and Stein, G.**, Reactions of the ferri-ferrocytochrome *c* system with superoxide/oxygen and CO_2^-/CO_2 studied by fast pulse radiolysis, *Biochim. Biophys. Acta*, 440, 573, 1976.
22. **Bielski, B. H. J. and Richter, H. W.**, A study of the superoxide radical chemistry by stopped-flow radiolysis and radiation induced oxygen consumption, *J. Am. Chem. Soc.*, 99, 3019, 1977.
23. **Sawada, Y. and Yamazaki, I.**, One-electron transfer reactions in biochemical systems. VIII. Kinetic study of superoxide dismutase, *Biochim. Biophys. Acta*, 327, 257, 1973.
24. **Farhataziz, and Ross, A. B.**, Selected specific rates of reactions of transients from water in aqueous solution. III. Hydroxyl radical and perhydroxyl radical and their radical ions, NSRDS-NBS Rep. 59, National Bureau of Standards, U.S. Department of Commerce, Washington, D.C., 1977.
25. **Halliwell, B.**, An atttempt to demonstrate a reaction between superoxide and hydrogen peroxide, *F.E.B.S. Lett.*, 72, 8, 1976.
26. **Rabani, J., Mulac, W. A., and Matheson, M. S.**, The pulse radiolysis of aqueous tetranitromethane. I. Rate constants and extinction coefficient of e_{aq}^- II. Oxygenated solutions, *J. Phys. Chem.*, 69, 53, 1965.
27. **Bors, W., Saran, M., and Michel, C.**, Pulse-radiolytic investigations of catechols and catecholamines. III. Adrenalone, *J. Phys. Chem.*, 83, 2447, 1979.
28. **Greenstock, C. L. and Ruddock, G. W.**, Determination of superoxide radical anion reaction rates using pulse radiolysis, *Int. J. Radiat. Phys. Chem.*, 8, 367, 1976.
29. **Adams, G. E. and Michael, B. D.**, Pulse radiolysis of benzoquinone and hydroquinone. Semiquinone formation by water elimination from trihydroxycyclohexadienyl radicals, *Trans. Faraday Soc.*, 63, 1171, 1967.
30. **Bors, W., Michel, C., Saran, M., and Lengfelder, E.**, Kinetic investigations of the autoxidation of adrenalin, *Z. Naturforsch.*, 33c, 891, 1978.
31. **McAdam, M. E.**, A consideration of the effects of added solutes on the activity of bovine superoxide dismutase, *Biochem. J.*, 161, 697, 1977.
32. **Asada, K. and Kanematsu, S.**, Reactivity of thiols with superoxide radicals, *Agric. Biol. Chem.*, 40, 1891, 1976.
33. **Gohn, M., Getoff, N., and Bjergbakke, E.**, Pulse radiolysis of adrenaline in acid aqueous solutions, *Int. J. Radiat. Phys. Chem.*, 8, 533, 1976.
34. **Gohn, M., Getoff, N., and Bjergbakke, E.**, Radiation chemistry studies on chemotherapeutic agents. I. Pulse radiolysis of adrenalin in aqueous solutions, *J.C.S. Faraday II*, 73, 406, 1977.
35. **Bors, W., Saran, M., and Michel, C.**, Pulse-radiolytic investigations of catechols and catecholamines. II. Reactions of Tiron with oxygen radical species, *Biochim. Biophys. Acta*, 582, 537, 1979.
36. **Greenstock, C. L. and Miller, R. W.**, The oxidation of Tiron by superoxide anion. Kinetics of the reaction in aqueous solution and in chloroplasts, *Biochim. Biophys. Acta*, 396, 11, 1975.
37. **Simic, M. and Hayon, E.**, Intermediates produced from the one-electron oxdation and reduction of hydroxylamines. Acid-base properties of the amino, hydroxyamino, and methoxyamino radicals, *J. Am. Chem. Soc.*, 93, 5982, 1971.
38. **Adams, G. E. and Boag, J. W.**, Spectroscopic studies of reactions of the ·OH radical, *Proc. Chem. Soc.*, 112, 1964.
39. **Bielski, B. H. J., Comstock, D. A., Haber, A., and Chan, P. C.**, Study of peroxidase mechanisms by pulse radiolysis. II. Reaction of horseradish peroxidase compound I with O_2^-, *Biochim. Biophys. Acta*, 350, 113, 1974.
40. **Bielski, B. H. J. and Gebicki, J. M.**, Study of peroxidase mechanisms by pulse radiolysis. III. The rate of reaction of O_2^- and $HO_2\cdot$ radicals with horseradish peroxidase compound I, *Biochim. Biophys. Acta*, 364, 233, 1974.
41. **Land, E. J. and Swallow, A. J.**, One-electron reactions in biochemical systems as studied by pulse radiolysis. IV. Oxidation of NADH, *Biochim. Biophys. Acta*, 234, 34, 1971.
42. **Nadezhdin, A. and Dunford, H. B.**, Oxidation of NADH by hydroperoxyl radical. A flash photolysis study, *J. Phys. Chem.*, 83, 1957, 1979.
43. **Bielski, B. H. J. and Chan, P. C.**, Kinetic study by pulse radiolysis of the lactate dehydrogenase-catalyzed chain oxidation of NADH by $HO_2\cdot$ and O_2^- radicals, *J. Biol. Chem.*, 250, 318, 1975.
44. **Chan, P. C. and Bielski, B. H. J.**, Glyceraldehyde-3-phosphate dehydrogenase-catalyzed chain oxidation of reduced nicotinamide adenine dinucleotide by perhydroxyl radicals, *J. Biol. Chem.*, 255, 874, 1980.

45. **Goldberg, B. and Stern, A.**, The role of the superoxide anion as a toxic species in the erythrocyte, *Arch. Biochem. Biophys.*, 178, 218, 1977.
46. **Halliwell, B.**, Generation of the superoxide radical during the peroxidatic oxidation of NADH by catalase at acid pH values, *F.E.B.S. Lett.*, 80, 291, 1977.
47. **Esnouf, M. P., Green, M. R., Hill, H. A. O., Irvine, G. B., and Walter, S. J.**, Dioxygen and the vitamin K-dependent synthesis of prothrombin, in *Oxygen Free Radicals and Tissue Damage, Ciba Found. Symp.*, 65 (new series), 1979, 187.
48. **Lown, J. W. and Weir, G.**, Studies related to antitumor antibiotics. XIV. Reactions of mitomycin B with DNA, *Can. J. Biochem.*, 56, 296, 1978.
49. **Hesse, S., Mezger, M., and Wolff, T.**, Activation of ^{14}C-chlorobiphenyls to protein-binding metabolites by rat liver microsomes, *Chem. Biol. Interact.*, 20, 355, 1978.
50. **Rotman, A., Daly, J. W., and Creveling, C. R.**, Oxygen-dependent reaction of 6-hydroxydopamine, 5,6-dihydroxytryptamine, and related compounds with proteins in vitro: a model for cytotoxicity, *Mol. Pharmacol.*, 12, 887, 1976.
51. **Dybing, E., Nelson, S. D., Mitchell, J. R., Sasame, H. A., and Gillette, J. R.**, Oxidation of α-methyldopa and other catechols by cytochrome P-450-generated superoxide anion: possible mechamism of methyldopa hepatitis, *Mol. Pharmacol.*, 12, 911, 1976.
52. **Chedekel, M. R., Smith, S. K., Post, P. W., Pokora, A., and Vessell, D. L.**, Photodestruction of pheomelanin: role of oxygen, *Proc. Natl. Acad. Sci. U.S.A.*, 75, 5395, 1978.
53. **Lynch, R. E. and Fridovich, I.**, Autoinactivation of xanthine oxidase. The role of superoxide radical and hydrogen peroxide, *Biochim. Biophys. Acta*, 571, 195, 1979.
54. **Sligar, S. G., Lipscomb, J. D., DeBrunner, P. G., and Gunsalus, I. C.**, Superoxide anion production by the autoxidation of cytochrome P-450_{cam}, *Biochem. Biophys. Res. Comm.*, 61, 290, 1974.
55. **Nilsson, R.**, Investigation of mixed function oxidation by means of chemiluminescence, *Biochim. Biophys. Acta*, 184, 237, 1969.
56. **Kameda, K., Ono, T., and Imai, Y.**, Participation of superoxide, hydrogen peroxide and hydroxyl radicals in NADPH-cytochrome P-450 reductase-catalyzed peroxidation of methyl linolenate, *Biochim. Biophys. Acta*, 572, 77, 1979.
57. **Goscin, S. A. and Fridovich, I.**, The role of superoxide radical in a non-enzymatic hydroxylation, *Arch. Biochem. Biophys.*, 153, 778, 1972.
58. **Bors, W., Michel, C., and Saran, M.**, On the nature of biochemically generated hydroxyl radicals. Studies using the bleaching of *p*-nitrosodimethylaniline as a direct assay method, *Eur. J. Biochem.*, 95, 621, 1979.
59. **Fee, J. A.**, On the question of superoxide toxicity and the biological function of superoxide dismutase, *Oxidases and Related Redox Systems*, Proc. 3rd Int. Symp., Albany, 1979
60. **Koen, Y. M., Peskin, A. V., Baru, V. A., Fedorchenko, V. V., Zbarskii, I. B., and Konstantinov, A. A.**, RNA transport from liver cell nuclei in vitro. Effect of superoxide dismutase on the release of rapidly labeled RNA from isolated nuclei, *Biochem./Biokhim.*, 43, 1673, 1978.
61. **Schubotz, L. M. and Weser, U.**, Are active site models of cuprein (Cu_2Zn_2-SOD) "superoxide dismutase"?, in *Metalloproteins*, Weser, U., Ed., Thieme Verlag, Stuttgart, 1979, 127.
62. **Younes, M. and Weser, U.**, Reactivity of superoxide dismutase-active Cu(II) complexes on the rate of adrenochrome formation, *F.E.B.S. Lett.*, 71, 87, 1976.
63. **Laurie, S. H. and Prime, D. M.**, The formation and nature of the mixed valence copper-D-penicillaminechloride cluster in aqueous solution and its relevance to the treatment of Wilson's disease, *J. Inorg. Biochem.*, 11, 229, 1979.
64. **Werringloer, J., Kawano, S., Chacos, N., and Estabrook, R. W.**, The interaction of divalent copper and the microsomal electron transport system. A re-examination of the effects of copper chelates on the function of cytochrome P-450, *J. Biol. Chem.*, 254, 11839, 1979.
65. **Miller, R. W. and MacDowall, F. D. H.**, The Tiron free radical as a sensitive indicator of chloroplastic photoautoxidation, *Biochim. Biophys. Acta*, 387, 176, 1975.
66. **Richter, C., Azzi, A., Weser, U., and Wendel, A.**, Hepatic microsomal dealkylations. Inhibition by a tyrosine-copper(II) complex provided with superoxide dismutase activity, *J. Biol. Chem.*, 252, 5061, 1977.
67. **McCord, J. M., Crapo, J. D., and Fridovich, I.**, Superoxide dismutase assays: a review of methodology, in *Superoxide and Superoxide Dismutases*, Michelson, A. M., McCord, J. M., and Fridovich, I., Eds., Academic Press, London, 1977, 11.
68. **Crapo, J. D., McCord, J. M., and Fridovich, I.**, Preparation and assay of superoxide dismutases, in *Methods in Enzymology 53, Part D, Biomembranes*, Fleischer, S. and Packer, L., Eds., Academic Press, New York, 382, 1978.
69. **Fridovich, I.**, Measuring the activity of superoxide dismutases; an embarrassment of riches, in *Superoxide Dismutase, Vol. I*, Oberley, L. W., Ed., CRC Press, Boca Raton, Fla., 1982, chap. 3.
70. **Heikkila, R. E. and Cabbat, F.**, A sensitive assay for superoxide dismutase based on the autoxidation of 6-hydroxydopamine, *Anal. Biochem.*, 75, 356, 1976.

71. **Marklund, S. and Marklund, G.**, Involvement of the superoxide anion radical in the autoxidation of pyrogallol and a convenient assay for superoxide dismutase, *Eur. J. Biochem.*, 47, 469, 1974.
72. **Cohen, G., Heikkila, R. E., and MacNamee, D.**, The generation of hydrogen peroxide, superoxide radical, and hydroxyl radical by 6-hydroxydopamine, dialuric acid and related cytotoxic agents, *J. Biol. Chem.*, 249, 2447, 1974.
73. **Asami, S. and Akazawa, T.**, Enzymic formation of glycolate in *Chromatium,* Role of superoxide radical in a transketolase-type mechanism, *Biochemistry,* 16, 2202, 1977.
74. **Auclair, C., de Prost, D., and Hakim, J.**, Superoxide anion production by liver microsomes from phenobarbital-treated rat, *Biochem. Pharmacol.*, 27, 355, 1978.
75. **Marshall, M. J. and Worsfold, M.**, Superoxide dismutase: a direct, continuous linear assay using the oxygen electrode, *Anal. Biochem.*, 86, 561, 1978.
76. **Markert, M., Allaz, M.-J., and Frei, J.**, Continuous monitoring of oxygen consumption and superoxide production by particles-stimulated human polymorphonuclear leukocytes, *F.E.B.S. Lett.*, 113, 225, 1980.
77. **Weser, U. and Voelcker, G.**, Evidence for accelerated H_2O_2 formation by erythrocuprein, *F.E.B.S. Lett.*, 22, 15, 1972.
78. **McCormick, J. P. and Thomason, T.**, Near-ultraviolet photooxidation of tryptophan. Proof of formation of superoxide ion, *J. Am. Chem. Soc.*, 100, 312, 1978.
79. **Heikkila, R. E., Cohen, G., and Manian, A. A.**, Reactivity of various phenothiazine derivatives with oxygen and oxygen radicals, *Biochem. Pharmacol.*, 24, 363, 1975.
80. **Misra, H. P. and Fridovich, I.**, Superoxide dismutase: "positive" spectrophotometric assays, *Anal. Biochem.*, 79, 553, 1977.
81. **Misra, H. P. and Fridovich, I.**, Superoxide dismutase: a photochemical augmentation assay, *Arch. Biochem. Biophys.*, 181, 308, 1977.
82. **Misra, H. P. and Fridovich, I.**, Superoxide dismutase and peroxidase: a positive activity stain applicable to polyacrylamide gel electropherograms, *Arch. Biochem. Biophys.*, 183, 511, 1977.
83. **Heikkila, R. E., Winston, B., Cohen, G., and Barden, H.**, Alloxan-induced diabetes — evidence for hydroxyl radical as a cytotoxic intermediate, *Biochem. Pharmacol.*, 25, 1085, 1976.
84. **Grankvist, K., Marklund, S., Sehlin, J., and Taeljedal, I. B.**, Superoxide dismutase, catalase and scavengers of hydroxyl radical protect against the toxic action of alloxan on pancreatic islet cells in vitro, *Biochem. J.*, 182, 17, 1979.
85. **Cederbaum, A. I., Dicker, E., and Cohen, G.**, Effect of hydroxyl radical scavengers on microsomal oxidation of alcohols and on associated microsomal reactions, *Biochemistry,* 17, 3058, 1978.
86. **Ohnishi, K. and Lieber, C. S.**, Respective role of superoxide and hydroxyl radical in the activity of the reconstituted microsomal ethanol-oxidizing system, *Arch. Biochem. Biophys.*, 191, 798, 1978.
87. **Mittal, C. K. and Murad, F.**, Activation of guanylate cyclase by superoxide dismutase and hydroxyl radical: a physiological regulator of guanosine 3′,5′-monophosphate formation, *Proc. Natl. Acad. Sci. U.S.A.*, 74, 4360, 1977.
88. **Salin, M. L. and McCord, J. M.**, Free radicals in leukocyte metabolism and inflammation, in *Superoxide and Superoxide Dismutases,* Michelson, A. M., McCord, J. M., and Fridovich, I., Eds., Academic Press, London, 1977, 257.
89. **Cohen, G.**, The generation of hydroxyl radicals in biologic systems: toxicological aspects, *Photochem. Photobiol.*, 28, 669, 1978.
90. **Auclair, C., Torres, M., Cramer, E., and Hakim, J.**, Cyanide-insensitive NADH oxidation by subcellular fractions isolated from human polymorphonuclear blood cells, *Enzyme,* 23, 225, 1978.
91. **Fendler, J. H.**, Radiation and photochemistry in membrane mimetic systems, *Radiat. Res.*, 74, 521, 1978.
92. **Patterson, L. K. and Hasegawa, K.**, Pulse radiolysis studies in model lipid systems. The influence of aggregation on kinetic behavior of OH induced radicals in aqueous sodium linoleate, *Ber. Bunsenges. Physik. Chem.*, 82, 951, 1978.
93. **Chapman, J. D. and Reuvers, A. P.**, The time-scale of radioprotection in mammalian cells, in *Radioprotection: Chemical Compounds — Biological Means, Experientia Suppl. 27,* Locker, A. and Fleming, K., Eds., Adler, Chicago, 1977, 9.
94. **Torres, M., Auclair, C., and Hakim, J.**, NADH oxidation and oxygen reductions by the granule-rich fraction isolated from human polymorphonuclear blood cells, *F.E.B.S. Lett.*, 97, 50, 1979.
95. **Cruz, F. S., Docampo, R., and Boveris, A.**, Generation of superoxide anions and hydrogen peroxide from β-lapachone in bacteria, *Antimicrob. Agents Chemother.*, 14, 630, 1978.
96. **Boveris, A., Cadenas, E., and Stoppani, A. O. M.**, Role of ubiquinone in the mitochondrial generation of hydrogen peroxide, *Biochem. J.*, 156, 435, 1976.
97. **Kono, Y., Takahashi, M. A., and Asada, K.**, Oxidation of manganous pyrophosphate by superoxide radicals and illuminated spinach chloroplasts, *Arch. Biochem. Biophys.*, 174, 454, 1976.
98. **Chedekel, M. R., Poh Agin, P., and Sayre, R. M.**, Photochemistry of pheomelanin: action spectrum for superoxide production, *Photochem. Photobiol.*, 31, 553, 1980.

99. **Bielski, B. H. J.**, Re-evaluation of the special and kinetic properties of $HO_2\cdot$ and O_2^- free radicals, *Photochem. Photobiol.*, 28, 645, 1978.
100. **Bors, W., Michel, C., and Saran, M.**, Superoxide anions do not react with hydroperoxides, *F.E.B.S. Lett.*, 107, 403, 1979.
101. **Marklund, S.**, Spectrophotometric study of spontaneous disproportionation of superoxide anion radical and sensitive direct assay for superoxide dismutase, *J. Biol. Chem.*, 251, 7504, 1976.
102. **McClune, G. J. and Fee, J. A.**, Stopped-flow spectrophotometric observation of superoxide dismutation in aqueous solution, *F.E.B.S. Lett.*, 67, 294, 1976.
103. **Kim, S., DiCosimo, R., and San Filippo, J.**, Spectrometric and chemical characterization of superoxide, *Anal. Chem.*, 51, 679, 1979.
104. **Valentine, J. S. and Curtis, A. B.**, A convenient preparation of solutions of superoxide anion and the reaction of superoxide anion with a copper(II) complex, *J. Am. Chem. Soc.*, 97, 224, 1975.
105. **Hoffman, B. M., Szymanski, T., and Basolo, F.**, Consideration of a report on the formulation of monomeric cobalt-dioxygen adducts. Continued support for Co(III)-O_2^-, *J. Am, Chem. Soc.*, 97, 673, 1975.
106. **Milligan, D. E. and Jacox, M. E.**, Infrared spectroscopic evidence for the species $HO_2\cdot$, *J. Chem. Phys.*, 38, 2627, 1963.
107. **Vorhaben, J. E. and Steele, R. H.**, Studies on the generation of electronic excitation states in a riboflavin-hydrogen peroxide-copper(II)-ascorbate redox system leading to chemiluminescence and/or aromatic hydroxylation, *Biochemistry*, 6, 1404, 1967.
108. **Arneson, R. M.**, Substrate-induced chemiluminescence of xanthine oxidase and aldehyde oxidase, *Arch. Biochem. Biophys.*, 136, 352, 1970.
109. **Finazzi Agro, A., Giovagnoli, C., de Sole, P., Calabrese, L., Rotilio, G., and Mondovi, B.**, Erythrocuprein and singlet oxygen, *F.E.B.S. Lett.*, 21, 183, 1972.
110. **Weser, U. and Paschen, W.**, Mode of singlet oxygen induced chemiluminescence in the presence of erythrocuprein, *F.E.B.S. Lett.*, 27, 248, 1972.
111. **Johnston, R. B., Keele, B. B., Misra, H. P., Webb, L. S., Lehmeyer, J. E., and Rajagopalan, K. V.**, Superoxide anion generation and phagocytic bactericidal activity, in *The Phagocytic Cell in Host Resistance*, Bellanti, J. A. and Dayton, D. H., Eds., Raven Press, New York, 1975, 61.
112. **Allen, R. C., Stjernholm, R. L., and Steele, R. H.**, Evidence for the generation of an electronic excitation state(s) in human polymorphonuclear leukocytes and its participation in bactericidal activity, *Biochem. Biophys. Res. Comm.*, 47, 679, 1972.
113. **Webb, L. S., Keele, B. B., and Johnston, R. B.**, Inhibition of phagocytosis-associated chemiluminescence by superoxide dismutase, *Infect. Immunity*, 9, 1051, 1974.
114. **Rosen, H. and Klebanoff, S. J.**, Chemiluminescence and superoxide production by myeloperoxidase deficient leukocytes, *J. Clin. Invest.*, 58, 50, 1976.
115. **Cheson, B. D., Christensen, R. L., Sperling, R., Kohler, B. E., and Babior, B. M.**, The origin of the chemiluminescence of phagocytosing granulocytes, *J. Clin. Invest.*, 58, 789, 1976.
116. **Tsan, M. F.**, Stimulation of the hexose monophosphate shunt independent of hydrogen peroxide and superoxide production in rabbit alveolar macrophages during phagocytosis, *Blood*, 50, 935, 1977.
117. **Harvath, L., Amirault, H. J., and Andersen, B. R.**, Chemiluminescence of human and canine polymorphonuclear leukocytes in the absence of phagocytosis, *J. Clin. Invest.*, 61, 1145, 1978.
118. **Shoaf, A. R. and Steele, R. H.**, Microsomal lipid peroxidation, drug oxidations and chemiluminescence: mechanisms, *Biochem. Biophys. Res. Comm.*, 61, 1363, 1974.
119. **Veldink, G. A., Garssen, G. J., Slappendel, S., Vliegenthart, J. F. G., and Boldingh, J.**, Chemiluminescence during lipoxygenase-catalysed oxygenation of linoleic acid, *Biochem. Biophys. Res. Comm.*, 78, 424, 1977.
120. **Cadenas, E., Boveris, A., and Chance, B.**, Low-level chemiluminescence of bovine heart submitochondrial particles, *Biochem. J.*, 186, 659, 1980.
121. **Cadenas, E., Boveris, A., and Chance, B.**, Low-level chemiluminescence of hydroperoxide-supplemented cytochrome *c*, *Biochem. J.*, 187, 131, 1980.
122. **Ballou, D., Palmer, G., and Massey, V.**, Direct demonstration of O_2^- production during the oxidation of reduced flavin and of its catalytic decomposition by erythrocuprein, *Biochem. Biophys. Res. Comm.*, 36, 898, 1969.
123. **Knowles, P. F., Gibson, J. F., Pick, F. M., and Bray, R. C.**, ESR evidence for enzymic reduction of oxygen to a free radical, the superoxide ion, *Biochem. J.*, 111, 53, 1969.
124. **Bray, R. C., Pick, F. M., and Samuel, D.**, Oxygen-17 hyperfine splitting in the EPR spectrum of enzymatically generated O_2^- *Eur. J. Biochem.*, 15, 352, 1970.
125. **Getz, D., Melamud, E., Silver, B. L., and Dori, Z.**, Electronic structure of dioxygen in Co(II) oxygen carriers, singlet oxygen or O_2^-?, *J. Am. Chem. Soc.*, 97, 3846, 1975.
126. **Tovrog, B. S. and Drago, R. S.**, Evidence against the O_2^- formation of Co(II) adducts to dioxygen, *J. Am. Chem. Soc.*, 96, 6765, 1974.

127. **Nishinaga, A.**, Reactions of cobalt-oxygen complexes with organic molecules, in *Biochemical and Medical Aspects of Active Oxygen*, Hayaishi, O. and Asada, K., Eds., University of Tokyo Press, Tokyo, 1977, 13.
128. **Aasa, R., Braenden, R., Deinum, J., Malmstroem, B. G., Reinhammar, B., and Vaenngard, T.**, A ^{17}O-effect on the EPR spectrum of the intermediate in the dioxygen-laccase reaction, *Biochem. Biophys. Res. Comm.*, 70, 1204, 1976.
129. **Eastland, G. W. and Symons, M. C. R.**, ESR studies of superoxide ions produced by radiolysis in alcoholic media, *J. Phys. Chem.*, 81, 1502, 1977.
130. **Green, M. R., Hill, H. A. O., and Tirner, D. R.**, The nature of the superoxide ion in dipolar aprotic solvents: the EPR spectra of the superoxide ion in *N,N*-dimethylformamide — evidence for hydrated forms, *F.E.B.S. Lett.*, 103, 176, 1979.
131. **Meisel, D., Czapski, G., and Samuni, A.**, Hydroperoxyl radical reactions. I. EPR study of the complexation of $HO_2\cdot$ with some metal ions, *J. Am. Chem. Soc.*, 95, 4148, 1973.
132. **Meisel, D., Levanon, H., and Czapski, G.**, Hydroperoxyl radical reactions. II. Cupric ions in modulated photolysis. EPR experiments, *J. Phys. Chem.*, 78, 779, 1974.
133. **Bains, M. S.**, An ESR study of Fenton-Ruff reagent, Fe^{3+}-H_2O_2 solution system, using Ti^{4+} ions as free radical detector ions, *J. Indian Chem. Soc.*, 53, 83, 1976.
134. **Meisel, D., Ilan, Y. A., and Czapski, G.**, Hydroperoxyl radical reactions. III. Pulse-radiolytic study of the reaction of the hydroperoxyl radical with some metal ions, *J. Phys. Chem.*, 78, 2330, 1974.
135. **Buettner, G. R.**, The spin trapping of superoxide and hydroxyl radicals, in *Superoxide Dismutase, Vol. II.*, Oberley, L. W., Ed., CRC Press, Boca Raton, Fla., 1982, chap. 4.
136. **Finkelstein, E., Rosen, G. M., Rauckman, E. J., and Paxton, J.**, Spin trapping of superoxide, *Mol. Pharmacol.*, 16, 676, 1979.
137. **Schmidt, K. H.**, Electrical conductivity techniques for studying the kinetics of radiation-induced chemical reactions in aqueous solutions, *Int. J. Radiat. Phys. Chem.*, 4, 439, 1972.
138. **Bothe, E. and Schulte-Frohlinde, D.**, The bimolecular decay of the α-hydroxy-methylperoxyl radicals in aqueous solution, *Z. Naturforsch.*, 33b, 786, 1978.
139. **Schuchmann, M. N. and von Sonntag, C.**, Hydroxyl radical-induced oxidation of 2-methyl-2-propanol in oxygenated aqueous solution. A product and pulse radiolysis study, *J. Phys. Chem.*, 83, 780, 1979.
140. **Peover, M. E. and White, B. S.**, Electrolytic reduction of oxygen in aprotic solvents: the superoxide ion, *Electrochim. Acta*, 11, 1061, 1966.
141. **Roberts, J. L., Morrison, M. M., and Sawyer, D.T.**, Base-induced generation of superoxide ion and hydroxyl radical from hydrogen peroxide, *J. Am. Chem. Soc.*, 100, 329, 1978.
142. **Rosenthal, I.**, The synthesis of ^{18}O-enriched KO_2, *J. Lab. Comp.*, 12, 317, 1976.
143. **Hayaishi, O., Hirata, F., Ohnishi, T., Henry, J. P., Rosenthal, I., and Katoh, A.**, Indoleamine 2,3-dioxygenase. Incorporation of $^{18}O_2^-$ and $^{18}O_2$ into the reaction products, *J. Biol. Chem.*, 252, 3548, 1977.
144. **Frimer, A. A. and Rosenthal, I.**, Chemical reactions of superoxide anion radical in aprotic solvents, *Photochem. Photobiol.*, 28, 711, 1978.
145. **Cahill, A. E. and Taube, H.**, The use of heavy oxygen in the study of reactions of hydrogen peroxide, *J. Am. Chem. Soc.*, 74, 2312, 1952.
146. **McCord, J. M. and Fridovich, I.**, Superoxide dismutase. An enzymic function for erythrocuprein (hemocuprein), *J. Biol. Chem.*, 244, 6049, 1969.
147. **Massey, V.**, The microestimation of succinate and the extinction coefficient of cytochrome *c*, *Biochim. Biophys. Acta*, 34, 255, 1959.
148. **van Gelder, B. F. and Slater, E. C.**, The extinction coefficient of cytochrome, *c*, *Biochim. Biophys. Acta*, 58, 593, 1962.
149. **Ilan, Y., Ilan, Y. A., and Czapski, G.**, Do copper ions influence the reduction of ferricytochrome *c* by $O_2^{\cdot-}$, *Biochim. Biophys. Acta*, 503, 399, 1978.
150. **Hodgson, E. K. and Fridovich, I.**, The accumulation of superoxide radical during the aerobic action of xanthine oxidase. A requiem for $H_2O_4^-$, *Biochim. Biophys. Acta*, 430, 182, 1976.
151. **Fridovich, I.**, Quantitative aspects of the production of O_2^- by milk xanthine oxidase, *J. Biol. Chem.*, 245, 4053, 1970.
152. **Cassell, R. H. and Fridovich, I.**, The role of superoxide radical in the autoxidation of cytochrome *c*, *Biochemistry*, 14, 1866, 1975.
153. **Butler, J., de Kok, J., de Weille, J. R., Koppenol, W. H., and Braams, R.**, Mechanism of the reaction of hydrated electrons with ferrocytochrome *c*, *Biochim. Biophys. Acta*, 459, 207, 1977.
154. **Simic, M. G. and Taub, I. A.**, Fast electron transfer processes in cytochrome *c* and related metalloproteins, *Biophys. J.*, 24, 285, 1978.
155. **Cohen, H. J. and Chovaniec, M. E.**, Superoxide generation by digitonin-stimulated guinea pig granulocytes. A basis for continuous assay for monitoring superoxide production and for the study of the activation of the generating system, *J. Clin. Invest.*, 61, 1081, 1978.

156. **Dri, P., Bellavite, P., Berton, G., and Rossi, F.,** Interrelationship between oxygen consumption, superoxide anion and hydrogen peroxide formation in phagocytosing guinea pig polymorphonuclear leucocytes, *Mol. Cell. Biochem.*, 23, 109, 1979.
157. **Lyakhovich, V. V., Mishin, V. M., and Pokrovskii, A. G.,** Interrelationship between the generation of radical-anions of oxygen and the reduction of artificial acceptors and cytochrome P-450 by NADPH-cytochrome *c* reductase, *Biochem./Biokhim.*, 42, 1033, 1977.
158. **Berman, M. C., Adnams, C. M., Ivanetich, K. M., and Kench, J. E.,** Autoxidation of soluble trypsin-cleaved microsomal ferrocytochrome b_5 and formation of superoxide radicals, *Biochem. J.*, 157, 237, 1976.
159. **Demma, L. S. and Salhany, J. M.,** Direct generation of superoxide anions by flash photolysis of human oxyhemoglobin, *J. Biol. Chem.*, 252, 1226, 1977.
160. **Azzi, A., Montecucco, C., and Richter, C.,** The use of acetylated ferricytochrome *c* for the detection of superoxide radicals produced in biological membranes, *Biochem. Biophys. Res. Comm.*, 65, 597, 1975.
161. **Kuthan, H., Tsuji, H., Graf, H., Ullrich, V., Werringloer, J., and Estabrook, R. W.,** Generation of superoxide anion as a source of hydrogen peroxide in a reconstituted monooxygenase system, *F.E.B.S. Lett.*, 91, 343, 1978.
162. **Buchanan, A. G. and Lees, H.,** The effects of pH and temperature on the assay of superoxide dismutase, *Can. J. Microbiol.*, 22, 1643, 1976.
163. **Simic, M. G. and Taub, I. A.,** Mechanisms of inter- and intramolecular electron transfer in cytochromes, *Faraday Disc.*, 63, 270, 1977.
164. **van Leeuwen, J. W., Raap, A., Koppenol, W. H., and Nauta, H.,** A tunnelling model to explain the reduction of ferricytochrome *c* by H and OH radicals, *Biochim. Biophys. Acta*, 503, 1, 1978.
165. **Altman, F. P.,** *Tetrazolium Salts and Formazans*, Springer-Verlag, Stuttgart, 1976.
166. **Baehner, R. L. and Nathan, D. G.,** Quantitative NBT test in chronic granulomatous disease, *N. Engl. J. Med.*, 278, 971, 1968.
167. **Baehner, R. L., Boxer, L. A., and Davis, J.,** The biochemical basis of nitroblue tetrazolium reduction in normal human and chronic granulomatous disease polymorphonuclear leukocytes, *Blood*, 48, 309, 1976.
168. **Roerig, D. L., Mascaro, L., and Aust, S. F.,** Microsomal electron transport: tetrazolium reduction by rat liver microsomal NADPH-cytochrome *c* reductase, *Arch. Biochem. Biophys.*, 153, 475, 1972.
169. **Rajagopalan, K. V. and Handler, P.,** Hepatic aldehyde oxidase. II. Differential inhibition of electron transfer to various electron acceptors, *J. Biol. Chem.*, 239, 2022, 1964.
170. **Miller, R. W.,** Reactions of superoxide anion, catechols, and cytochrome *c*, *Can. J. Biochem.*, 48, 935, 1970.
171. **Rosenkrantz, H.,** The reactivity of catechol amines, 5-hydroxy indoles, and phenols, with blue tetrazolium, *Arch. Biochem. Biophys.*, 81, 194, 1959.
172. **Bielski, B. H. J., Shiue, G. G., and Bajuk, S.,** Reduction of nitroblue tetrazolium by CO_2^- and O_2^- radicals, *J. Phys. Chem.*, 84, 830, 1980.
173. **Nishikimi, M.,** The generation of superoxide anion in the reaction of tetrahydropteridines with molecular oxygen, *Arch. Biochem. Biophys.*, 166, 273, 1975.
174. **Takabe, T., Miyakawa, M., and Nikai, S.,** The micellar acceleration of the reduction of nitroblue tetrazolium by superoxide anion radicals during the autoxidation of polyphenols, *Bull. Chem. Soc. Jpn.*, 51, 321, 1978.
175. **Amano, D., Kagosaki, Y., Usui, T., Yamamoto, S., and Hayaishi, O.,** Inhibitory effects of superoxide dismutase and various other proteins on the nitroblue tetrazolium reduction by phagocytizing guinea pig polymorphonuclear leukocytes, *Biochem. Biophys. Res. Comm.*, 66, 272, 1975.
176. **Wilkinson, R. W., Powars, D. R., and Hochstein, P.,** New evidence for the role of NADH oxidase in phagocytosis by human granulocytes, *Biochem. Med.*, 13, 83, 1975.
177. **Woolf, J. H. and Gregory, E. M.,** Superoxide anion independent reduction of nitroblue tetrazolium by human renal tissue, *Clin. Biochem.*, 9, 241, 1976.
178. **Auclair, C., Torres, M., and Hakim, J.,** Superoxide anion involvement in NBT reduction catalyzed by NADPH-cytochrome P-450 reductase: a pitfall, *F.E.B.S. Lett.*, 89, 26, 1978.
179. **Nishikimi, M., Rao, N. A., and Yagi, K.,** The occurrence of superoxide in the reaction of reduced phenazine methosulfate and molecular oxygen, *Biochem. Biophys. Res. Comm.*, 46, 849, 1972.
180. **Richter, H. W. and Waddell, W. H.,** The active oxygen species in the phenazine methosulfate/NADH/O_2 hydroxylating system, in *Oxygen and Oxy-Radicals in Chemistry and Biology*, Rogers, M. A. J. and Powers, E. L., Eds., Academic Press, New York, 1981, 711.
181. **Ponti, V., Dianzani, M. U., Cheeseman, K., and Slater, T. F.,** Studies on the reduction of nitroblue tetrazoliumchloride mediated through the action of NADH and phenazine methosulphate, *Chem. Biol. Interact.*, 23, 281, 1978.

182. **Peterson, D. A., Gerrard, J. M., Rao, G. H. R., Krick, R. P., and White, J. G.,** Ferrous iron mediated oxidation of arachidonic acid: studies employing nitroblue tetrazolium (NBT), *Prostagland. Med.*, 1, 304, 1978.
183. **Rao, G. H. R., Gerrard, J. M., Eaton, J. W., and White, J. G.,** Arachidonic acid peroxidation, prostaglandin synthesis and platelet function, *Photochem. Photobiol.*, 28, 845, 1978.
184. **Yamashoji, S., Yoshida, H., and Kajimoto, G.,** Photooxidation of linoleic acid by ultraviolet light and effect of superoxide anion quencher, *Agric. Biol. Chem.*, 43, 1249, 1979.
185. **Zhuravleva, O. S. and Berdnikov, V. M.,** The mechanism of the consumption of hydrogen peroxide in the iron(III)ion-catechol-hydrogen peroxide-anisole system (the Hamilton system), *Russ. J. Phys. Chem.*, 50, 1369, 1976.
186. **Hodgson, E. K. and Fridovich, I.,** Reversal of the superoxide dismutase reaction, *Biochem. Biophys. Res. Comm.*, 54, 270, 1973.
187. **Miller, R. W., Kerr, C. T., and Curry, J. R.,** Mammalian dihydroorotate-ubiquinone reductase complex, *Can. J. Biochem.*, 46, 1099, 1968.
188. **Taylor, W. H., Taylor, C. D., and Taylor, M. L.,** Biosynthetic dihydroorotate dehydrogenase from *Lactobacillus bulgaricus* partial characterization of the enzyme, *J. Bacteriol.*, 119, 98, 1974.
189. **Forman, H. J. and Kennedy, J.,** Dihydroorotate-dependent superoxide production in rat brain and liver, *Arch. Biochem. Biophys.*, 173, 219, 1976.
190. **Mishin, V. M., Pokrovskii, A. G., and Lyakhovich, V. V.,** Interaction of various acceptors with anion radicals of oxygen in liver microsomes, *Biochem./Biokhim.*, 41, 627, 1976.
191. **Ikediobi, C. O.,** Bleaching of dichlorophenol indophenol by a coupled oxidation with linoleate catalyzed by soybean lipoxygenase, *Agric. Biol. Chem.*, 41, 2369, 1977.
192. **Greenstock, C. L. and Wiebe, R. H.,** Kinetic studies of peroxides and peroxy radicals, and their reactions with biological molecules, in *Oxygen and Oxy-Radicals in Chemistry and Biology*, Rogers, M. A. J. and Powers, E. L., Eds., Academic Press, New York, 1981, 119.
193. **Hawley, M. D., Tatawawadi, S. V., Piekarski, S., and Adams, R. N.,** Electrochemical studies of the oxidation pathways of catecholamines, *J. Am. Chem. Soc.*, 89, 447, 1967.
194. **Sternson, A. W., McCreery, R., Feinberg, B., and Adams, R. N.,** Electrochemical studies of adrenergic neurotransmitters and related compounds, *J. Electroanal. Chem.*, 46, 313, 1973.
195. **Swan, G. A.,** Structure, chemistry, and biosynthesis of the melanins, *Fortschritte der Chemie Organischer Naturstoffe*, Zechmeister, L., Ed., Springer-Verlag, Wien, 31, 521, 1974.
196. **Misra, H. P. and Fridovich, I.,** The role of superoxide radicals in the autoxidation of epinephrine and a simple assay for superoxide dismutase, *J. Biol. Chem.*, 247, 3170, 1972.
197. **Sobotka, H., Barsel, N., and Chanley, J. D.,** The aminochromes, in *Fortschritte der Chemie Organischer Naturstoffe*, Zechmeister, L., Ed., Springer-Verlag Wien, 14, 217, 1957.
198. **Bors, W., Michel, C., Saran, M., and Lengfelder, E.,** The involvement of oxygen radicals during the autoxidation of adrenalin, *Biochim. Biophys. Acta*, 540, 162, 1978.
199. **Sun, M. and Zigman, S.,** An improved spectrophotometric assay for superoxide dismutase based on epinephrine autoxidation, *Anal. Biochem.*, 90, 81, 1978.
200. **Sun, M. and Zigman, S.,** A new fluorometric assay for superoxide dismutase and prostaglandin synthetase in ocular tissues, *Fed. Proc.*, 37, 1781, 1978.
201. **Sun, M., Paxhia, T., and Zigman, S.,** Fluorescence assay for superoxide dismutase based on epinephrine oxidation: role of metal ions on superoxide formation, *Fed. Proc.*, 38, 572, 1979.
202. **Bors, W., Saran, M., Michel, C., Lengfelder, E., Fuchs, C., and Spöttl, R.,** Pulse-radiolytic investigations of catechols and catecholamines. I. Adrenaline and adrenochrome, *Int. J. Radiat. Biol.*, 28, 353, 1975.
203. **Harrison, W. H., Whisler, W. W., and Hill, B. J.,** Catecholamine oxidation and ionization properties indicated from the H^+-release, tritium exchange, and spectral changes which occur during ferricyanide oxidation, *Biochemistry*, 7, 3089, 1968.
204. **Graham, D. G.,** Oxidative pathways for catecholamines in the genesis of neuromelanin and cytotoxic quinones, *Mol. Pharmacol.*, 14, 633, 1978.
205. **Uemura, T., Chiesara, E., and Cova, D.,** Interaction of epinephrine metabolites with the liver microsomal electron transport system, *Mol. Pharmacol.*, 13, 196, 1977.
206. **Powis, G.,** Hepatic microsomal metabolism of epinephrine and adrenochrome by superoxide-dependent and independent pathways, *Biochem. Pharmacol.*, 28, 83, 1979.
207. **Schenkman, J. B., Jansson, I., Powis, G., and Kappus, H.,** Active oxygen in liver microsomes: mechanism of epinephrine oxidation, *Mol. Pharmacol.*, 15, 428, 1979.
208. **Weser, U. and Schubotz, L. M.,** Conformational aspects of superoxide dismutase-active copper chelates and catechols, *Bioinorg. Chem.*, 9, 505, 1978.
209. **Rapp, U., Adams, W. C., and Miller, R. W.** Purification of superoxide dismutase from fungi and characterization of the reaction of the enzyme with catechols by EPR spectroscopy, *Can. J. Biochem.*, 51, 158, 1973.

210. **Winterbourn, C. C., French, J. K., and Claridge, R. F. C.,** Superoxide dismutase as an inhibitor of reactions of semiquinone radicals, *F.E.B.S. Lett.,* 94, 269, 1978.
211. **Miller, R. W. and Rapp, U.,** The oxidation of catechols by reduced flavins and dehydrogenases. An ESR study of the kinetics and initial products of oxidation, *J. Biol. Chem.,* 248, 6084, 1973.
212. **Fridovich, I. and Handler, P.,** Xanthine oxidase. V. Differential inhibition of the reduction of various electron acceptors, *J. Biol. Chem.,* 237, 916, 1962.
213. **Miller, R. W. and Massey, V.,** Dihydroorotic dehydrogenase. II. Oxidation and reduction of cytochrome *c*, *J. Biol. Chem.,* 240, 1466, 1965.
214. **Hirata, F. and Hayaishi, O.,** Studies on indoleamine 2,3-dioxygenase. I. Superoxide anion as substrate, *J. Biol. Chem.,* 250, 5960, 1975.
215. **Staudinger, H. J. and Ullrich, V.,** Zum Reaktionsmechanismus der nicht-enzymatischen Hydroxylierung. Zugleich Untersuchungen über die Aktivierung von Sauerstoff, *Z. Naturforsch.,* 19b, 409, 1964.
216. **Ullrich, V. and Staudinger, H. J.,** Aktivierung von Sauerstoff in Modellsystemen, in *Biochemie des Sauerstoffs,* Hess, B. and Staudinger, Hj., Eds., Springer-Verlag, Berlin, 1968, 229.
217. **Norman, R. O. C. and Lindsay Smith, J. R.,** Mechanisms of aromatic hydroxylation and ring-opening reactions, in *Oxidases and Related Redox Systems, Vol. I.,* King, T. E., Mason, H. S., and Morrison, M., Eds., John Wiley & Sons, New York, 1965, 131.
218. **Halliwell, B.,** Superoxide and hydroxylation reactions, in *Superoxide and Superoxide Dismutases,* Michelson, A. M., McCord, J. M., and Fridovich, I., Eds., Academic Press, London, 1977, 335.
219. **Ravindranath, S. D., Kumar, A. A., Kumar, R. P., Vaidyanathan, C. S., and Rao, N. A.,** Mechanism of aromatic hydroxylation. Properties of a model for pyridine nucleotide-dependent flavoprotein hydroxylase, *Arch. Biochem. Biophys.,* 165, 478, 1974.
220. **Kumar, R. P., Ravindranath, S. D., Vaidyanathan, C. S., and Rao, N. A.,** Mechanism of hydroxylation of aromatic compounds. II. Evidence for the involvement of O_2^- in enzymatic hydroxylations, *Biochem. Biophys. Res. Comm.,* 49, 1422, 1972.
221. **Kumar, A. A., Vaidyanathan, C. S., and Rao, N. A.,** Mechanism of aromatic hydroxylation: the kinetic mechanism and the involvement of superoxide anions in the reaction catalyzed by m-hydroxybenzoate-4-hydroxylase from *Aspergillus niger* (UBC-814), *Indian J. Biochem. Biophys.,* 15, 5, 1978.
222. **Massey, V., Strickland, S., Mayhew, S. G., Howell, L. G., Engel, P. C., Matthews, R. G., Schuman, M., and Sullivan, P. A.,** The production of superoxide radicals in the reaction of reduced flavins and flavorproteins with molecular oxygen, *Biochem. Biophys. Res. Comm.,* 36, 891, 1969.
223. **Halliwell, B.,** Hydroxylation of *p*-coumaric acid by illuminated chloroplasts. The role of superoxide, *Eur. J. Biochem.,* 55, 355, 1975.
224. **Halliwell, B.,** Superoxide-dependent formation of hydroxyl radicals in the presence of iron chelates, *F.E.B.S. Lett.,* 92, 321, 1978.
225. **Halliwell, B. and de Rycker, J.,** Superoxide and peroxidase-catalysed reactions. Oxidation of dihy droxyfumarate, NADH and dithiothreitol by horseradish peroxidase, *Photochem. Photobiol.,* 28, 757, 1978.
226. **Adams, P. A., Berman, M. C., and Baldwin, D. A.,** The participation of both O_2^- and HO_2^- species in the haemin-catalyzed *para*-hydroxylation of aniline, *J.C.S. Chem. Comm.,* 856, 1979.
227. **Henry, J. P. and Michelson, A. M.,** Light emissions involving the superoxide anion, in *Biochemical and Medical Aspects of Active Oxygen,* Hayaishi, O. and Asada, K., Eds., University of Tokyo Press, Tokyo, 1977, 135.
228. **Totter, J. R., Castro de Dugros, E., and Riveiro, C.,** The use of chemiluminescent compounds as possible indicators of radical production during xanthine oxidase action, *J. Biol. Chem.,* 235, 1839, 1960.
229. **Greenlee, L., Fridovich, I., and Handler, P.,** Chemiluminescence induced by operation of iron-flavoproteins, *Biochemistry,* 1, 779, 1962.
230. **Puget, K. and Michelson, A. M.,** Iron-containing superoxide dismutase from luminous bacteria, *Biochimie,* 56, 1255, 1974.
231. **Lavelle, F., Michelson, A. M., and Dimitrijevic, L.,** Biological protection by superoxide dismutase, *Biochem. Biophys. Res. Comm.,* 55, 350, 1973.
232. **Albrecht, H. O.,** Über die Chemilumineszenz des Aminophthalsäurehydrazids, *Z. Physik. Chem. (Leipzig),* 136, 321, 1928.
233. **White, E. H., Zafiriou, O., Kaegi, H. H., and Hill, J. H. M.,** Chemiluminescence of luminol: the chemical reaction, *J. Am. Chem. Soc.,* 86, 940, 1964.
234. **White, E. H. and Bursey, M. M.,** Chemiluminescence of luminol and related hydrazides: the light emission step, *J. Am. Chem. Soc.,* 86, 941, 1964.
235. **Lee, J. and Seliger, H. H.,** Spectral characteristics of the excited states of the product of the chemiluminescence of luminol, *Photochem. Photobiol.,* 11, 247, 1970.
236. **Baxendale, J. H.,** Pulse radiolysis study of the chemiluminescence from luminol, *J.C.S. Faraday I,* 69, 1665, 1973.

237. **Wuerzberg, E. and Haas, Y.**, A pulse radiolysis study of the chemiluminescence of some luninol-like molecules, *J. Phys. Chem.*, 83, 2687, 1979.
238. **Quickenden, T. I.**, Chemiluminescence in liquids, *J. N. Z. Inst. Chem.*, 28, 10, 1964.
239. **Gundermann, K. D.**, Chemiluminescence of diazaquinones and related compounds, in *Chemiluminescence and Bioluminescence*, Cormier, M. J., Hercules, D. M., and Lee, J., Eds., Plenum Press, New York, 1973, 209.
240. **Henry, J. P. and Michelson, A. M.**, Superoxide and chemiluminescence, in *Superoxide and Superoxide Dismutases*, Michelson, A. M., McCord, J. M., and Fridovich, I., Eds., Academic Press, London, 1977, 283.
241. **Takahama, U. and Nishimura, M.**, Light-induced chemiluminescence of luminol in spinach chloroplasts fragments: reaction of O_2^- with electron transfer components, *Plant Cell Physiol.*, 18, 1139, 1977.
242. **Isacsson, U. and Wettermark, G.**, Chemiluminescence in analytical chemistry, *Anal. Chim. Acta*, 68, 339, 1974.
243. **Schroeder, H. R. and Yeager, F. M.**, Chemiluminescence yields and detection limits of some isoluminol derivatives in various oxidation systems, *Anal. Chem.*, 50, 1114, 1978.
244. **Yurow, H. W. and Sass, S.**, Detection of various α-substituted ketones via chemiluminescence of 5-amino-2,3-dihydro-1,4-phthalazinedione (luminol), *Anal. Chim. Acta*, 68, 203, 1974.
245. **Yurow, H. W. and Sass, S.**, Detection of various α-substituted nitriles and *gem* halonitroalkanes by chemiluminescence, *Anal. Chim. Acta.*, 77, 324, 1975.
246. **Michelson, A. M.**, Production of superoxide by metal ions, in *Superoxide and Superoxide Dismutases*, Michelson, A. M., McCord, J. M., and Fridovich, I., Eds., Academic Press, London, 1977, 77.
247. **Henry, J. P., Isambert, M. F., and Michelson, A. M.**, Studies in bioluminescence. III. The *Pholas dactylus* system, *Biochim. Biophys. Acta*, 205, 437, 1970.
248. **Michelson, A. M. and Isambert, M. F.**, Studies in bioluminescence. XI. Further studies on the *Pholas dactylus* system. Mechanism of luciferase, *Biochimie*, 55, 619, 1973.
249. Unpublished observations from our laboratory, 1977.
250. **Elstner, E. F., Heupel, A., and Vaklinova, S.**, Über die Oxidation von Hydroxylamin durch isolierte Chloroplasten und die mögliche Funktion einer Peroxidase aus Spinatblättern bei der Oxidation von Ascorbin- und Glycolsäure, *Z. Pflanzenphysiol.*, 62, 184, 1970.
251. **Elstner, E. F., Stoffer, C., and Heupel, A.**, Determination of superoxide free radical ion and hydrogen peroxide as products of photosynthetic oxygen reduction, *Z. Naturforsch.*, 30c, 53, 1975.
252. **Schmid, R.**, Deactivation of superoxide dismutase on EDTA-treated chloroplasts, *F.E.B.S. Lett.*, 60, 98, 1975.
253. **Bors, W., Lengfelder, E., Saran, M., Michel, C., Fuchs, C., and Frenzel, C.**, Oxidation of hydroxylamine to nitrite as an assay for the combined presence of superoxide anions and hydroxyl radicals, *Biochem. Biophys. Res. Comm.*, 75, 973, 1977.
254. **Kono, Y.**, Generation of superoxide radical during autoxidation of hydroxylamine and an assay for superoxide dismutase, *Arch. Biochem. Biophys.*, 186, 189, 1978.
255. **Takabe, T., Miyakawa, M., and Nikai, S.**, Effects of Cu^{2+} and detergents on the autoxidation of hydroxylamine involving superoxide anion radicals, *Agric. Biol. Chem.*, 43, 613, 1979.
256. **Fridovich, I. and Handler, P.**, Detection of free radicals in illuminated dye solutions by the initiation of sulfite oxidation, *J. Biol. Chem.*, 235, 1835, 1960.
257. **McCord, J. M. and Fridovich, I.**, The utility of superoxide dismutase in studying free radical reactions. I. Radicals generated by the interaction of sulfite, dimethylsulfoxide and oxygen, *J. Biol. Chem.*, 244, 6056, 1969.
258. **Tyler, D. D.**, Polarographic assay and intracellular distribution of superoxide dismutase in rat liver, *Biochem. J.*, 147, 493, 1975.
259. **Yang, S. F.**, Sulfoxide formation from methionine or its sulfide analogs during aerobic oxidation of sulfite, *Biochemistry*, 9, 5009, 1970.
260. **Tuazon, P. T. and Johnson, S. L.**, Free radical and ionic reaction of bisulfite with NADH and its analogues, *Biochemistry*, 16, 1183, 1977.
261. **Inouye, B., Ikeda, M., Ishida, T., Ogata, M., Akiyama, J., and Utsumi, K.**, Participation of superoxide free radical and Mn^{2+} in sulfite oxidation, *Toxicol. Appl. Pharmacol.*, 46, 29, 1978.
262. **Peiser, G. D. and Yang, S. F.**, Chlorophyll destruction in the presence of bisulfite and linoleic acid hydroperoxide, *Phytochemistry*, 17, 79, 1978.
263. **Peiser, G. D. and Yang, S. F.**, Sulfite-mediated destruction of β-carotene, *J. Agric. Food Chem.*, 27, 446, 1979.
264. **Yamazaki, I. and Yokota, K. N.**, Oxidation states of peroxidase, *Mol. Cell. Biochem.*, 2, 39, 1973.
265. **Yamazaki, I.**, Peroxidase, in *Molecular Mechanisms of Oxygen Activation*, Hayaishi, O., Ed., Academic Press, New York, 1974, 535.

266. **Bielski, B. H. J. and Chan, P. C.,** Re-evaluation of the kinetics of lactate dehydrogenase-catalyzed chain oxidation of NADH by superoxide radicals in the presence of EDTA, *J. Biol. Chem.,* 251, 3841, 1976.
267. **Hirata, F. and Hayaishi, O.,** Superoxide anion as an intermediate or a substrate for certain oxygenases, in *Superoxide and Superoxide Dismutases,* Michelson, A. M., McCord, J. M., and Fridovich, I., Eds., Academic Press, London, 1977, 395.
268. **Kido, T., Soda, K., and Asada, K.,** Properties of 2-nitropropane dioxygenase of *Hansenula mrakii.* Formation and participation of superoxide, *J. Biol. Chem.,* 253, 226, 1978.
269. **Nakamura, S. and Yamazaki, I.,** One-electron transfer reactions in biochemical systems. IV. A mixed mechanism in the reaction of milk xanthine oxidase with electron acceptors, *Biochim. Biophys. Acta,* 189, 29, 1969.
270. **Odajima, T. and Yamazaki, I.,** Myeloperoxidase of the leucocyte of normal blood. III. The reaction of ferric myeloperoxidase with superoxide, *Biochim. Biophys. Acta,* 284, 355, 1972.
271. **Debey, P. and Balny, C.,** Production of superoxide ions in rat liver microsomes, *Biochimie,* 55, 329, 1973.
272. **Bielski, B. H. J. and Chan, P. C.,** Enzyme-catalyzed free radical reactions with NADH. I. Lactate dehydrogenase-catalyzed chain oxidation of bound NADH by superoxide radicals, *Arch. Biochem. Biophys.,* 159, 873, 1973.
273. **Bielski, B. H. J. and Chan, P. C.,** Studies on free and enzyme-bound NAD· free radicals, *J. Am. Chem. Soc.,* 102, 1713, 1980.
274. **Halliwell, B.,** Lignin synthesis: the generation of hydrogen peroxide and superoxide by horseradish peroxidase and its stimulation by Mn(II) and phenols, *Planta,* 140, 81, 1978.
275. **Oyanagui, Y.,** Inhibition of superoxide anion production in nonstimulated guinea pig peritoneal exudate cells by anti-inflammatory drugs, *Biochem. Pharmacol.,* 27, 777, 1978.
276. **Gross, G. G., Janse, C., and Elstner, E. F.,** Involvement of malate, monophenols, and superoxide in hydrogen peroxide formation by isolated cell walls from horseradish *(Armoracia lapathifolia* Gilib), *Planta,* 136, 271, 1977.
277. **Anbar, M. and Neta, P.,** A compilation of specific bimolecular rate constants for the reactions of hydrated electrons, hydrogen atoms and hydroxyl radicals with inorganic and organic compounds in aqueous solution, *Int. J. Appl. Radiat. Isot.,* 18, 493, 1967.
278. **Bors, W., Lengfelder, E., Saran, M., Fuchs, C., and Michel, C.,** Reactions of oxygen radical species with methional: a pulse radiolysis study, *Biochem. Biophys. Res. Comm.,* 70, 81, 1976.
279. **Bors, W., Saran, M., and Michel, C.,** Radical intermediates involved in the bleaching of the carotenoid crocin. Hydroxyl radicals, superoxide anions and hydrated electrons, *Int. J. Radiat. Biol.,* in press.
280. **Heikkila, R. E. and Cabbat, F. S.,** Protection against alloxan-induced diabetes in mice by the hydroxyl radical scavenger dimethylurea, *Eur. J. Pharmacol.,* 52, 57, 1978.
281. **Cederbaum, A. I., Dicker, E., Rubin, E., and Cohen, G.,** Effect of thiourea on microsomal oxidation of alcohols and associated microsomal functions, *Biochemistry,* 18, 1187, 1979.
282. **Beauchamp, C. O. and Fridovich, I.,** A mechanism for the production of ethylene from methional. The generation of the hydroxyl radical by xanthine oxidase, *J. Biol. Chem.,* 245, 4641, 1970.
283. **Koppenol, W. H., Butler, J., and van Leeuwen, J. W.,** The Haber-Weiss cycle, *Photochem. Photobiol.,* 28, 655, 1978.
284. **Weinstein, J. and Bielski, B. H. J.,** Kinetics of the interaction of $HO_2\cdot$ and O_2^- radicals with hydrogen peroxide. The Haber-Weiss reaction, *J. Am. Chem. Soc.,* 101, 58, 1979.
285. **Gilbert, B. C., Hodgeman, D. K. C., and Norman, R. O. C.,** ESR studies. XXXVIII. Evidence for the formation of dimeric radical-cations, $[R_2S\text{-}SR_2]^+$, in the one-electron oxidation of sulfides, *J. C. S. Perkin II,* 1748, 1973.
286. **Bonifacic, M., Moeckel, H., Bahnemann, D., and Asmus, K. D.,** Formation of positive ions and other primary species in the oxidation of sulphides by hydroxyl radicals, *J. C. S. Perkin II,* 675, 1975.
287. **Asmus, K. D.,** Stabilisation of oxidized sulfur centers in organic sulfides. Radical cations and odd-electron sulfur-sulfur bonds, *Acc. Chem. Res.,* 12, 436, 1979.
288. **Saran, M., Bors, W., Michel, C., and Elstner, E. F.,** Formation of ethylene from methionine. Reactivity of radiolytically produced oxygen radicals and effect of substrate activation, *Int. J. Radiat. Biol.,* 37, 521, 1980.
289. **Pryor, W. A. and Tang, R. H.,** Ethylene formation from methional, *Biochem. Biophys. Res. Comm.,* 81, 498, 1978.
290. **Cederbaum, A. I., Miwa, G., Cohen, G., and Lu, A. Y. H.,** Production of hydroxyl radicals and their role in the oxidation of ethanol by a reconstituted microsomal system containing cytochrome P-450 purified from phenobarbital-treated rats, *Biochem. Biophys. Res. Comm.,* 91, 747, 1979.
291. **Cohen, G. and Cederbaum, A. I.,** Chemical evidence for production of hydroxyl radicals during microsomal electron transfer, *Science,* 204, 66, 1979.

292. **Cohen, G. and Cederbaum, A. I.,** Microsomal metabolism of hydroxyl radical scavenging agents: relationship to the microsomal oxidation of alcohols, *Arch. Biochem. Biophys.,* 199, 438, 1980.
293. **Konze, J. R. and Elstner, E. F.,** Pyridoxalphosphate-dependent ethylene production from methionine by isolated chloroplasts, *F.E.B.S. Lett.,* 66, 8, 1976.
294. **Elstner, E. F., Saran, M., Bors, W., and Lengfelder, E.,** Oxygen activation in isolated chloroplasts. Mechanism of ferredoxin-dependent ethylene formation from methionine, *Eur. J. Biochem.,* 89, 61, 1978.
295. **Hiller, K.-O., Göbl, M., Masloch, B., and Asmus, K.-D.,** On the oxidation mechanism of methionine in aqueous solutions, in *Radiation Biology and Chemistry Research Developments,* Edwards, H. E., Navarathnan, S., Parson, W. J, and Phillips, G. O., Eds., Elsevier, Amsterdam, 1979, 73.
296. **Hiller, K.-O., Masloch, B., Göbl, M., and Asmus, K.-D.,** Mechanism of the ·OH radical-induced oxidation of methionine in aqueous solutions, *J. Am. Chem. Soc.,* 103, 2734, 1981.
297. **Repine, J. E., Eaton, J. W., Anders, M. W., Hoidal, J. R., and Fox, R. B.,** Generation of hydroxyl radical by enzymes, chemicals and human phagocytes in vitro. Detection with the anti-inflammatory agent, dimethyl sulfoxide, *J. Clin. Invest.,* 64, 1642, 1979.
298. **Gilbert, B. C., Norman, R. O. C., and Sealy, R. C.,** ESR studies. XLIII. Reaction of dimethylsulfoxide with the hydroxyl radical, *J.C.S. Perkin II,* 303, 1975.
299. **Meissner, G., Henglein, A., and Beck, G.,** Pulsradiolytische Untersuchung von Dimethylthioäther und Dimethylsulfoxid in wässriger Lösung, *Z. Naturforsch.,* 22b, 13, 1967.
300. **Veltwisch, D., Janata, E., and Asmus, K. D.,** Primary processes in the reaction of OH-radicals with sulphoxides, *J. C. S. Perkin II,* 146, 1980.
301. **Klein, S. M., Cohen, G., and Cederbaum, A. I.,** The interaction of hydroxyl radicals with dimethylsulfoxide produces formaldehyde, *F.E.B.S. Lett.,* 116, 220, 1980.
302. **Loeff, I. and Stein, G.,** The radiation and photochemistry of aqueous solutions of benzene, *J. Chem. Soc.,* 2623, 1963.
303. **Willson, R. L., Greenstock, C. L., Adams, G. E., Wageman, R., and Dorfman, L. M.,** The standardization of ·OH-radical rate data from radiation chemistry, *Int. J. Radiat. Phys. Chem.,* 3, 211, 1971.
304. **Srinivasan, T. K. K., Balakrishnan, I., and Reddy, M. P.,** On the nature of the products of γ-radiolysis of aerated aqueous solutions of benzene, *J. Phys. Chem.,* 73, 2071, 1969.
305. **Bhatia, K.,** Reactions of the radiolytically generated hydroxycyclohexadienyl radical in aqueous benzene systems, *Radiat. Res.,* 59, 537, 1974.
306. **Walling, C. and Johnson, R. A.,** Fenton's reagent. V. Hydroxylation and side-chain cleavage of aromatics, *J. Am. Chem. Soc.,* 97, 363, 1975.
307. **Jefcoate, C. R. E., Lindsay Smith, J. R., and Norman, R. O. C.,** Hydroxylation. IV. Oxidation of some benzenoid compounds by Fenton's reagent and the ultraviolet irradiation of hydrogen peroxide, *J. Chem. Soc. B.,* 1013, 1969.
308. **Bhatia, K.,** Determination of trace phenol in aqueous solution by aqueous liquid chromatography, *Anal. Chem.,* 45, 1344, 1973.
309. **Rigo, A., Stevanato, R., Finazzi Agro, A., and Rotilio, G.,** An attempt to evaluate the rate of the Haber-Weiss reaction by using ·OH radical scavenger, *F.E.B.S. Lett.,* 80, 130, 1977.
310. **Jacob, N., Balakrishnan, I., and Reddy, M. P.,** Characterization of the hydroxyl radical in some photochemical reactions, *J. Phys. Chem.,* 81, 17, 1977.
311. **Lindsay Smith, J. R., Shaw, B. A. J., Foulkes, D. M., Jeffrey, A. M., and Jerina, D. M.,** Aromatic hydroxylation. VI. Oxidation of naphthalene by dioxygen in the presence of iron(II) salts, *J. C. S. Perkin II,* 1583, 1977.
312. **Zevos, N. and Sehested, K.,** Pulse radiolysis of aqueous naphthalene solutions, *J. Phys. Chem.,* 82, 138, 1978.
313. **O'Neill, P., Steenken, S., van der Linde, H., and Schulte-Frohlinde, D.,** Reactions on OH radicals with nitrophenols in aqueous solution, *Radiat. Phys. Chem.,* 12, 13, 1978.
314. **Nofre, C., Cier, A., and Lefier, A.,** Activation de l'oxygène moléculaire en solution aqueuse par les ions métalliques et leurs chélates, *Bull. Soc. Chim. Fr.,* 530, 1961.
315. **Melhuish, W. H. and Sutton, H. C.,** Study of the Haber-Weiss reaction using a sensitive method for detection of ·OH radicals, *J. C. S. Chem. Comm.,* 970, 1978.
316. **Matthews, R. W. and Sangster, D. F.,** Measurement by benzoate radiolytic decarboxylation of relative rate constants for hydroxyl radical reactions, *J. Phys. Chem.,* 69, 1938, 1965.
317. **Sagone, A. L., Decker, M. A., Wells, R. M., and Democko, C.,** A new method for the detection of hydroxyl radical production by phagocytic cells, *Biochim. Biophys. Acta,* 628, 90, 1980.
318. **Halliwell, B.,** Generation of hydrogen peroxide, superoxide and hydroxyl radicals during the oxidation of dihydroxyfumaric acid by peroxidase, *Biochem. J.,* 163, 441, 1977.
319. **Halliwell, B. and Ahluwalia, S.,** Hydroxylation of *p*-coumaric acid by horseradish peroxidase. The role of superoxide and hydroxyl radicals, *Biochem. J.,* 153, 513, 1976.

320. **Mill, T., Hendry, D. G., and Richardson, H.,** Free-radical oxidants in natural waters, *Science,* 207, 886, 1980.
321. **Norman, R. O. C. and Radda, G. K.,** Aromatic hydroxylation: the electrophilic character of the hydroxyl radical and its significance in biological hydroxylation, *Proc. Chem. Soc.,* 138, 1962.
322. **Dizdaroglu, M. and Simic, M. G.,** Radiation-induced conversion of phenylalanines to tyrosines, *Radiat. Res.,* 83, 437, 1980.
323. **Hamilton, G. A.,** Oxidation by molecular oxygen. II. The oxygen atom transfer mechanism for mixed-function oxidases and the model for mixed-function oxidases, *J. Am. Chem. Soc.,* 86, 3391, 1964.
324. **Hamilton, G. A., Friedman, J. P., and Campbell, P. M.,** The hydroxylation of anisole by hydrogen peroxide in the presence of catalytic amounts of ferric ion and catechol. Scope, requirements, and kinetic studies, *J. Am. Chem. Soc.,* 88, 5266, 1966.
325. **Cadet, J. and Teoule, R.,** Comparative study of oxidation of nucleic acid components by hydroxyl radicals, singlet oxygen and superoxide anion radicals, *Photochem. Photobiol.,* 28, 661, 1978.
326. **Hariharan, P. V., Courtney, J., and Eleczko, S.,** Production of hydroxyl radicals in cell systems exposed to haematoporphyrin and red light, *Int. J. Radiat. Biol.,* 37, 691, 1980.
327. **Hatada, M., Kraljic, I., El Samahy, A., and Trumbore, C. N.,** Radiolysis and photolysis of the hydrogen peroxide-*p*-nitrosodimethylaniline-oxygen system, *J. Phys. Chem.,* 78, 888, 1974.
328. **Kraljic, I. and Trumbore, C. N.,** *p*-Nitrosodimethylaniline as an ·OH radical scavenger in radiation chemistry, *J. Am. Chem. Soc.,* 87, 2547, 1965.
329. **Dainton, F. S. and Wiseall, B.,** Reactions of nitrosodimethylaniline with free radicals, *Trans. Faraday Soc.,* 64, 694, 1967.
330. **Shah, S., Trumbore, C. N., Giessner, B., and Park, W.,** Studies in the radiation and photochemistry of aqueous *p*-nitrosodimethylaniline, in *Radiation Chemistry, Vol. I,* Adv. Chem. Ser. 81, Gould, R. F., Ed., ACS Publishing, Washington, D.C., 1968, 321.
331. **Baxendale, J. H. and Khan, A. A.,** The pulse radiolysis of *p*-nitrosodimethylaniline in aqueous solutions, *Int. J. Radiat. Phys. Chem.,* 1, 11, 1969.
332. **Floyd, R. A., Soong, L. M., Stuart, M. A., and Reigh, D. L.,** Free radicals and carcinogenesis. Some properties of the nitroxyl free radicals produced by covalent binding of 2-nitrosofluorene to unsaturated lipids of membranes, *Arch. Biochem. Biophys.,* 185, 450, 1978.
333. **Elstner, E. F. and Zeller, H.,** Bleaching of *p*-nitrosodimethylaniline by photosystem I of chloroplast lamellae, *Plant Sci. Lett.,* 13, 15, 1978.
334. **Kraljic, I., El Mohsni, S., and Arvis, M.,** A general method for the identification of primary reactions in sensitized photooxidations, *Photochem. Photobiol.,* 27, 531, 1978.
335. **Friend, J. and Mayer, A. M.** The enzymic destruction of carotenoids by isolated chloroplasts, *Biochim. Biophys. Acta,* 41, 422, 1960.
336. **Weber, F., Laskawy, G., and Grosch, W.,** Co-oxidation of carotene and crocin by soyabean lipoxygenase isoenzymes, *Z. Lebensm. Unters. Forsch.,* 155, 142, 1974.
337. **Weber, F. and Grosch, W.,** Co-oxidation of a carotenoid by the enzyme lipoxygenase: influence on the formation of linoleic acid hydroperoxides, *Z. Lebensm. Unters. Forsch.,* 161, 223, 1976.
338. **Dicks, J. W. and Friend, J.,** The oxidation of carotenoids by mitochondria from sugar beet leaves. II. Crocin destruction by an unsaturated fat oxidase system in Triton extracts of mitochondria, *Phytochemistry,* 6, 1193, 1967.
339. **Dicks, J. W. and Friend, J.,** The oxidation of carotenoids by mitochondria from sugar beet leaves. III. Crocin oxidation by a peroxidase system, *Phytochemistry,* 7, 1933, 1968.
340. **Dicks, J. W.,** The participation of peroxidase and oxalic acid in crocin destruction by a particulate system from sugar beet leaves, *Phytochemistry,* 9, 1433, 1970.
341. **Montalbini, P., Koch, F., Burba, M., and Elstner, E. F.,** Increase in lipid-dependent carotene destruction as compared to ethylene formation and chlorophyllase activity following mixed infection of sugar beet (*Beta vulgaris* L.) with beet yellows virus and beet mild yellowing virus, *Physiol. Plant Pathol.,* 12, 211, 1978.
342. **Bors, W., Michel, C., and Saran, M.,** Organic oxygen radicals in biology: generation and detection, in *Oxygen and Oxy-Radicals in Chemistry and Biology,* Rogers, M. A. J. and Powers, E. L., Eds., Academic Press, New York, 1981, 79.
343. **Saran, M., Michel, C., and Bors, W.,** The bleaching of crocin by oxygen radicals, in *Chemical and Biochemical Aspects of Superoxide and Superoxide Dismutase,* Bannister, J. V. and Hill, H. A. O., Eds., Elsevier, Amsterdam, 1980, 38.
344. **McCord, J. M. and Day, E. D.,** Superoxide-dependent production of hydroxyl radical catalyzed by iron-EDTA complex, *F.E.B.S. Lett.,* 86, 139, 1978.
345. **Iddon, B., Phillips, G. O., Robbins, K. E., and Davies, J. V.,** Radiation chemistry of aqueous solutions of indole and its derivatives, *J. Chem. Soc. B,* 1887, 1971.
346. **Armstrong, R. C. and Swallow, A. J.,** Pulse- and γ-radiolysis of aqueous solutions of tryptophan, *Radiat. Res.,* 40, 563, 1969.

347. **Bors, W., Michel, C., and Saran, M.,** Generation and reactions of various types of oxygen radicals, *Bull. Eur. Physiopath. Resp.,* 17 (suppl.), 13, 1981.
348. **Goldstein, I. M., Kaplan, H. G., Edelson, H. S., and Weissmann, G.,** Ceruloplasmin. A scavenger of superoxide anion radicals, *J. Biol. Chem.,* 254, 4040, 1979.
349. **Henry, M.-F. and Vignais, P. M.,** Production of superoxide anions in *Paracoccus denitrificans, Arch. Biochem. Biophys.,* 203, 365, 1980.
350. **Yamashoji, S., Yoshida, H., and Kajimoto, G.,** Evidence for generation of O_2^- or an O_2^--like factor in the decomposition of linoleic acid hydroperoxide, *Agric. Biol. Chem.,* 43, 665, 1979.
351. **Egan, R. W., Paxton, J., and Kuehl, F. A.,** Mechanism for irreversible self-deactivation of prostaglandin synthetase, *J. Biol. Chem.,* 251, 7329, 1976.
352. **Kuehl, F. A., Egan, R. W., Humes, J. L., Beveridge, G. C., and van Arman, C. G.,** Evidence for a pivotal role of the endoperoxide, PGG_2, in inflammatory processes, in *Biochemical Aspects of Prostaglandins and Thromboxanes,* Kharasch, N. and Fried, J., Eds., Academic Press, New York, 1977, 55.
353. **Egan, R. W., Gale, P. H., Beveridge, G. C., Phillips, G. B., and Marnett, L. J.,** Radical scavenging as the mechanism for stimulation of prostaglandin cyclo-oxygenase and depression of inflammation by lipoic acid and sodium iodide, *Prostaglandins,* 16, 861, 1978.
354. **Ullrich, V.,** Dioxygen activation by cytochrome P-450, in *Oxygen and Oxy-Radicals in Chemistry and Biology,* Rogers, M. A. J. and Powers, E. L., Eds., Academic Press, New York, 1981, 497.
355. **Sligar, S. G., Shasty, B. S., and Gunsalus, I. C.,** Oxygen reactions of the P-450 heme protein, in *Microsomes and Drug Oxidations,* Ullrich, V., Roots, I., Hildebrandt, A., Estabrook, R. W., and Conney, A. H., Eds., Pergamon Press, London, 1977, 202.
356. **Yamamoto, H., Takei, H., Yamamoto, T., and Kimura, M.,** Ferric ion catalyzed decomposition of hydrogen peroxide in acetonitrile-water mediums, *Chem. Pharm. Bull.,* 27, 789, 1979.
357. **Ingold, K. U.,** Peroxy radicals, *Acc. Chem. Res.,* 2, 1, 1969.
358. **Kochi, J. K.,** Oxygen radicals, in *Free Radicals,* Vol. 2, Kochi, J. K., Ed., Interscience, New York, 1973, 665.
359. **Simic, M.,** The chemistry of peroxy radicals and its implication to radiation biology, in *Fast Processes in Radiation Chemistry and Biology,* Adams, G. E., Fielden, E. M., and Michael, B. D., Eds., John Wiley & Sons, New York, 1975, 162.

Chapter 4

THE SPIN TRAPPING OF SUPEROXIDE AND HYDROXYL RADICALS

Garry R. Buettner

TABLE OF CONTENTS

I. INTRODUCTION TO SPIN TRAPPING

A. The Experiment

The spin trapping technique involves the addition reaction of a very reactive short-lived free radical to a diamagnetic compound, spin trap, to produce a relatively long-lived free radical product, spin adduct, which can be easily studied by electron spin resonance, ESR. In favorable cases, the resulting ESR spectrum allows the identification of the original reactive radical, or if no unique assignment is feasible, it is still possible to learn something about the nature of R·, the short-lived free radical, and thus still probe the free radical chemistry of interest.

B. Historical Aspects of Spin Trapping

The spin trapping technique was proposed independently by five different research groups in the late 1960s. Janzen's group examined the capabilities of nitrones, in particular α-phenyl-*tert*-butyl nitrone (PBN), to undergo free radical addition reactions, and it was this group that introdued the term "spin trapping".[1,2] Forshult et al.,[3] Chalfont et al.,[4] and Leaver and Ramsay[5-7] introduced the use of 2-nitroso-2-methylpropane (NtB) as a radical trap. The use of nitrosobenzene and its derivatives as spin traps was introduced by Terabe and Konaka.[8] This initial work demonstrated that both nitrone and nitroso compounds are capable of undergoing addition reactions with reactive free radicals to form more stable radical products, spin adducts.

Several excellent reviews are now available on various aspects of spin trapping. C. A. Evans has an excellent overview of spin trapping that can serve as a starting point for the uninitiated.[9] An early review by Janzen focuses on the chemical aspects of spin trapping,[10] while his later review provides us with a critical look at spin trapping in biological systems.[11] McCay et al. have summarized some aspects of the use of spin traps in enzyme systems that produce free radicals.[12] Other excellent early reviews by Langercrantz[13] and Perkins[14] should also be consulted. A recent review focusing on the practical aspects of the spin trapping of superoxide and hydroxyl radical presents a brief summary of the many applications of spin trapping in the study of these radicals.[15]

II. SPIN TRAPS

A. Nitroso Spin Traps

Nitroso compounds, such as NtB, can provide considerably more information that can be used to identify the trapped free radical than can nitrones. The trapped radical adds directly to the nitroso nitrogen (see Figure 1), so that in the spin adduct, the group added is in the immediate vicinity of the nitroxide moiety. This close proximity of the added group to the nitroxide center results in the hyperfine splitting constants being quite sensitive to the nature of the trapped radical. In addition, elements of the trapped radical may lead to additional hyperfine splitting being present in the spin adduct spectrum. Thus, nitroso compounds can in general provide an excellent fingerprint of the trapped radical.[16-20] However, nitroso compounds are both thermally and photochemically unstable[10,21-23] and, in addition, can have a tendency to dimerize[10] making their use difficult in some applications. Ohkuma et al.,[23] while attempting to investigate the dissociation rate and equilibrium constant for the reaction of NtB dimer going to NtB monomer, observed the occurrence of side reactions that resulted in the loss of spin trapping agent. Warming the solution results in higher monomer concentration; however, decomposition and the extreme volatility of NtB became difficult problems to cope with.

Allowances must also be made for the visible absorption by nitroso compounds

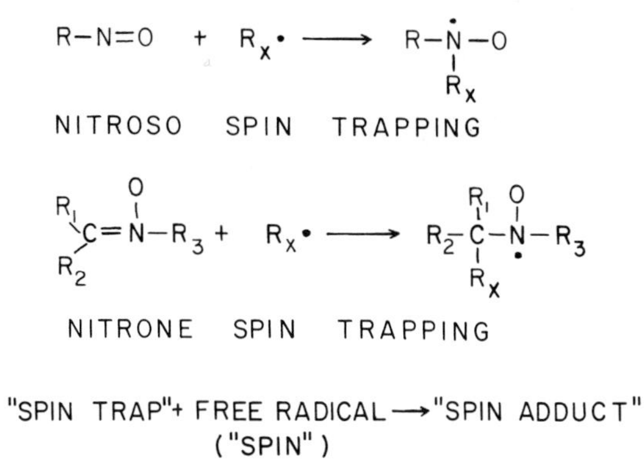

FIGURE 1. The nitroso and nitrone spin-trapping reactions.

when employing the spin trapping technique in photochemical studies. For example, the UV-irradiation of an aqueous solution of NtB will give rise to an intense triplet signal due to the di-*tert*-butylnitroxide radical, a result of the spin trapping of the *tert*-butyl radical produced by the photolysis of NtB and the subsequent spin trapping of this radical by NtB.[23]

The NtB spin adducts of oxygen-centered radicals, such as the hydroxyl radical, superoxide radical (perhydroxyl radical), and alkoxyl radicals, appear to be unstable.[11,24-26] For example, the hydroxyl radical spin adduct of NtB decomposes to yield *t*-butyl radicals,[24] thus yielding radicals that were not part of the original free radical chemistry being probed. The apparent unsuitability of nitroso compounds to probe the chemistry of oxygen-centered radicals leaves us with the nitrones as a tool in the spin trapping technique to probe free radical chemistry involving these species.

B. Nitrone Spin Traps

The two most popular nitrone spin traps in use are PBN(α-phenyl-*tert*-butyl nitrone) and DMPO (5,5-dimethyl-pyrroline-*N*-oxide or 5,5-dimethylpyrroline-1-oxide). PBN has good shelf stability and produces relatively stable spin adducts with both carbon-centered and oxygen-centered radicals. Unfortunately, the hyperfine splitting constants of the nitroxides resulting from the addition reaction of free radicals to nitrones are not nearly as sensitive to the nature of R· as the nitroxides derived from nitroso spin traps. However, DMPO, as introduced by Janzen,[27,28] is considerably more sensitive to the nature of R· than PBN and provides reasonable variations in the hyperfine splittings to discriminate between free radicals. For example, as seen in Table 1, the hyperfine splitting constants for ·OH with DMPO are $a_N = a_\beta^H$ = 14.9 gauss[29-31] while the splitting constants with PBN are a_N = 15.3 gauss and a_β^H = 2.75 gauss.[32] For the superoxide (perhydroxyl) radical spin adduct, the splitting constants are a_N = 14.3 gauss, a_β^H = 11.7 gauss, and a_γ^H = 1.25 gauss for DMPO,[32] while a_N = 14.8 gauss and a_β^H = 2.75 gauss for PBN. Thus, from the hyperfine splitting constants, the superoxide and hydroxyl radical spin adducts are easily distinguished with DMPO, while it is more difficult to discriminate between these same spin adducts with PBN as a spin trapping agent.

Additional nitrone spin trapping agents that may be of use are TMPO (2,5,5-trimethyl-1-pyrroline-*N*-oxide),[33] 4-POBN(α-4-pyridyl-*N*-oxide *tert*-butyl nitrone),[34] 4-MePyBN, (4-(*N*-methyl pyridinion) *tert*-butyl nitrone), or a variation of 4-MePyBN, such as 3-MePyBN or 2-MePyBN.[35,36]

TMPO lacks the β-hydrogen that is present in DMPO, and thus considerably less

Table 1
ESR PARAMETERS FOR VARIOUS SPIN ADDUCTS[a]

Spin adduct	Source	g	A_N/G	A_β^H/G	A_γ^H/G	Other	Ref.
DMPO/OOH	30% H_2O_2 + hυ	2.0061	14.3	11.7	1.25	—	32
DMPO/OOH	Chlorophyll + hυ	2.0061	14.3	11.7	1.25	—	72
DMPO/OOH	CdS + hυ	2.0061	14.1	11.3	1.25	—	86
DMPO/OOH	Xanthine + xanthine oxidase	2.0060	14.2	11.2	1.3	—	44
DMPO/OH	1% H_2O_2 + hυ	2.0060	15.3	15.3	—	—	32
DMPO/OH	Hematoporphyrin + hυ	2.0060	15.0	15.0	—	—	58
DMPO/OH	ADP-Fe(III) + H_2O_2	—	14.83	14.83	—	—	31
DMPO/OH	1% H_2O_2 + hυ	2.0062	15.0	15.	—	—	68
DMPO/OH	Chlorophyll + hυ	2.0061	14.9	14.9	—	—	30
DMPO/OH	$(NH_4)_2S_2O_8$	—	14.77	14.77	—	—	31
DMPO/OH	Radiolysis of H_2O	—	15.0	15.0	—	—	23
DMPO/OH	OH^- + $S_2O_8^{2-}$ + hυ	2.0057	14.90	14.90	—	—	41
DMPO/OH	1% H_2O_2 + hυ	—	14.9	14.9	—	—	72
DMPO/H	Radiolysis of H_2O (e^-)	—	16.7	22.4[b]	—	—	23
DMPO/H	SO_3^{2-} + hυ	2.0054	16.58	22.50[b]	—	—	41
DMPO/D (^2H)	Radiolysis of D_2O	—	—	—	—	$a_D = 3.3$	23
DMPO/CH_2OH	Radiolysis of H_2O + CH_3OH	—	16.0	22.7	—	—	23
DMPO/CH_2OH	$S_2O_8^{2-}$ + CH_3OH + hυ	2.0053	15.87	22.57	—	—	41
DMPO/CH_3CHOH	Microsomes + EtOH	2.0068	15.8	22.8	—	—	68
DMPO/CH_3CHOH	Hematoporphyrin + EtOH + hυ	2.0067	15.8	22.9	—	—	58
DMPO/$\cdot CO_2^-$	Chlorophyll + HCO_2^- + hυ	—	15.6	18.7	—	—	30
DMPO/$\cdot CO_2^-$	$S_2O_8^{2-}$ + HCO_2^- + hυ	2.0054	15.79	18.97	—	—	41
DMPO/CH_3	$S_2O_8^{2-}$ + $CH_3CO_2^-$ + hυ	2.0052	16.33	23.24	—	—	41
DMPO/$\cdot C_6H_5$	$S_2O_8^{2-}$ + $C_6H_5CO_2^-$ + hυ	2.0053	15.97	24.30	—	—	41
DMPO/$\cdot NH_2$	$S_2O_8^{2-}$ + NH_3 + hυ	2.0054	15.85	19.03	—	$a_{N_2} = 1.71$	41
DMPO/N_3 (^{14}N)	1% H_2O_2 + N_3^- + hυ	—	14.7	14.7	—	$a_{N_2} = 3.0$	85
DMPO/N_3 (^{15}N)	1% H_2O_2 + N_3^- + hυ	—	14.7	14.7	—	$a_{N_2} = 4.5$	85
DMPO/N_3	Hematoporphyrin + N_3^- + hυ	2.006	14.9	14.9	—	$a_{N_2} = 3.0$	58
DMPO/$SO_3^{\cdot -}$	SO_3^{2-} + hυ	2.0055	14.55	16.66	—	—	41
DMPO/$SO_4^{\cdot -}$	$S_2O_8^{2-}$ + hυ	2.0059	13.82	10.10	1.42 and 0.83	—	41
TMPO/$O_2^{\cdot -}$	Xanthine + xanthine oxidase	—	15.6	—	—	—	40
TMPO/OH	H_2O_2 + hυ	—	15.7	—	—	—	40
PBN/OOH	30% H_2O_2 + hυ	2.0057	14.8	2.75	—	—	32

PBN/OH	1% H_2O_2 + hυ	2.0057	15.3	2.75	—	—	32
PBN/OH	1% H_2O_2 + hυ	—	15.6	2.7	—	—	72
PBN/OH	1% H_2O_2 + hυ	2.0061	15.5	2.75	—	—	68
PBN/OH	$Na_2S_2O_8$	—	15.13	2.63	—	—	31
PBN/OH	$(NH_4)_4S_2O_8$	—	15.23	2.60	—	—	31
PBN/OH	Radiolysis of H_2O	—	15.6	2.7	—	—	23
PBN/H	Radiolysis of H_2O	—	16.8	10.9[b]	—	—	23
PBN/CH_2OH	Microsomal NADPH + EtOH	2.0056	16.00	3.74	—	—	67
PBN/CH_2OH	1% H_2O_2 + CH_3OH + hυ	—	16.07	3.86	—	—	67
PBN/CH_3CHOH	1% H_2O_2 + EtOH + hυ	—	15.94	3.34	—	—	67
PBN/CH_3CHOH	Microsomal-NADPH + EtOH	2.0056	16.10	3.35	—	—	67
PBN/TRIS	Microsomal-NADPH-Tris	—	16.00	3.75	—	—	67
PBN/DMSO-A	Microsomal-NADPH	2.0056	16.46	3.60	—	—	67
PBN/DMSO-B	Microsomal-NADPH	2.0058	15.10	3.42	—	—	67
PBN/linoleic acid-A	Microsomal-NADPH	—	16.54	3.35	—	—	67
PBN/linoleic acid-B	Microsomal-NADPH	—	15.50	3.75	—	—	67
PBN/N_3	1% H_2O_2 + N_3^- + hυ	—	15.05	2.06	—	$a_N = 2.06$	85
PBN/N_3	1% H_2O_2 + N_3^- + hυ	—	14.9	2.1	—	$a_{N_2} = 2.1$	87
PBN/N_3	H_2O_2 + Fe(III) + ADP + N_3^-	—	14.9	2.1	—	$a_{N_2} = 2.1$	87
4-MePyBN/$O_2^{\bar{\cdot}}$	Riboflavin + hυ	2.0091	13.78	1.65	—	—	36
4-MePyBN/OH	1% H_2O_2 + hυ	—	15.0	1.9	—	—	72
4-MePyBN/OH	H_2O_2 + hυ	—	14.70	1.45	0.38	—	35
4-MePyBN/OD	D_2O_2 + hυ	—	14.76	1.43	—	—	35
4-MePyBN/H	$S_2O_8^{2-}$	—	15.51	6.24[b]	—	—	35
4-MePyBN/N_3	1% H_2O_2 + N_3^- + hυ	—	14.68	1.95	—	$a_{N_2} = 1.95$	85
3-MePyBN/OH	H_2O_2 + hυ	—	14.80	1.42	0.32	—	35
3-MePyBN/OD	D_2O_2 + hυ	—	14.78	1.38	—	—	35
3-MePyBN/H	$S_2O_8^{2-}$	—	15.40	6.21[b]	—	—	35
2-MePyBN/OH	H_2O_2 + hυ	—	14.95	3.90	—	—	35
2-MePyBN/OD	D_2O_2 + hυ	—	14.96	4.10	—	—	35
2-MePyBN/H	$S_2O_8^{2-}$	—	15.42	6.22[b]	—	—	35
4-POBN/$O_2^{\bar{\cdot}}$	Xanthine + xanthine oxidase	—	14.16	1.75	—	—	40
4-POBN/OH[c]	1% H_2O_2 + hυ	—	14.97	1.68	0.34	—	34
4-POBN/CH_3CHOH	H_2O_2 + EtOH + hυ	—	15.56	2.59	—	—	15

[a] A selected set of spin adducts.

[b] Two protons.

[c] A study of splitting *vs.* pH is also presented by Janzen et al;[34] the values quoted are averages.

information is available for the characterization of the various spin adducts. 4-POBN is a water soluble nitrone which is unique in its suitability for the detection of hydroxyl radical at pH 6 and 7.[31,34] It produces unusually sharp lines that show splitting not only from the nitrogen and β-hydrogen but also from a γ-hydrogen. In addition it is quite resistent to hydrolysis at pH 6 or 7; while at pH 2 the hydrolysis is quite rapid ($t_{1/2}$ = 13.8 min) yielding the hydroxy spin adduct.[34] 4-MePyBN is also water soluble and produces unique spin adducts with both the hydroxyl free radical[35] and the superoxide radical.[36]

Being Schiff bases, nitrones can undergo hydrolysis to yield aldehydes and hydroxylamines

$$R - CH = N - R' + H_2O \rightarrow RCHO + R'NHOH \qquad (1)$$

Obviously, it is desirable that these reactions be slow for spin-trapping studies in aqueous solutions. Ohkuma et al.[23] have examined the rate of these reactions for DMPO, TMPO, and PBN inaqueous solutions of varying pH. Both DMPO and TMPO showed no significant change in their UV spectra after 20 hr in the pH range of 1.0 to 12.0. However, PBN did undergo hydrolysis in acid solution, but in the pH range 7.2 to 11.8 no change in the UV spectrum was observed after 72 hr. The hydrolysis of PBN was found to be first order with respect to both PBN and hydrogen ion concentration with a second order rate constant of 90.0 $M^{-1}hr^{-1}$. Thus, the rate of hydrolysis of nitrones appears to be dependent upon the structure, with acyclic nitrones being the most susceptible, aryl nitrones less so, and cyclic nitrones offering stiff resistance to hydrolysis.[37] It would appear that in the pH range of biological interest all three spin traps present no large problems from hydrolysis. However, the possibility should be kept in mind.

One additional consideration is the possible oxidation of the spin-trapping agent. Floyd and Soong observed the apparent oxidation of DMPO to form DMPOX (see Figure 2) in a hematin and cumene hydroperoxide biochemical system.[38] Finkelstein et al.[15] have examined this process and have found that other oxidizing agents such as lead dioxide can lead to this species. The DMPOX spectrum shows a solvent effect with a_N = 7.2 gauss, a_γ^H = $a_{\gamma 1}^H$ = 4.1 gauss in water.

C. Spin Adduct Characteristics

Often a major difficulty in studying free radicals in both chemical and biological systems with ESR spectroscopy is being able to produce a sufficient quantity of radicals for observation. If the radicals are very short-lived, their concentration may be too low for detection; this is especially true in biological systems. The spin-trapping technique allows the accumulation of free radicals in the form of relatively stable, long-lived spin adducts to a concentration that is suitable for ESR detection. A great many spin adducts are considerably more stable and longer-lived than the original free radicals, thus permitting a study of the original free radical. However, not all spin adducts have a sufficiently long lifetime for study. For example, the hydroxyl spin adduct of nitroso spin traps are too short-lived for detection, with the hydroxyl spin adduct undergoing disproportionation to yield diamagnetic products.[11,24-26,39] Thus, the lifetimes of the spin adducts are very important considerations for the success of spin trapping experiments, i.e., the longer the lifetime, the easier it is to accumulate a sufficient number of spin adduct molecules for detection and study by ESR spectroscopy.

1. ·OH Spin Adducts

Hydroxyl radicals react rapidly with nitrones to produce spin adducts that are relatively long-lived. However, the lifetime of the hydroxyl spin adduct of nitrones varies

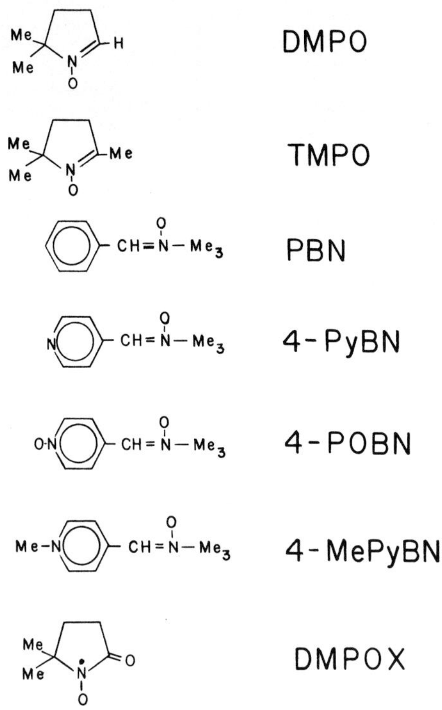

FIGURE 2. Some nitrone spin traps that have been used in the study of oxygen radical chemistry.

considerably. The hydroxyl spin adducts of cyclic nitrones, such as DMPO and TMPO, are in general more stable than the spin adducts of aryl nitrones such as PBN and 4-POBN (see Figure 3). For example, Finkelstein and co-workers[15,40] have observed that the half-lives of DMPO/OH and 4-POBN/OH are 2.6 hr and 23 sec, respectively, when in the presence of 0.6% hydrogen peroxide at pH 7.4. Recently, Kirino et al.[41] observed the half-life of DMPO/OH to be less than 5 sec in a spin-trapping experiment consisting of the photolysis of peroxydisulfate and DMPO in base. Thus, it is apparent that the spin adduct lifetime varies in relation to the chemical environment of the spin-trapping system.

The peroxydisulfate, a strong oxidizing agent with $E^\circ_{S.H.R.E.} = +2.01$ volts,[42] present in the spin-trapping solution could well have oxidized the nitroxide formed in the spin-trapping reaction. It must be remembered that the nitroxides that result from spin-trapping experiments are free radicals that can easily participate in oxidation or reduction reactions when appropriate reagents are present. For example, both oxidation and reduction are facile; with di-*tert*-butyl nitroxide, the one-electron potentials are +0.55 V and −1.63 V as measured polarographically in acetonitrile.[43] We have observed varying lifetimes for spin adducts of DMPO generated with the Fenton reaction

$$Fe(II) + H_2O_2 \rightarrow Fe(III) + OH^- + \cdot OH \qquad (2)$$

depending on the H_2O_2 and iron salt concentration of the spin-trapping mixture.[44] Janzen and co-workers have observed that MePyBN spin adducts are in general more stable than the spin adducts of PBN or 4-POBN;[35] however, no indication of the half-lives were given. The apparent stability of the hydroxyl radical spin adduct of DMPO

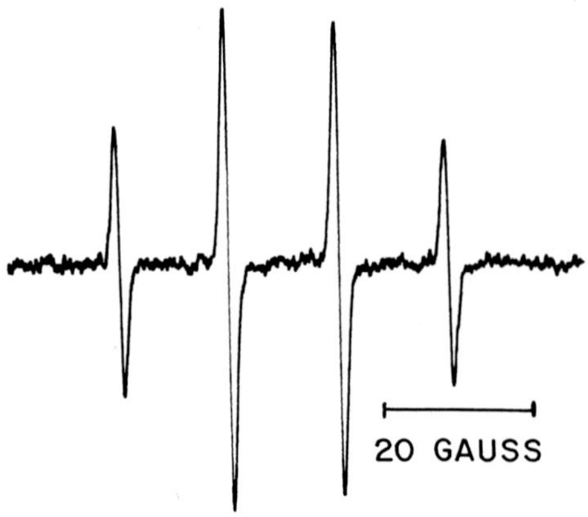

FIGURE 3. The hydroxyl spin adduct spectrum of DMPO produced by a solution of 50 mM DMPO, 100 μM H_2O_2, and 100 μM Fe(II) as ferrousammonium sulfate all in distilled water. The spectrum was obtained on a Varian E-4 equipped with an aqueous sample cell. Time constant = 0.5 sec, modulation amplitude = 0.8 gauss, gain = 5×10^3, power = 20 mW at a scan rate of 25 gauss/min. This 1:2:2:1 pattern results from the accidental equivalency of a_N and a_β^H; $a_N = a_\beta^H = 14.9$ gauss, g = 2.0061.

makes it quite useful as an agent for the trapping of this radical; PBN and MePyBN are also effective traps, but the lifetime of their ·OH adducts appears not to be as favorable as DMPO/OH.

Unfortunately, caution is in order when assigning an observed hydroxyl spin adduct of a nitrone spin trap. This species may arise from some source other than hydroxyl radical trapping. As mentioned above, the hydrolysis of 4-POBN in acid will yield an addition product identical to the hydroxyl spin adduct of 4-POBN when in the presence of H_2O_2.[34] Here, an intermediate in the hydrolysis reaction is oxidized to produce an ESR active compound identical to the ·OH radical spin adduct. In addition, the DMPO/OOH spin adduct is quite unstable and decomposes into DMPO/OH.[40,48] Further discussion of this observation is reserved for the next section.

The highly reactive nature of the hydroxyl radical can be used to assist in the verification that an observed hydroxyl radical spin adduct actually arose from the spin trapping of·OH, as opposed to some other chemistry of the spin trap itself. Compounds (scavengers) may be introduced into the hydroxyl radical generating system that are known to undergo specific reactions to produce new radicals. These new radicals can react with spin-trapping agents to produce a characteristic spin adduct. If the scavenger concentration is "low" compared to spin trap, the hydroxyl spin adduct signal will be dominant. However, as the scavenger concentration is raised, while holding the spin trap concentration constant, the hydroxyl spin adduct signal will decrease in intensity while the signal from the scavenger radical will increase in intensity. One could also alter the relative amounts of hydroxyl adduct and scavenger adduct by varying the concentration of spin trap while keeping the scavenger concentration constant. At "high" spin trap concentrations most of the hydroxyl radicals generated will react with the trap. But at "low" spin trap concentrations the hydroxyl radical will more readily react with the scavenger, producing more scavenger radical and subsequently more scavenger radical spin adduct and less of the hydroxyl radical adduct (see Figure 4).

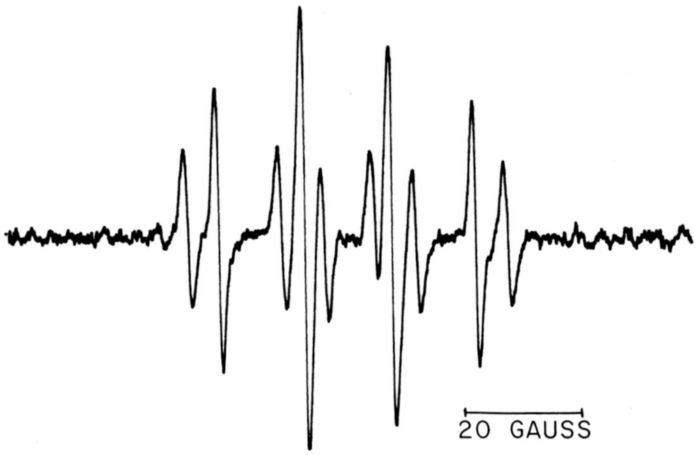

FIGURE 4. The hydroxyethyl and hydroxyl radical spin adducts of DMPO. Note the 1:2:2:1 pattern of the hydroxyl radical and the larger splitting of $a_\beta{}^H$ from the carbon-centered hydroxyethyl radical. The spectrum was produced by a solution of 50 m*M* DMPO, 100 μM Fe(II) from ferrousammonium sulfate, 100 μM H_2O_2 in the presence of 5.5 *M* EtOH. If the EtOH concentration is raised or the DMPO concentration is lowered, the hydroxyl radical adduct signal intensity decreases and the intensity of the hydroxylethyl radical increases. Instrument and settings are the same as in Figure 4. For the hydroxyethyl radical adduct of DMPO, a_N = 15.8 gauss, $a_\beta{}^H$ = 22.9 gauss, g = 2.0067.

Typical scavengers that may be employed are ethanol, formate ion, azide ion, or propanol.

$\cdot OH + CH_3CH_2OH \rightarrow CH_3\dot{C}HOH + H_2O$ $k = 1.1 \times 10^9$*   (Ref. 45,46)   (3)

$\cdot OH + HCO_2^- \rightarrow H_2O + \cdot CO_2^-$ $k = 3 \times 10^9$ (Ref. 46)   (4)

$\cdot OH + N_3^- \rightarrow OH^- + N_3\cdot$ $k = 1 \times 10^{10}$ (Ref. 46)   (5)

$\cdot OH + (CH_3)_2CHOH \rightarrow H_2O + (CH_3)_2\dot{C}OH$ $k = 2.2 \times 10^9$ (Ref. 46)   (6)

Each scavenger radical gives a distinct spin adduct with DMPO that can easily be distinguished from the hydroxyl radical adduct (see Table 1).

Kirino et al.[41] have also observed that the stability of the spin adducts of DMPO is greater when a neutral or bulky group is added to the spin trap. However, this may be an artifact of the presence of peroxydisulfate in their spin-trapping mixtures, although neutral and bulky groups may provide good protection of the nitroxide moiety through steric hinderance. Finkelstein et al.[40] have pointed out that spin traps lacking a β-hydrogen, such as TMPO, are less susceptible to H· abstraction and, therefore, should produce more stable spin adducts. Unfortunately, information is lost that may be helpful in the identification of particular spin adducts.

2. $O_2^{\overline{\cdot}}/HO_2\cdot$ Spin Adducts

Superoxide will add to the double bond of nitrones albeit at a much slower rate than the hydroxyl free radical.[33] The resulting spin adduct with DMPO is distinctive in that splitting from a γ-hydrogen is present in the spectrum (see Table 1 and Figure 1).

* Units of k are $dm^3mol^{-1}s^{-1}$.

However, other spin adducts of DMPO also present similar spectra such as the benzyloxy radical adduct.[15] Thus, it is necessary to make careful determinations of g-values and splitting constants when appropriate.

The spin adducts of $O_2^{\overline{\cdot}}/HO_2\cdot$ are much less stable than the corresponding spin adducts with $\cdot OH$.[15,36,40,47,48] For example, Buettner and Oberley have observed that the half-life of the DMPO/OOH spin adduct formed in a light-riboflavin-DETAPAC (diethylenetriaminepentaacetic acid) system is pH dependent. The half-life of DMPO/OOH ranged from 27 sec at pH 9 to 91 sec at pH 5 with a value of 58 sec at pH 7.[48] Finkelstein et al.[40] have observed a half-life approximately 100 sec for DMPO/OOH generated by a system of DMPO, and tetraammonium superoxide in *N,N*-dimethylformamide mixed in pH 7.8 phosphate buffer.

Using a riboflavin system, Sanderson and Chedekel have determined the half-life of the 4-MePyBN/OOH spin adduct.[36] Again, the half-life was found to be pH dependent with values of 83 sec at pH 5.5, 78 sec at pH 7.0, and 65 sec at pH 8.0, a stability similar to the DMPO/OOH spin adduct.

In studies on the spin trapping of superoxide, Finkelstein et al.[40] have determined that the DMPO/OOH spin adduct can decompose to yield DMPO/OH, as an increase in DMPO/OH signal intensity correlated inversely to a decrease in the DMPO/OOH signal intensity. In control experiments they found that superoxide dismutase depressed both the DMPO/OOH and DMPO/OH signals while catalase was without effect, ruling out the possibility of the Fenton reaction[49,50] making a significant contribution to the DMPO/OH signal intensity.

In contrast to DMPO, it was found that the superoxide spin adduct of TMPO is quite stable with no significant change in signal intensity from TMPO/OOH for a period of 4 min.[40] TMPO/OOH was also found to yield a unique spin adduct spectrum that can be distinguished from TMPO/OH. Unfortunately TMPO appears to have poor shelf stability and must be synthesized not long before its desired use as a spin trapping agent.[51]

The observation of $O_2^{\overline{\cdot}}/HO_2\cdot$ spin adducts can be a problem if the flux of superoxide is low. In addition, it has been noted that metal ions, even in trace amounts, can greatly affect the nature and time course of the spin adducts observed in superoxide generating systems(Figure 5).[40,47,48] Two possibilities present themselves: the metal ions present can compete with the spin trapping agent for the superoxide and/or the metal ions can react with the $O_2^{\overline{\cdot}}/HO_2\cdot$ spin adducts. Buettner et al.[47] have found that DETAPAC will greatly enhance the DMPO/OOH signal observed in superoxide generating spin trapping systems. Ethylenediaminitetraacetic acid (EDTA) was without beneficial effect in their experiments. They also observed that DETAPAC can eliminate the interfering effect of metals, even iron, on the xanthine oxidase-nitroblue-tetrazolium assay for superoxide dismutase, while EDTA would not eliminate this interference from iron. Thus, the competing reactions of metals with superoxide can be eliminated with DETAPAC. Finkelstein et al.[40] have noted that DETAPAC is as effective in eliminating interference from metals as the removal of these trace amounts of metals by passing all water and buffers through a chelex 100 column.

In addition, metals are known to accelerate the decomposition of hydroperoxides.[52] We have observed that the chelating agents DETAPAC and EDTA can slow the decomposition of benzyl peroxide by iron salts using a DMPO spin-trapping system.[53] One should be reminded that $O_2^{\overline{\cdot}}/HO_2\cdot$ spin adducts of the various spin-trapping agents are peroxides and as such their inherent instability and chemistry as peroxides needs to be considered. An extremely useful tool is the use of superoxide dismutase to inhibit the production of the $O_2^{\overline{\cdot}}/HO_2\cdot$ spin adduct in superoxide spin-trapping system. The specificity of superoxide dismutase provides a unique means of determining the involvement of superoxide in the production of spin adducts observed in spin-trapping experiments.

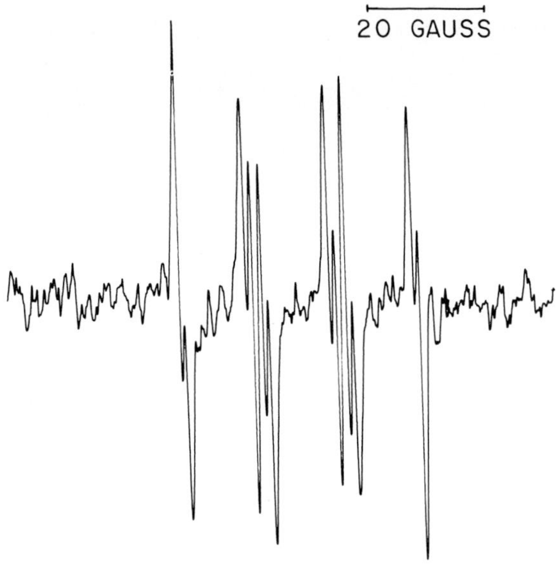

FIGURE 5. The superoxide spin adduct spectrum of DMPO produced by a solution of 0.1 m*M* xanthine, 0.15 m*M* DETAPAC, 70 m*M* DMPO, and 0.012 unit/m*ℓ* of xanthine oxidase in 50 m*M* phosphate buffer, pH 7.8. The spectrum was obtained on a Varian E-109 equipped with the E-238 aqueous solution cavity. Time constant = 0.5 sec, modulation amplitude = 0.8 gauss, gain = 5×10^4, power = 20 mW at room temperature, scan rate = 25 gauss/min, a_N = 14.2 gauss, a_β^H = 11.2 gauss, a_γ^H = 1.3 gauss, g = 2.0060.

3. Kinetics of $\cdot OH$ and $O_2^{\overline{\cdot}}/HO_2\cdot$ Addition to Spin Traps

a. *$O_2^{\overline{\cdot}}/HO_2\cdot$*

Using kinetic competition procedures Finkelstein et al. have examined the kinetics of the addition reaction of $O_2^{\overline{\cdot}}/HO_2\cdot$ to DMPO and TMPO.[33] They found the reaction to be first order with respect to both $O_2^{\overline{\cdot}}/HO_2\cdot$ and DMPO. The apparent rate constant was found to be pH dependent and could be explained by assuming that DMPO can react with $HO_2\cdot$ and $O_2^{\overline{\cdot}}$ at different rates. For DMPO, $k_{HO_2^{\cdot}} = 6600$ $M^{-1}s^{-1}$, $k_{O_2} = 10\ M^{-1}s^{-1}$ and

$$k_{app} = \frac{k_{HO_2^{\cdot}} + X\,k_{O_2^{\overline{\cdot}}}}{1 + X} \qquad (7)$$

where $X = Ka/[H^+] = 10^{(pH-pKa)}$. The pKa of superoxide is taken as 4.88.[54] This model predicts that the reaction between DMPO and $HO_2\cdot$ dominates below pH 7.7, while the reaction of DMPO with $O_2^{\overline{\cdot}}$ is dominant above pH 7.7. TMPO was found to have an apparent rate constant of 3.2 $M^{-1}s^{-1}$ for the addition of $O_2^{\overline{\cdot}}/HO_2\cdot$.

The pH dependence of the apparent rate constant for superoxide spin trapping by DMPO presents a dilemma. At low pH, the trapping reaction is most efficient; however, the dismutation reaction is very much faster for $HO_2\cdot$ than $O_2^{\overline{\cdot}}$.[54]

$$HO_2^{\cdot} + HO_2^{\cdot} \longrightarrow H_2O_2 + O_2 \qquad k = 7.6 \times 10^{5*} \qquad \text{(Ref. 54)} \qquad (8)$$

$$HO_2^{\cdot} + O_2^{-} \xrightarrow{H_2O} H_2O_2 + O_2 + OH^- \qquad k = 8.5 \times 10^{7} \qquad \text{(Ref. 54)} \qquad (9)$$

* Units of k are $dm^3mol^{-1}s^{-1}$.

$$O_2^{\bar{\cdot}} + O_2^{\bar{\cdot}} \xrightarrow{2H_2O} H_2O_2 + O_2 + 2\,OH^- \quad k \approx 10^{0} \text{ very slow} \quad \text{(Ref. 54)} \qquad (10)$$

while

$$\text{DMPO} + O_2^{\bar{\cdot}} \xrightarrow{H_2O} \text{DMPO/OOH} + OH^- \quad k = 10 \quad \text{(Ref. 33)} \qquad (11)$$

$$\text{DMPO} + HO_2^{\cdot} \longrightarrow \text{DMPO/OOH} \quad k = 6.6 \times 10^3 \quad \text{(Ref. 33)} \qquad (12)$$

From these rate constants we see that the steady-state concentration of $HO_2\cdot$ in acid solution will be much lower than the steady-state concentration of $O_2^{\bar{\cdot}}$ in alkaline solution. Consequently, at low pH the advantage of a higher rate of trapping is offset by the increased rate of dismutation. Thus, contrary to first impressions, DMPO can best compete with spontaneous dismutation at high pH.

b. ·OH

The hydroxyl radical is perhaps the most reactive free radical that arises in biological systems.

$$\cdot OH + e^- + H^+ \rightarrow H_2O, E^{\circ\prime} = 2.33 \text{ V at pH } 7.0 \quad \text{(Ref. 55)} \qquad (13)$$

It can react exceedingly fast with solutes by three basic mechanisms: hydrogen abstraction ($k = 10^7 - 10^9\ M^{-1}s^{-1}$),[46] electron transfer ($k = 10^7 - 10^9\ M^{-1}s^{-1}$),[46] and addition reactions ($k = 10^8 - 10^{10}\ M^{-1}s^{-1}$).[46] Thus, one would predict that the addition reaction of hydroxyl radical to the nitrone double bond should also be quite rapid. Indeed, using a kinetic competition approach, Finkelstein et al.[33] have determined the rate for the addition reaction of hydroxyl radical to DMPO, TMPO, 4-POBN, and CDMPO (2-carboxy-5,5-dimethyl-1-pyrroline-*N*-oxide).[33] Using ethanol as a competitive inhibitor they found

$$\text{DMPO} + \cdot OH \rightarrow \text{DMPO/OH} \quad k = 3.4 \times 10^9\ M^{-1}s^{-1} \text{ or } k = 2.1 \times 10^9\ M^{-1}s^{-1} \qquad (14)$$

$$\text{TMPO} + \cdot OH \rightarrow \text{TMPO/OH} \quad k = 3.8 \times 10^9\ M^{-1}s^{-1} \qquad (15)$$

$$\text{4-POBN} + \cdot OH \rightarrow \text{4-POBN/OH} \quad k = 1.9 \times 10^9\ M^{-1}s^{-1} \qquad (16)$$

$$\text{CDMPO} + \cdot OH \rightarrow \text{CDMPO/OH} \quad k = 1.55 \text{ to } 2.68 \times 10^9\ M^{-1}s^{-1} \qquad (17)$$

In the determination of the rate of reaction of ·OH with DMPO, two different hydroxyl radical generating systems were used, the photolysis of H_2O_2 and a Fenton system. The lower rate constant observed with the Fenton system may be a result of the presence of Fe^{2+} in the Fenton system, as oxidation of Fe^{2+} by alkyl radicals such as the α-hydroxethyl radical is possible.[49]

4. Variations in Hyperfine Splitting

The hyperfine splittings used to characterize the various spin adducts arise from an interaction of the unpaired electron of the nitroxide moiety with the nuclei that possess magnetic moments such as $^{14}N(I = 1)$, $^{15}N(I = \frac{1}{2})$, $^{1}H(I = \frac{1}{2})$, $^{2}H(I = 1)$, etc. The greater the probability of the unpaired electron being at a particular nucleus, the stronger that interaction and subsequently the larger the associated hyperfine splitting constant.[56] In addition,the geometry of the spin adduct can influence the magnitude of the hyperfine interactions.[9,27] In their examination of the hyperfine splitting of various spin adducts of DMPO in benzene solution, Janzen and Liu[27] found that hydroxylalkyl radical adducts produced large values for a_N and a_β^H probably due to an increase in spin density on the nitrogen and a small dihedral angle for the β-CH caused by intramolecular hydrogen bonding. In addition, the nature of the solvent can influence the hyperfine splitting constant. For example, the 1-hydroxyethyl radical ($CH_3\dot{C}HOH$) spin adduct of DMPO has been found to have hyperfine splitting constants of a_N = 15.03 gauss and a_β^H = 22.53 gauss in benzene,[27] while a_N = 15.8 gauss and a_β^H = 22.9 gauss[56-58] in aqueous solution.

In addition, the same spin adducts may have different hyperfine splitting constants in the same solvent (see Table 1). This variation is perhaps attributable to the nature of the additional components of the reaction mixture. Janzen et al.[31] have studied these variations for the hydroxyl and hydroperoxyl radicals. They observed a correlation in these variations by making plots of a_β^H vs. a_N. They found that the data produced straight lines that could be fitted empirically by the following equations:

$$a_\beta^H = 0.604\, a_N - 6.53 \text{ for PBN/OH} \tag{18}$$

$$a_\beta^H = 1.26\, a_N - 15.7 \text{ for PBN/OOH} \tag{19}$$

These correlations allow one to be more confident when making spin adduct assignments, even though previously published data from a different spin-trapping system on the same spin adduct may be somewhat at variance.

5. Singlet Oxygen and Spin Trapping with Nitrones

Often in the study of active oxygen species in biological systems, the question of the involvement of singlet oxygen is raised. Using a photochemical system, Harbour et al.[59] have found that indeed singlet oxygen can react with some nitrones. They observed that PBN ($k = 1.4 \times 10^8\ M^{-1}s^{-1}$), 4-PyBN ($k = 1.4 \times 10^8\ M^{-1}s^{-1}$), 4-POBN ($k = 1.2 \times 10^8\ M^{-1}s^{-1}$), 4-MePyBN ($k = 8 \times 10^7\ M^{-1}s^{-1}$), and DMPO ($k = 1.8 \times 10^7\ M^{-1}s^{-1}$) all reacted with singlet oxygen. However, only 4-POBN, 4-MePyBN, and DMPO yielded paramagnetic products. The hyperfine splitting constants observed corresponded to the ·OH spin adduct in each case. However, only very small quantities of spin adduct were detected in these experiments. Thus, the observation of ·OH radical adduct at low concentration may indicate singlet oxygen generation rather than hydroxyl radical production. An examination of the rate constants suggests that DMPO may be the best spin trap to use in probing free radical reactions in systems that may generate singlet oxygen. It should be kept in mind that these reactions with singlet oxygen will deplete both the spin trap and the oxygen in the system and thus may effect the chemistry of the system.

An additional consideration brought to light by Peters and Rogers[60] and Saito et al.[61] is that under appropriate conditions singlet oxygen can easily be reduced to superoxide. Thus, it may be that in some experiments singlet oxygen could lead to the free radical chemistry of superoxide and hydroxyl radical.

III. A FEW APPLICATIONS

A. Spin Trapping and Phagocytes

The formation of active oxygen species by phagocytosing cells is a subject of current interest as their role in microbicidal activity is not understood.[62] The use of spin trapping by Green et al.[63] to observe the production of oxygen radicals by stimulated neutrophils may be the first use of spin trapping in intact cells. Using DMPO, they were able to observe both DMPO/OH and DMPO/OOH formation by stimulated neutrophils. Both signals were abolished by superoxide dismutase; however, in the presence of catalase the DMPO/OH adduct was observed. Thus, considering their experimental conditions, it seems reasonable that the majority of the DMPO/OH spin adduct observed was the result of the decomposition of the superoxide spin adduct, DMPO/OOH, to form the DMPO/OH spin adduct rather than the result of ·OH radical trapping. The DMPO/OOH adduct was observed in the absence of superoxide dismutase and catalase.

Rosen and Klebanoff[64] were able to detect only the DMPO/OH spin adduct in a spin-trapping system with stimulated human polymorphonuclear leukocytes. Again, superoxide dismutase markedly suppressed the observed signal while catalase brought about an approximate 20% decline in signal intensity. Again, this is consistent with the initial trapping of $O_2^{\overline{\cdot}}/HO_2\cdot$ by DMPO to form DMPO/OOH and the subsequent breakdown of DMPO/OOH to DMPO/OH. These results suggest that only a small amount of DMPO/OH came from the trapping of ·OH itself. These examples demonstrate the importance of using superoxide dismutase and catalase as tools to interpret experimental results in $O_2^{\overline{\cdot}}$, H_2O_2, and ·OH generating systems.

B. Lipid Peroxidation

The role of the superoxide and hydroxyl free radicals in lipid peroxidation has been of great interest since the realization that oxygen radicals exist in biological systems. Pederson and Aust have demonstrated the production of superoxide by purified NADPH-cytochrome *c* reductase systems from rat liver microsomes that are believed to be involved in lipid peroxidation.[65] Fong et al.[66] have proposed that the hydroxyl radical initiates lipid peroxidation in liver microsomes. Piette's group has employed spin trapping in an effort to study free radical production in these processes.[57,67-69] They have gathered evidence that appears to be consistent with the role of the hydroxyl free radical as an initiator of microsomal lipid peroxidation during reduced nicotinamide adenine dinucleotide phosphate (NADPH) oxidation. They believe that cytochrome P450 reductase may be the enzyme responsible for this NADPH-dependent lipid peroxidation in liver microsomes.[67] They were able to observe the PBN/OH and DMPO/OH adducts in their microsomal system. In the presence of hydroxyl radical scavengers they detected the appropriate scavenger radical spin adduct of their spin trap. The hydroxyl radical is thought to arise from the Fenton reaction, as catalase inhibited[68] the formation of DMPO/OH while superoxide dismutase was without effect.[69] The presence of DMPO inhibited the formation of malondialdehyde (MDA). This is consistent with DMPO being an excellent scavenger of hydroxyl radical. Cederbaum and Cohen found DMPO to be an excellent inhibitor of microsomal oxidation of ethanol and 1-butanol as well as the metabolism of 2-keto-4-thiomethylbutyrate.[70] They were able to exclude the possible toxic action of DMPO on microsomes, and as such were able to confirm the involvement of hydroxyl radical in these processes.

Tien et al.[71] have used spin trapping in their studies of superoxide-dependent lipid peroxidation. As expected, Fenton's reagent readily stimulates lipid peroxidation. This peroxidation is inhibited by catalase and hydroxyl radical scavengers but not by superoxide dismutase. However, if adenosine 5′-diphosphate (ADP) is introduced into Fenton's reagent, a type of lipid peroxidation results which is not inhibited by hydroxyl radical scavengers and, as seen by the use of DMPO, the amount of hydroxyl radical is greatly reduced.

Xanthine-oxidase-promoted lipid peroxidation is not inhibited by catalase and is greatly stimulated by ADP.[71] Hydroxyl radical scavengers are without effect. In addition, xanthine oxidase-promoted lipid peroxidation is very effectively inhibited by superoxide dismutase but not by catalase. The roles of superoxide and iron are of great interest in these processes and spin trapping can play a role in elucidating the mechanism of these events.

C. Photochemical Systems

The technique of spin trapping has been used to observe the photosensitized production of the superoxide and hydroxyl free radicals. The production of superoxide by the illumination of spinach chloroplasts was the first application of spin trapping to biological systems.[72] The superoxide spin adduct of DMPO was observed upon the illumination with red light of a spinach chloroplast suspension. Methyl viologen was found to enhance the production of DMPO/OOH, thus supporting the hypothesis that methyl viologen accepts electrons from the primary acceptor of photosystem I and then subsequently transfers these electrons to O_2 producing $O_2^{\cdot -}$.

The use of DMPO as a spin trap in conjunction with hydroxyl radical scavengers such as formate and 2-propanol have allowed Harbour and Bolton[30] to observe the photoproduction of the hydroxyl free radical during the photooxidation of chlorophyll both in vitro and in vivo. Their experiments suggest that the hydroxyl radical may play a key role in the destruction of chlorophyll. Ginkel and Raison,[73] using DMPO, have also observed the production of DMPO/OOH and DMPO/OH in chloroplasts, photosystem I particles, or chlorophyll-*a* lipid vesicles.

The production of superoxide by melanin photoreactions has been studied by the spin-trapping technique.[36,73,74] The structure and reactivity of melanins is the subject of an extensive review by Sealy et al.[76]

The detection of superoxide from illuminated riboflavin by spin trapping was first observed by Buettner and Oberley.[48] The riboflavin-DETAPAC-light system has since been used by others in the studying of various aspects of spin trapping.[15,33,36,40]

We have also observed the apparent photosensitized production of superoxide and hydroxyl radicals by protoporphyrin[77] and hematoporphyrin,[58] both of which are of potential use in the treatment of cancer.

D. Free Radical Production by Drugs

The role of superoxide and superoxide dismutase in cancer and cancer treatment is of great current interest.[78-80] The use of spin trapping to detect the production of active oxygen species by antitumor drugs has provided considerable insight into their possible modes of action. Lown and co-workers[81] have observed the formation of the PBN/OH adduct with PBN in the presence of reduced mitomycin C, mitomycin B, or streptonigrin. The presence of catalase or superoxide dismutase in the spin trapping mixture inhibits the generation of the PBN/OH adduct, implying the involvement of both superoxide and hydrogen peroxide in its formation. In somewhat similar experiments, the production of the hydroxyl and superoxide radicals from bleomycin was observed in the presence of iron (II) and not iron (III).[82-84] These studies reinforced the thinking on the active role of metal ions in bleomycin toxicity while providing additional insights on the mechanism of action of these drugs.

IV. CONCLUSION

Spin trapping can provide a means of probing free radical chemistry in biochemical and biological systems that was never before possible because of the extremely low steady-state concentration of the free radicals of interest. It is not without its pitfalls and shortcomings, but with a judicious choice of control experiments and an understanding of the chemistry of spin traps, considerable new insights can be gained in the involvement of free radicals in biology.

ACKNOWLEDGMENTS

The assistance of Maxine Dreyer in manuscript preparation is most greatly acknowledged. This work was supported in part by the Haines Research Fund, the Research Corporation, and NSF Grant #PRM-8109079.

REFERENCES

1. **Janzen, E. G. and Blackburn, B. J.,** Detection and identification of short-lived free radicals by an electron spin resonance trapping technique, *J. Am. Chem. Soc.,* 90, 5909, 1968.
2. **Janzen, E. G. and Blackburn, B. J.,** Detection and identification of short-lived free radicals by electron spin resonance trapping techniques (spin trapping). Photolysis of organolead, -tin, and -mercury compounds, *J. Am. Chem. Soc.,* 91, 4481, 1969.
3. **Forshult, S., Lagercrantz, C., and Torssell, K.,** Use of nitroso compounds as scavengers for the study of short-lived free radicals in organic reactions, *Acta Chem. Scand.,* 23, 522, 1969.
4. **Chalfont, G. R., Perkins, M. J., and Horsfield, A.,** A probe for homolytic reactions in solution. II. The polymerization of styrene, *J. Am. Chem. Soc.,* 90, 7141, 1968.
5. **Leaver, I. H. and Ramsay, G. C.,** Trapping of radical intermediates in the photoreduction of benzophenone, *Tetrahedron,* 25, 5669, 1969.
6. **Leaver, I. H. and Ramsay, G. C.,** ESR of nitroxide radicals. I. Photolysis of indoles, phenols, disulphides, and thiols, *Aust. J. Chem.,* 22, 1891, 1969.
7. **Leaver, I. H. and Ramsay, G. C.,** ESR of nitroxide radicals. II. Photosensitization of indoles, phenols and thiols by a diphenylpyrazoline, *Aust. J. Chem.,* 22, 1899, 1969.
8. **Terabe, S. and Konaka,R.,** Electron spin resonance studies on oxidation with nickel peroxide. Spin trapping of free radical intermediates, *J. Am. Chem. Soc.,* 91, 5655, 1969.
9. **Evans, C. A.,** Spin trapping, *Aldrichimica Acta,* 12, 23, 1979.
10. **Janzen, E. G.,** Spin trapping, *Acc. Chem. Res.,* 4, 31, 1971.
11. **Janzen, E. G.,** A critical review of spin trapping in biological systems, in *Free Radicals in Biology,* Vol. 4, Pryor, W. A., Ed., Academic Press, New York, 1980, chap. 4.
12. **McCay, P. B., Noguchi, T., Fong, K.-L., Lai, E. K., and Poyer, J. L.,** Production of radicals from enzyme systems and the use of spin traps, in *Free Radicals in Biology,* Vol. 4, Pryor, W. A., Ed., Academic Press, New York, 1980, chap. 5.
13. **Langercrantz, C.,** Spin trapping of some short-lived radicals by the nitroxide method, *J. Phys. Chem.,* 75, 3466, 1971.
14. **Perkins, M. J.,** The trapping of free radicals by diamagnetic scavengers, in *Essays on Free-Radical Chemistry,* Spec. Publ. No. 24, Chemical Society, London, 1970, 97.
15. **Finkelstein, E., Rosen, G. M., and Rauckman, E. J.,** Spin trapping of superoxide and hydroxyl radical: practical aspects, *Arch. Biochem. Biophys.,* 200, 1, 1980.
16. **Joshi, A., Rustgi, S., and Riesz, P.,** E.S.R. of spin-trapped radicals in gamma-irradiated aqueous solutions of nucleic acids and their constituents, *Int. J. Radiat. Biol.,* 30, 151, 1976.
17. **Rustgi, S., Joshi, A., Moss, H., and Riesz, P.,** ESR of spin-trapped radicals in aqueous solutions of amino acids, reaction of the hydroxyl radical, *Int. J. Radiat. Biol.,* 31, 415, 1977.
18. **Rustgi, S., Joshi, A., Riesz, P., and Friedberg, F.,** ESR of spin-trapped radicals in aqueous solutions of amino acids, reaction of the hydrated electron, *Int. J. Radiat. Biol.,* 32, 533, 1977.
19. **Joshi, A., Rustgi, S., Moss, H., and Riesz, P.,** ESR of spin-trapped radicals in aqueous solutions of peptides, reactions of the hydroxyl radical, *Int. J. Radiat. Biol.,* 33, 205, 1978.
20. **Joshi, A., Moss, H., and Riesz, P.,** ESR study of the post-radiolysis growth of spin-trapped radicals in γ-irradiated aqueous solutions of thymine, *Int. J. Radiat. Biol.,* 34, 165, 1978.
21. **Terabes, S., Kuruma, K., and Konaka, R.,** Spin trapping by use of nitroso-compounds. VI. Nitrosodurene and other nitrosobenzene derivatives, *J. Chem. Soc. Perkin Trans.,* 2, 1252, 1973.
22. **Macker, A., Wajer, Th. A. J. W., and DeBoer, Th. J.,** C-nitroso compounds. VI. Acyl-alkyl-nitroxides from acyl radicals and nitroso compounds as studied by ESR, *Tetrahedron,* 24, 1623, 1968.
23. **Ohkuma, T., Kirino, Y., and Ohkuma, T.,** Some physiochemical properties of 2-methyl-2-nitrosoprane, phenyl N-*tert*-butyl nitrone, 5,5-dimethylpyrroline-N-oxide and 2,5,5-trimethyl-pyrroline-N-oxide and the feasibility of their use as spin traps in aqueous solution, *Chem. Pharm. Bull.,* 29, 25, 1981.

24. **Sargent, F. P. and Gardy, E. M.,** Spin trapping of radicals formed during radiolysis of aqueous solutions. Direct electron spin resonance observations, *Can. J. Chem.,* 54, 275, 1976.
25. **Wargon, J. A. and Williams, F.,** Electron spin resonance studies of radical trapping in the radiolysis of organic liquids. I. Evidence for the primary formation of the methoxy radical in methanol, *J. Am. Chem. Soc.,* 94, 7917, 1972.
26. **Mackor, A., Wajer, Th. A. J., DeBoer, Th. J., and Von Voorst, J. D. W.,** C-nitroso compounds. III. Alkoxy-alkyl-nitroxides as intermediates in the reaction of alkoxy-radicals with nitroso compounds, *Tetrahedron Lett.,* p. 385, 1967.
27. **Janzen, E. G. and Liu, J. I.-P.,** Radical addition reaction of 5,5-dimethyl-1-pyrroline-1-oxide. ESR spin trapping with a cyclic nitrone, *J. Mag. Resonance,* 9, 510, 1973.
28. **Janzen, E. G., Evans, C. A., and Liu, J. I.-P.,** Factors influencing hyperfine splitting in the ESR spectra of five-membered ring nitroxides, *J. Mag. Resonance,* 9, 513, 1973.
29. **Floyd, R. A., Soong, L. M., Stuart, M. A., and Reigh, D. L.,** Spin trapping of free radicals produced from nitrosoamine carcinogens, *Photochem. Photobiol.,* 28, 857, 1978.
30. **Harbour, J. R. and Bolton, J. R.,** The involvement of the hydroxyl radical in the destructive photooxidation of chlorophylls in vivo and in vitro, *Photochem. Photobiol.,* 28, 231, 1978.
31. **Janzen, E. G., Nutter, D. E., Jr., David, E. R., Blackburn, B. J., Poyer, J. L., and McCay, P. B.,** On spin trapping hydroxyl and hydroperoxyl radicals, *Can. J. Chem.,* 56, 2237, 1978.
32. **Harbour, J. R., Chow, V., and Bolton, J. R.,** An electron spin resonance study of the spin adducts of OH and HO_2 radicals with nitrones in the ultraviolet photolysis of aqueous hydrogen peroxide solutions, *Can. J. Chem.,* 52, 3549, 1974.
33. **Finkelstein, E., Rosen, G. M., and Rauckman, E. J.,** Spin trapping. Kinetics of the reaction of superoxide and hydroxyl radicals with nitrones, *J. Am. Chem. Soc.,* 102, 4994, 1980.
34. **Janzen, E. G., Wang, Y. Y., and Shetty, R.V.,** Spin trapping with α-pyridyl 1-oxide N-*tert*-butyl nitrones in aqueous solutions. A unique electron spin resonance spectrum for the hydroxyl radical adduct, *J. Am. Chem. Soc.,* 100, 2923, 1978.
35. **Janzen, E. G., Dudley, R. L., and Shetty, R. V.,** Synthesis and electron spin resonance chemistry of nitronyl labels for spin trapping. α-Phenyl N-[5-(5-Methyl-2,2-dialkyl-1,3-dioxanyl)] nitrones and α-(*N*-alkylpyridinium) *N*-(*tert*-butyl) nitrones, *J. Am. Chem. Soc.,* 101, 243, 1979.
36. **Sanderson, D. G. and Chedekel, M. R.,** Spin trapping of the superoxide radical by 4-(*N*-methylpyridinium) *t*-butyl nitrone, *Photochem. Photobiol.,* 32, 573, 1980.
37. **Hamer, J. and Macaluso, A.,** Nitrones, *Chem. Rev.,* 64, 473, 1964.
38. **Floyd, R. A. and Soong, L. A.,** Spin trapping in biological systems. Oxidation of 5,5-dimethylpyrroline-1-oxide by a hydroperoxide-hematin system, *Biochem. Biophys. Res. Commun.,* 74, 79, 1977.
39. **Asmus, K. D., Wigger, A., and Henglein, A.,** Pulsradiolytische Untersuchung einiger elementar Prozesse der Nitrobenzolreducktion, *Ber. Bunsenges, Physik. Chem.,* 70, 862, 1966.
40. **Finkelstein, E., Rosen, G. M., and Rauckman, E. J.,** Spin trapping of superoxide, *Mol. Pharmacol.,* 16, 676, 1979.
41. **Kirino, Y., Ohkuma, T., and Kwan, T.,** Spin trapping with 5,5-dimethylpyrroline-*N*-oxide in aqueous solution, *Chem. Pharm. Bull.,* 29, 29, 1981.
42. **Latimer, W. M.,** *Oxidation Potentials,* Prentice-Hall, New York, 1952.
43. **Hoffman, A. K. and Henderson, A. T.,** A new stable free radical: di-*t*-butylnitroxide, *J. Am. Chem. Soc.,* 83, 4671, 1961.
44. **Buettner, G. R.,** unpublished observations, 1980.
45. **Rabani, J. and Slein, G.,** The reactivity of OH radicals with ferro-ferricyanide, formate, ethanol and amino acids in irradiated solutions, *Trans. Faraday Soc.,* 58, 2150, 1962.
46. **Farhataziz and Ross, A. B.,** Selected specific rates of reactions of transients from water in aqueous solution. III. Hydroxyl radical and perhydroxyl radical and their radical ions, *Natl. Stand. Ref. Data Soc. Natl. Bur. Stand. (U.S.),* 59, 1977.
47. **Buettner, G. R., Oberley, L. W., and Leuthauser, S. W. H. C.,** The effect of iron on the distribution of superoxide and hydroxyl radicals as seen by spin trapping and on the superoxide dismutase assay, *Photochem. Photobiol.,* 28, 693, 1978.
48. **Buettner, G. R. and Oberley, L. W.,** Considerations in the spin trapping of superoxide and hydroxyl radical in aqueous systems using 5,5-dimethylpyrroline-1-oxide, *Biochem. Biophys. Res. Commun.,* 83, 69, 1978.
49. **Walling, C.,** Fenton's reagent revisited, *Acc. Chem. Res.,* 8, 125, 1975.
50. **Cohen, G. and Sinet, P. M.,** Fenton's reagent — once more revisited, in *Chemical and Biochemical Aspects of Superoxide and Superoxide Dismutase,* Bannister, J. V. and Hill, H. A. O., Eds., Elsevier, Amsterdam, 1980, 27.
51. **Janzen, E. G.,** personal communication, 1980.
52. **Sosnovsky, G. and Rawlinson, D. J.,** The chemistry of hydroperoxides in the presence of metal ions, in *Organic Peroxides,* Vol. 2, Swern, D., Ed., Interscience, New York, 1971, 153.

53. **Buettner, G. R. and Oberley, L. W.**, unpublished observations, 1978.
54. **Behar, D., Czapski, G., Rabani, J., Dorfman, L. M., and Schwartz, H. A.**, The acid dissociation constant and decay kinetics of the perhydroxyl radical, *J. Phys. Chem.*, 74, 3209, 1970.
55. **Fee, J. A. and Valentine, J. S.**, Chemical and physical properties of superoxide, in *Superoxide and Superoxide Dismutases*, Michelson, A. M., McCord, J. M., and Fridovich, I., Eds., Academic Press, New York, 1977, 26.
56. **Wertz, J. E. and Bolton, J. R.**, *Electron Spin Resonance*, McGraw-Hill, New York, 1972, chap. 3.
57. **Lai, C.-S. and Piette, L. H.**, Hydroxyl radical production involved in lipid peroxidation of rat liver microsomes, *Biochem. Biophys. Res. Commun.*, 78, 51, 1977.
58. **Buettner, G. R. and Oberley, L. W.**, The apparent production of superoxide and hydroxyl radicals by hematoporphyrin and light as seen by spin-trapping, *F.E.B.S. Lett.*, 121, 161, 1980.
59. **Harbour, J. R., Issler, S. L., and Hair, M. L.**, Singlet oxygen and spin trapping with nitrones, *J. Am. Chem. Soc.*, 102, 7778, 1980.
60. **Peters, G. and Rogers, M. A. J.**, On the feasibility of electron transfer to singlet oxygen from mitochondrial components, *Biochem. Biophys. Res. Commun.*, 96, 770, 1980.
61. **Saito, I., Matsuura, T., and Inoue, K.**, Formation of superoxide ion from singlet oxygen. On the use of a water-soluble singlet oxygen source, *J. Am. Chem. Soc.*, 103, 188, 1981.
62. **Klebanoff, S. J. and Clark, R. A.**, *The Neutrophil: Function and Clinical Disorders*, Elsevier, Amsterdam, 1978.
63. **Green, M. R., Hill, H. A. O., Okolow-Zubkowska, M. J., and Segal, A. W.**, The production of hydroxyl and superoxide radicals by stimulated human neutrophils-measurements by EPR spectroscopy, *F.E.B.S. Lett.*, 100, 23, 1979.
64. **Rosen, H. and Klebanoff, S. J.**, Hydroxyl radical generation by polymorphonuclear leukocytes measured by electron spin resonance spectroscopy, *J. Clin. Invest.*, 64, 1725, 1979.
65. **Pederson, T. C. and Aust, S. D.**, NADPH-dependent lipid peroxidation catalyzed by purified NADPH-cytochrome c reductase from rat liver microsomes, *Biochem. Biophys. Res. Commun.*, 48, 789, 1972.
66. **Fong, K.-L., McCay, P. B., and Poyer, J. L.**, Evidence that peroxidation of lysosomal membranes is initiated by hydroxyl free radicals produced during flavin enzyme activity, *J. Biol. Chem.*, 248, 7792, 1973.
67. **Saprin, A. N. and Piette, L. H.**, Spin trapping and its application in the study of lipid peroxidation and free radical production with liver microsomes, *Arch. Biochem. Biophys.*, 180, 480, 1977.
68. **Lai, C.-S. and Piette, L. H.**, Spin-trapping studies of hydroxyl radical production involved in lipid peroxidation, *Arch. Biochem. Biophys.*, 190, 27, 1978.
69. **Lai, C.-S., Grover, T. A., and Piette, L. H.**, Hydroxyl radical production in a purified NADPH-cytochrome c (P-450) reductase system, *Arch. Biochem. Biophys.*, 193, 373, 1979.
70. **Cederbaum, A. I. and Cohen, G.**, Inhibition of the microsomal oxidation of ethanol and 1-butanol by the free radical, spin-trapping agent 5,5-dimethyl-1-pyrroline-1-oxide, *Arch. Biochem. Biophys.*, 204, 397, 1980.
71. **Tien, M., Svingen, B. A., and Aust, S.D.**, Superoxide dependent lipid peroxidation, *Fed. Proc.*, 40, 179, 1981.
72. **Harbour, J. R. and Bolton, J. R.**, Superoxide formation in spinach chloroplasts: electron spin resonance detection by spin trapping, *Biochem. Biophys. Res. Commun.*, 64, 803, 1975.
73. **Ginkel, G. V. and Raison, J. K.**, Light-induced formation of $O_2^{\cdot-}$ oxygen radicals in systems containing chlorophyll, *Photochem. Photobiol.*, 32, 793, 1980.
74. **Chedekel, M. R., Smith, S. K., Post, P. W., and Vessell, D. L.**, Photodestruction of pheomelanin: role of oxygen, *Proc. Natl. Acad. Sci. U.S.A.*, 75, 5395, 1978.
75. **Felix, C. C., Hyde, J. S., Sarna, T., and Sealy, R. C.**, Melanin photoreactions in aerated media: electron spin resonance evidence for production of superoxide and hydrogen peroxide, *Biochem. Biophys. Res. Commun.*, 84, 335, 1978.
76. **Sealy, R. C., Felix, C. C., Hyde, J. S., and Swartz, H. M.**, Structure and reactivity of melanins: influence of free radicals and metal ions, in *Free Radicals in Biology IV*, Pryor, W. A., Ed., Academic Press, New York, 1980, chap. 7.
77. **Buettner, G. R. and Oberley, L. W.**, Superoxide formation by protoporphyrin as seen by spin trapping, *F.E.B.S. Lett.*, 98, 18, 1979.
78. **Oberley, L. W. and Buettner, G. R.**, Role of superoxide dismutase in cancer: a review, *Cancer Res.*, 39, 1141, 1979.
79. **Buettner, G. R. and Oberley, L. W.**, Oxygen, oxy radicals and superoxide dismutase in cancer and cancer treatment, in *Oxygen and Oxy-Radicals in Chemistry and Biology*, Rodgers, M. A. J. and Powers, E. L., Eds., Academic Press, New York, 1981, 606.
80. **Oberley, L. W.**, Superoxide dismutase and cancer, in *Superoxide Dismutase*, Vol. 2, Oberley, L. W., Ed., CRC Press, Boca Raton, Fla., 1982, chap. 6.

81. **Lown, J. W., Sim, S.-K., and Chen, H.-H.,** Hydroxyl radical production by free and DNA-bound aminoquinone antibiotics and its role in DNA degradation. Electron spin resonance detection of hydroxyl radicals by spin trapping, *Can. J. Biochem.*, 56, 1042, 1978.
82. **Sugiura, Y. and Kikuchi, T.,** Formation of superoxide and hydroxy radicals in iron(II)-bleomycin-oxygen system: electron spin resonance detection by spin trapping, *J. Antibiot.*, 31, 1310, 1978.
83. **Oberley, L. W. and Buettner, G. R.,** The production of hydroxyl radical by bleomycin and iron(II), *F.E.B.S. Lett.*, 97, 47, 1979.
84. **Sugiura, Y.,** Bleomycin-iron complexes. Electron spin resonance study, ligand effect, and implication for action mechanism, *J. Am. Chem. Soc.*, 102, 5208, 1980.
85. **Kremers, W. and Singh, A.,** Electron spin resonance study of spin-trapped azide radicals in aqueous solutions, *Can. J. Chem.*, 58, 1592, 1980.
86. **Harbour, J. R. and Hair, M. L.,** Superoxide generation in the phototysis of aqueous cadmium sulfide dispersions. Detection by spin trapping, *J. Phys. Chem.*, 81, 1791, 1977.
87. **Janzen, E. G., Stonks, H. J., Nutter, D. E., Davis, E. R., Blount, H. N., Poyer, J. L., and McCay, P. B.,** Spin trapping azidyl ($N_3\cdot$), cyanatyl ($OCN\cdot$), cyanyl ($\cdot CN$) radicals, and clorine atom ($Cl\cdot$), *Can. J. Chem.*, 58, 1596, 1980.

Chapter 5

THE ORGANIC CHEMISTRY OF SUPEROXIDE ANION RADICAL

Aryeh A. Frimer

TABLE OF CONTENTS

I. INTRODUCTION

July 4, 1776 marks not only the birthday of a nation but also the approximate year of the discovery of oxygen by Scheele, Priestly, or Lavoisier.[1] During the past 200 years this life-supporting molecule has been characterized, scrutinized, and utilized in manifold and assorted ways. Nevertheless, even today little is understood regarding the exact mechanism of its metabolism in living organisms, or how its chemical activity is moderated and controlled during biological oxidations.

Some clues are available and are best understood by quickly considering the electronic nature of molecular oxygen. In the ground state, it contains two unpaired electrons and is paramagnetic; hence, its interaction with an external electron pair would be expected to face a spin restriction. Indeed, oxygen in the ground state tends to act like a biradical and most of its oxidation reactions proceed by one-electron steps involving free radical intermediates. If we consider, as well, oxygen's electron affinity, some involvement of superoxide anion radicals ($O_2^{\cdot-}$) in oxidations of biological significance might be anticipated. One might have also suspected the existence of a mechanism of defense against any potential toxicity of these radicals within aerobic cells. Indeed, the experimental detection of $O_2^{\cdot-}$ as a product of enzymatic reduction of molecular oxygen, as well as the recently recognized function of erythrocuprein (superoxide dismutase) as the enzymatic catalyst for its disproportionation to molecular oxygen and hydrogen peroxide, have catapulted this radical anion to the forefront of biochemical research. Most recently $O_2^{\cdot-}$ has been implicated as playing a pivotal role in both aging and cancer. Physical chemists have also shown an interest in this species due to its omnipresence in irradiated solutions.

While significant progress has been achieved in the understanding of the ways to generate and disproportionate $O_2^{\cdot-}$ over the past decade, the major research efforts have been directed toward the investigation of this species in aqueous media. Nevertheless, the use of organic solvents may be essential to the further development of the chemistry of $O_2^{\cdot-}$. Investigation into the exact role of $O_2^{\cdot-}$ has been seriously hampered because of the fact that its protonation in protic solvents leads to the formation of several other radical oxidants, in particular the hydroperoxy radical $HO_2\cdot$, hydrogen peroxide, and the highly reactive hydroxyl radical $\cdot OH$.[2] It is clear that these side reactions can be avoided by the use of aprotic solvents. In addition, it is likely that aprotic media correspond more closely than aqueous solutions to the hydrophobic conditions in the active center of the biological systems which generate the superoxide anion.

II. GENERATION OF SUPEROXIDE ANION RADICAL IN APROTIC MEDIA

The question now becomes: how exactly does one generate $O_2^{\cdot-}$ in aprotic media? One approach is electrolytic reduction of molecular oxygen which has been carried out in dimethylsulfoxide (DMSO), dimethylformamide (DMF), pyridine, acetone, and methylene chloride.[3,4] The advantage of this method is that $O_2^{\cdot-}$ can be generated pure and in a controlled fashion. Furthermore, cyclic voltametry allows one to follow the course of the reaction and uncover intermediates. However, the electrochemical equipment required for such work is not widely available. What's more, the concentrations of $O_2^{\cdot-}$ resulting from electrochemical generation are too small ($<10^{-2}$ *M*) for efficient product studies.[5]

Superoxide salts, such as the commercially available potassium superoxide (KO_2), can conveniently serve as a well-defined source of $O_2^{\cdot-}$. KO_2 is slightly soluble in the more polar organic solvents such as DMSO. However, the complexation of the superoxide salts with crown ethers[6] bypasses their solubility restrictions in aprotic media and solutions of KO_2 can be readily prepared in benzene, toluene, acetonitrile, DMSO, pyridine, etc. The principal advantage of this method is its convenience and suitability for synthetic procedures. One caveat is, nevertheless, in order: this approach may yield solutions that are contaminated with varying amounts of K_2O_2 and KOH since the commercially available KO_2 is only 96% pure. The slow decomposition of crown ethers in these solutions has also been noted.[7]

Finally, the tetramethylammonium superoxide salt can be prepared from tetramethylammonium hydroxide and KO_2 in a solid phase reaction.[8,9] This salt is readily soluble in a number of aprotic solvents producing highly stable solutions of superoxide in the absence of crown ethers. Nevertheless, the whole preparation procedure is quite cumbersome and, although this procedure has been carried out several times successfully, a violent explosion has been reported in one instance.[10]

Oxygen-18 labeled KO_2 has been synthesized by Rosenthal[11] and used for mechanistic studies on the mode of $O_2^{\overline{\cdot}}$ attack.[10] Valentine[12] has noted, however, that here, too, precautions against explosion should be taken during the preparation of $K^{18}O_2$.

III. MODES OF $O_2^{\overline{\cdot}}$ ACTION

Before we proceed to describe the various organic reactions of $O_2^{\overline{\cdot}}$, it will be useful to discuss the five properties which control its modes of reaction. For as its name indicates, superoxide anion radical should have a dual nature. On the one hand, it is a negatively charged species. We will therefore be concerned with its *Bronstead basicity* and *nucleophilicity*. On the other hand, $O_2^{\overline{\cdot}}$ bears an unpaired electron. Thus, we will have to consider its capability of inducing characteristic *free radical reactions* (hydrogen atom abstraction, olefin addition, and radical coupling) and how well it functions as either a one-*electron reducing agent* or *oxidant*. Let us then briefly focus on each of these properties and consider how they might control the various mechanisms of $O_2^{\overline{\cdot}}$ action in protic and aprotic media.

A. Basicity

We can get some insight as to the relative Bronstead basicity of $O_2^{\overline{\cdot}}$ by knowing the acid strength of its conjugate acid, the hydroperoxy radical (HOO·) (Equation 1).

$$HB + O_2^{\overline{\cdot}} \rightleftarrows B^- + HOO\cdot \qquad (1)$$

In a 1970 pulse radiolysis study in aqueous solution, Behar and co-workers[13] calculated a pK_a value of 4.88 ± 0.10 for HOO·. More recently, Bielski and Allen,[14] using ultrapure reagents, determined that the actual number is slightly lower and should be 4.75 $\pm$ 0.08. In any case, these values suggest that $O_2^{\overline{\cdot}}$ in aqueous media is a relatively weak base (on the order of acetate).

Valentine notes[12] that the low pK_a value of HOO· in water should not in fact be terribly surprising. After all, $O_2^{\overline{\cdot}}$, like fluoride, is a small nonpolarizable anion which is expected to be particularly stabilized in protic media by a tightly bound solvation sphere of hydrogen-bonded solvent. Indeed, in the gas phase both $O_2^{\overline{\cdot}}$ and F^- have large hydration affinities.[15] In the case of fluoride the solvent envelope renders it a weak base (pK_a of HF = 3.2) and a relatively unreactive nucleophile.[16-19]

In aprotic solvents, however, the situation is expected to be radically different. For with the hydrogen bonds now absent, the dissolved anions are essentially "naked".[20] Fluoride, for example, proves to be strongly nucleophilic[16,17,20] and quite basic[20] in aprotic solvents such as DMF, acetonitrile, DMSO, and benzene. Similarly, the basicity of another small nonpolarized anion, the hydroxide ion, is 14 orders of magnitude greater in dry DMSO than it is in water.[21a] Thus, while the basicity of $O_2^{-\cdot}$ in aprotic solvents has yet to be determined quantitatively, it is expected to be quite substantial. Indeed, gas phase studies[21b] suggest that $O_2^{-\cdot}$, in the absence of solvent stabilizing influences, is substantially more basic than acetate, though it does not deprotonate phenol.

Sawyer[3,4,22] has furthermore argued that Equation 1 is merely half the story. For once the hydroperoxy radical is formed, it reacts further with $O_2^{-\cdot}$ in a second rapid ($k = 8.86 \pm 0.43 \times 10^7\ M^{-1}\ s^{-1}$)[14] and exothermic[3,22] dismutation process (Equation 2):

$$HO_2\cdot + O_2^{-\cdot} \longrightarrow HO_2^- + O_2 \qquad (2)$$

It is this dismutation reaction which drives the initial deprotonation step (Equation 1) to the right, increasing the *effective* basicity of $O_2^{-\cdot}$, i.e., its ability to effect proton removal from substrates. The *effective* basicity of $O_2^{-\cdot}$ will be clearly greater than its Bronstead basicity, i.e., its ability to remove protons from substrate as indicated by the pK_a value for its conjugate acid.

If we now combine Equations 1 and 2 we obtain Equation 3.

$$2O_2^{-\cdot} + HB \rightleftharpoons O_2 + HO_2^- + B^- \qquad (3)$$

This equation summarizes the acid catalyzed disproportionation of superoxide which has an equilibrium constant[3,4,22] of $K = 2.5 \times 10^8$. Clearly $O_2^{-\cdot}$ will have only a fleeting existence in water and a short lifetime in other protic solvents.

This conclusion is of extreme importance because it indicates that *superoxide activity in aqueous solvents will be limited mostly by kinetic rather than by thermodynamic factors.*[12] Put another way, only those modes of reaction that can compete with rapid dismutation will be observed in protic media. In aprotic media, however, the lifetime of $O_2^{-\cdot}$ is expected to be long because *solvent induced* acid catalyzed disproportionation (Equation 3) is precluded, while simple disproportionation (Equation 4) is energetically unfavorable[23] (see Section III.E).

$$2O_2^{-\cdot} \longrightarrow {}^3O_2 + O_2^{-2} \quad \Delta G > 28\ \frac{kcal}{mol} \qquad (4)$$

Returning now to Equation 3, Sawyer[3,22] further notes that, assuming of a pK_a of 15 for water, this equilibrium constant for the acid catalyzed disproportion of $O_2^{-\cdot}$ indicates that superoxide can promote proton transfer from substrates and solvents to an extent equivalent to that of a conjugate base of an acid with a pK_a of approximately 23. This renders $O_2^{-\cdot}$ eight orders of magnitude more effective than hydroxide ion. Exactly how aprotic media will affect these values quantitatively requires clarification. Based on our previous discussion, one would expect Equation 3 to be shifted even more to the right, increasing the effective basicity of $O_2^{-\cdot}$ all the more. In any case, it is clear that the possibility of deprotonation must be considered in any reaction of $O_2^{-\cdot}$ with substrates containing labile acidic hydrogens. As we shall see (Section IV.B), recent studies have established that a number of weakly acidic organic compounds are deprotonated efficiently in aprotic solvents in the presence of superoxide ion.

B. Nucleophilicity

Many of the considerations we have just mentioned in discussing the basicity of $O_2^{-\cdot}$ are relevant as well to the question of its nucleophilicity. In protic media, $O_2^{-\cdot}$ like F^- is expected to be a poor nucleophile because of a strong hydrogen-bonding interaction with the solvent.[15-18] Furthermore, we must remember that in protic solutions any nucleophilic attack by $O_2^{-\cdot}$ must compete with rapid disproportionation (Equation 3). It should not be surprising, then, that in aqueous solution little evidence for nucleophilic reactivity has been found.

In aprotic solvents, on the other hand, disproportionation is essentially absent (see Section III.E) and the lifetime of $O_2^{\cdot-}$ is quite long. In the absence of hydrogen bonding with solvent, the "naked" $O_2^{\cdot-}$ anion should now show a substantial degree of nucleophilic reactivity as does F^-.[16-18] What is surprising, however, is the kinetic data which suggest that $O_2^{\cdot-}$ is one of the most potent S_N2 nucleophiles yet studied.[24,25] Thus the second order rate constant for the reaction of KO_2 with alkyl bromides in DMSO is on the order of 10^2 $M^{-1}sec^{-1}$, while for nearly all other nucleophiles this value lies between 10^{-2} to 10^1 $M^{-1}sec^{-1}$. The question is "why"?

Of the variety of factors which determine the relative reactivity of nucleophilic reagents,[26] there seem to be three which are cited most often as the sources for high nucleophilicity. These are polarizability, basicity, and the alpha effect, i.e., the presence of unshared pairs of electrons on the atom adjacent to the nucleophilic atom.[27,28]

Regarding the first of these three, clearly $O_2^{\cdot-}$, which is comprised of two electronegative second row elements, is not expected to be particularly polarizable. However, it should be pointed out that while polarizability enhances nucleophilicity in protic solvents, it has the opposite effect in dipolar aprotic solvents.[18] This is because dipolar aprotic solvents are themselves much more polarizable than protic solvents and interact strongly with the nucleophile retarding its action. Nevertheless, while the lack of polarization does not detract from the nucleophilicity of $O_2^{\cdot-}$ in aprotic solvent, it still does not explain why superoxide is so much more reactive than the nonpolarizable fluoride anion. Thus, Parker[16] and Liotta[29] report that F^- and Cl^- have essentially the same nucleophilicity in aprotic solvents, while superoxide is 1000 times more potent a nucleophile than Cl^-.[24]

Turning now to basicity, though we have indeed suggested that the Bronstead basicity of $O_2^{\cdot-}$ should increase in aprotic solvents, gas phase studies[21] indicate that $O_2^{\cdot-}$ is less basic than phenoxide and certainly methoxide. Yet, $O_2^{\cdot-}$ is at least 50 times more nucleophilic than methoxide (in DMSO) toward methyl bromide.[24] More importantly, recent studies in the gas phase[30] and aprotic media[31,32] reveal no correlation between the basicity of a "naked" ion and its nucleophilicity.

Superoxide does qualify as an α-effect nucleophile and perhaps the α-effect is manifest in the exceptional nucleophilicity displayed by $O_2^{\cdot-}$.[24,25] If this hypothesis is correct, then other α-effect nucleophiles such as peroxy anions[28] ought to reveal exceptional nucleophilicity in the appropriate aprotic media. Indeed, Johnson et al.,[33] in their concluding comments to a study of the nucleophilic substitution of $O_2^{\cdot-}$ on alkyl halides, note that the solubility of KO_2 in DMSO and DMF containing 18-crown-6 is quite high. Nevertheless, primary peroxy anions react more rapidly with alkyl halides than does superoxide. Similarly, the second order rate constant for the reaction of phenyl benzoate with electrogenerated $O_2^{\cdot-}$ in either DMF or pyridine is 3 $M^{-1}sec^{-1}$ and for chlorophenyl benzoate 25 $M^{-1}sec^{-1}$. By comparison Curci and DiFuria[34] have found that the second order rate constant for the reaction of potassium *t*-butylperoxide with *p*-nitrophenyl benzoate in toluene solution containing crown ether is 7.6×10^3 $M^{-1}sec^{-1}$. This again suggests that peroxy anions are highly nucleophilic. Clearly an extensive kinetic study of the α-effect in aprotic solvents is called for.

Danen[24,25] has suggested that, in the particular case of $O_2^{\cdot-}$, a further factor may be at play. As we shall shortly see, $O_2^{\cdot-}$ is a good electron donor. It is plausible, then, that the supernucleophilicity of $O_2^{\cdot-}$ can be at least partially attributed to a significant electron transfer contribution in the transition state (Equation 5).

$$O_2^{\cdot-} + RX \longrightarrow (O_2^{\cdot-} \cdots\cdot R \cdots\cdot X \longleftrightarrow O_2 \cdots\cdot \dot{R} \cdots\cdot X^-) \longrightarrow RO_2\cdot + X^- \qquad (5)$$

Although such contributions can be drawn for all nucleophiles, it may be particularly important for $O_2^{\overline{\cdot}}$ because of the inherent stability of molecular oxygen.

C. Free Radical Character

Free radicals characteristically undergo two types of reactions: (1) addition to double bonds; and (2) abstraction of labile hydrogen atoms. Interestingly, superoxide undergoes the above reactions reluctantly if at all. For example, simple olefins such as tetramethyl- and trimethylethylene, 2-methyl-2-pentene, cyclohexene and its 1,2-dimethyl analog are totally unreactive to $O_2^{\overline{\cdot}}$.[35,36] Bors et al. report that the water soluble carotenoid Crocin, although bleached by radiolytically generated $HO_2\cdot$, is inert to $O_2^{\overline{\cdot}}$.[37] The same is true for aromatic hydrocarbons such as benzene, naphthalene, anthracene, and rubrene.[35] Tetraphenylethylene,[35]2,5-diphenylfuran,[35] and styrene[38-40] yielded only minute amounts of polar compounds.

These observations are really not that surprising from a thermodynamic perspective. The addition of $O_2^{\overline{\cdot}}$ to an olefin would involve as the first step the process shown in Equation 6.

$$O_2^{\overline{\cdot}} + CH_2 = CH_2 \longrightarrow \cdot CH_2CH_2OO^- \qquad (6)$$

The heat of reaction for this process is merely the negative value of the bond dissociation energy of the C–O bond in the product, $D(\cdot CH_2CH_2{-}OO^-)$. The latter can be calculated using the following cycle.

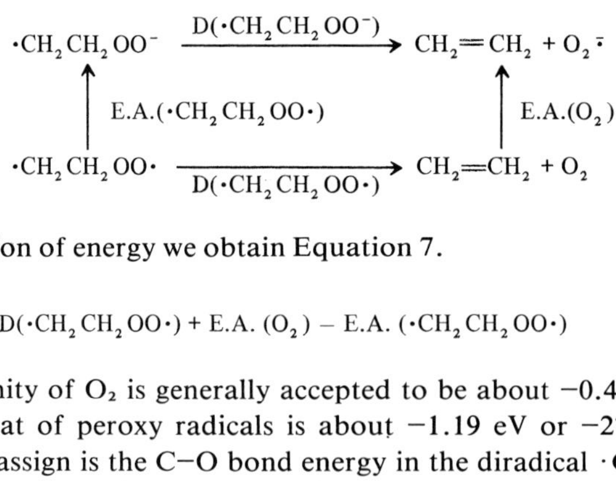

From the conversion of energy we obtain Equation 7.

$$D(\cdot CH_2CH_2OO^-) = D(\cdot CH_2CH_2OO\cdot) + E.A.\,(O_2) - E.A.\,(\cdot CH_2CH_2OO\cdot) \qquad (7)$$

The electron affinity of O_2 is generally accepted to be about −0.44 eV or −10.1 kcal/mol,[41-44] while that of peroxy radicals is about −1.19 eV or −27.4 kcal/mol.[45] The difficult value to assign is the C–O bond energy in the diradical $\cdot CH_2CH_2OO\cdot$. However, we can approximate its value from its monoradical analog $CH_3CH_2OO\cdot$. Like other peroxy radicals, C–O bond cleavage in the latter requires 27 kcal/mol,[46] and yields ethyl radical and molecular oxygen (Equation 8).

$$CH_3CH_2OO\cdot \longrightarrow CH_3CH_2\cdot + O_2 \quad D = 27\ \frac{\text{kcal}}{\text{mol}} \qquad (8)$$

By comparison C–O cleavage of the diradical $\cdot CH_2CH_2OO\cdot$ leads to the formation of ethylene and O_2 (Equation 9).

$$\cdot CH_2CH_2OO\cdot \longrightarrow CH_2 = CH_2 + O_2 \qquad (9)$$

As Benson notes[47] in the related case of C–H dissociation in CH_3CH_2-H vs. $\cdot CH_2CH_2$-H, the difference in bond dissociation energies should be the bond energy of the π bond, H_π (C=C), in ethylene which is generally accepted[48] to be around 63 kcal/mol. This leads to a value of −36 kcal/mol for $D(\cdot CH_2CH_2OO\cdot)$. If we now introduce all these values into Equation 7, we obtain that $D(\cdot CH_2CH_2OO^-)$ is equal to −18.6 kcal and consequently for Equation 6, ΔH is endothermic by approximately 18.6 kcal. Of course, whether a reaction proceeds spontaneously also depends on the entropy change, which in our case decreases. Hence, the free energy for Equation 6 should be even more endothermic.

Let us turn now to the abstraction of an allylic hydrogen (e.g., from propene) as outlined in Equation 10. The heat of reaction for this process can be estimated from the net bond energy difference between the bonds broken and those formed[49] as shown in Equation 11.

$$CH_2{=}CH{-}CH_3 + O_2^{\overline{\cdot}} \longrightarrow CH_2{=}CH{-}CH_2\cdot + HOO^- \quad (10)$$

$$\Delta H_{10} = D(CH_2{=}CHCH_2{-}H) - D(H{-}OO^-) \quad (11)$$

The allylic C–H bond dissociation energy of propene is 88.6 kcal/mol,[50] while that of the H–O bond in HOO^- has been recently calculated[51] to be 63.4 kcal/mol. Thus ΔH_{10} is endothermic by 25.2 kcal/mol.

The reader is reminded that the above calculations are for the gas phase. In solution a variety of other considerations come into play. Nevertheless, these results do suggest that neither hydrogen abstraction nor olefin addition by $O_2^{\overline{\cdot}}$ are expected to be primary reaction pathways with simple olefins, and this should be kept in mind in drawing up reaction mechanisms. On the other hand, the above results suggest that if the bond energy of the labile hydrogen drops below 70 kcal/mol, hydrogen atom abstraction by $O_2^{\overline{\cdot}}$ may become an energetically favorable process.

Valentine[12] also notes that the relative reactivity of a radical as a hydrogen abstractor is known to be related to the bond energy of the bond that is formed. The H–O bond energy (63.4 kcal/mol)[51] lies between the bond energies of $HO_2\cdot$ (47 kcal/mol)[46] and HI (71 kcal/mol).[48] Since neither atomic iodine nor molecular oxygen are reactive hydrogen abstractors, it is expected that $O_2^{\overline{\cdot}}$ should be reactive via this free radical mode only with exceptionally labile hydrogens.

There remains one further reaction that is quite typical of free radicals (although not necessarily of radical anions),[52] and that is radical combination (Equation 12).

$$R\cdot + O_2^{\overline{\cdot}} \longrightarrow RO_2^- \quad (12)$$

This process is of course the essence of the "termination" step in free radical chain reactions and the rate of such radical couplings approach the diffusion controlled limit.[53] Indeed a mechanism involving the combination of $O_2^{\overline{\cdot}}$ with organic radicals has been postulated for the oxidation of phenols,[54-56] carbanions,[57-59] and diketones.[60] In these cases, however, coupling of the generated radical with molecular oxygen is equally plausible (Equations 13 and 14).

$$R^- + O_2 \longrightarrow R\cdot + O_2^{\overline{\cdot}} \quad (13)$$

$$R\cdot + O_2 \longrightarrow RO_2\cdot \quad (14)$$

Merenyi and Lind[61] have recently suggested that $O_2^{\overline{\cdot}}$ couples with luminol radical to form the crucial peroxide adduct. They maintain that in their system $O_2^{\overline{\cdot}}$ and H_2O_2 are present but not oxygen. However, as we shall see (Section IV.A.1), $O_2^{\overline{\cdot}}$ rapidly dismutates to O_2 in the presence of hydrogen peroxide. Thus, in this case too, the involvement of molecular oxygen is quite plausible.

There are several documented cases, in fact, where it is clear that $O_2^{\overline{\cdot}}$ does not react with free radicals by coupling. The first is the reaction of $O_2^{\overline{\cdot}}$ with the hydroperoxy

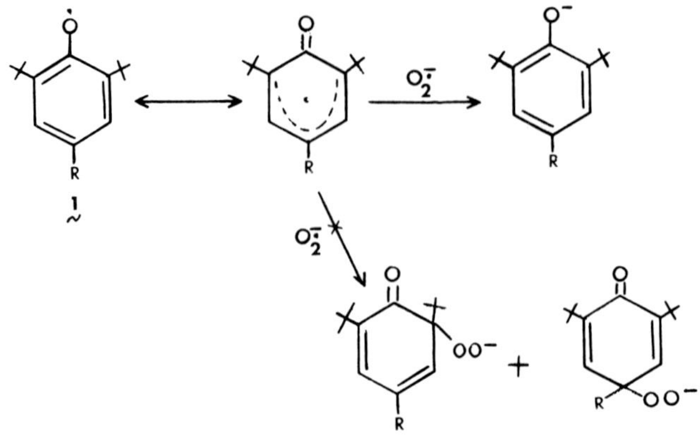

FIGURE 1. Reaction of $O_2^{\overline{\cdot}}$ with phenoxy radicals.

radical during the acid catalyzed disproportionation of $O_2^{\overline{\cdot}}$. As we have noted previously this process yields hydroperoxy anion and molecular oxygen (Equation 2) and it is generally assumed that it involves an electron transfer. The possibility of radical coupling does not seem to have been seriously considered (Equation 15).

$$HOO\cdot + O_2^{\overline{\cdot}} \longrightarrow HOOOO^- \longrightarrow HOO^- + O_2 \qquad (15)$$

Indeed, based on the analogous decomposition of $ClOO^-$ to singlet oxygen ($^1O_2$),[62] $HOOOO^-$ should likewise decompose to $^1O_2$ almost exclusively. However, $^1O_2$, if formed, is produced in no more than 0.2%[63] (see discussion in Section IV.H).

The second case is discussed by Le Berre and Berguer,[38,39] who report that $O_2^{\overline{\cdot}}$ reacts only extremely sluggishly with triphenylmethyl radical. Since the triphenylmethylperoxy anion is stable,[38] this sluggishness cannot be attributed to the reversibility of the reaction.

Nishinaga et al.[64,65] have found that $O_2^{\overline{\cdot}}$ does not couple with phenoxy radicals **1**, but reduces them to the corresponding phenolates (Figure 1). Similarly the radical cations of 1,3-diphenylisobenzofuran,[66] ferrocene,[66] and 2,5-dialkyl-1,4-dimethoxybenzenes[64] do not couple with $O_2^{\overline{\cdot}}$ but are reduced by it (see Section IV.G).

Radical-radical coupling has indeed been observed in the reaction of $O_2^{\overline{\cdot}}$ with the cation radicals of methyl viologen[67] and thianthrene[68] (see Section IV.G). This coupling, however, is "gimmicked" in that it is, in fact, not a pure radical reaction but rather an anion-cation coupling. Thus, it would seem that the radical-radical combination may be ruled out as a possible mechanism in $O_2^{\overline{\cdot}}$ reactions.

D. One-Electron Reductant

Considering the stability of molecular oxygen, electron transfer from $O_2^{\overline{\cdot}}$ to an appropriate reducible substrate (Equation 16) should be an energetically desirable process.

$$A + O_2^{\overline{\cdot}} \longrightarrow A^- + O_2 \qquad (16)$$

(A = anion, cation, radical, or neutral species)

Indeed it is electron transfer which is the essence of the dismutation reaction (Equation 2). Interestingly, however, the reduction potential for oxygen is sensitive to the nature of the solvent in which the measurement is made. In aprotic solvents the E°′ (vs. normal hydrogen electrode [NHE]) has an average value[69] of −0.57 V, which is nearly 0.25 V less than that observed for water (−0.33 V). This gap cannot be attributed to differences in the dielectric constants of the media, since the reduction potential of oxygen is relatively insensitive to the differing dielectric constants of a variety of aprotic solvents (CH_2Cl_2, acetone, acetonitrile, DMF, and DMSO). Rather it is probably due to the increased solvation (and hence stability) of $O_2^{\overline{\cdot}}$ in aqueous media.[70,71] As we have noted previously, the data suggest that $O_2^{\overline{\cdot}}$ is expected to be a much stronger reducing agent in aprotic solvents than it is in protic media.

E. One-Electron Oxidant

Equation 17 [69] indicates that the reduction of $O_2^{\overline{\cdot}}$ to hydrogen peroxide is a thermodynamically favorable process.

$$O_2^{\overline{\cdot}} + e^- + 2H^+ \rightleftharpoons H_2O_2\,,\ E^{\circ\prime} = +0.87 \qquad (17)$$

However, one has to be cautious about making any generalizations about the effectiveness of superoxide ion as an oxidizing agent. What Equation 17 does clearly tell us is that $O_2^{\overline{\cdot}}$ can be a strong oxidant when protons are readily accessible. Indeed, Sawyer[4] reports that the electrochemical reduction of superoxide ion in aprotic solvents occurs at −1.75 V vs. NHE; but even then, it occurs only when there is a concomitant abstraction of a proton from the solvent (Equation 18).

$$O_2^{\overline{\cdot}} + e^- + \text{H–solvent} \longrightarrow HO_2^- + \text{sol}^- \qquad (18)$$

The electrochemical generation of O_2^{-2} per se has not been observed. Obviously its formation is energetically no mean task and requires some positive species such as protons to increase its stabilization. In this light, it is not surprising that $O_2^{\overline{\cdot}}$ will not disproportionate to O_2 and O_2^{-2} in aprotic media (Equation 19)[23]

$$2\,O_2^{\overline{\cdot}} \longrightarrow {}^3O_2 + O_2^{-2},\ \Delta G > 28\ \frac{\text{kcal}}{\text{mol}} \qquad (19)$$

as it does so rapidly in protic solvents (Equation 20).[72]

$$2\,O_2^{\overline{\cdot}} + 2H^+ \longrightarrow H_2O_2 + {}^3O_2, \Delta G = -29.2\ \frac{\text{kcal}}{\text{mol}} \qquad (20)$$

We see, then, that in the absence of protons, $O_2^{\overline{\cdot}}$ is a poor electron acceptor (roughly equivalent to Na^+).[22] But we have also previously argued that $O_2^{\overline{\cdot}}$ is unlikely to induce oxidation of a substrate via hydrogen abstraction except for the most labile hydrogens. Thus in aprotic solvents we can in general rule out mechanistic possibilities which require any of the formal oxidation steps shown in Equations 21 to 24.

$$A^- + O_2^{\overline{\cdot}} \longrightarrow A + O_2^{-2} \qquad (21)$$

$$A\cdot + O_2^{\overline{\cdot}} \longrightarrow A^+ + O_2^{-2} \qquad (22)$$

$$A + O_2^{\overline{\cdot}} \longrightarrow A^{\dot{+}} + O_2^{-2} \qquad (23)$$

$$AH + O_2^{\overline{\cdot}} \longrightarrow A\cdot + HOO^- \qquad (24)$$

It follows that the repeated observation of oxidized products in $O_2^{\overline{\cdot}}$ reactions (vide infra) may require the agency of some other oxidants (e.g., HOO·, ·OH, O_2, etc.) produced in the course of the reaction.

IV. ORGANIC CHEMISTRY OF $O_2^{\overline{\cdot}}$

Having discussed the possible modes of $O_2^{\overline{\cdot}}$ action, let us now survey the reactions superoxide undergoes with the different functional groups and see how well our predictions hold up.

A. Reaction of Superoxide with Peroxides

Even a cursory examination of the recent literature of $O_2^{\overline{\cdot}}$ reactions quickly reveals the omnipresent role of peroxides as intermediates or final products. It is clear, therefore, that an understanding of the reactions of $O_2^{\overline{\cdot}}$ with this oxygen-oxygen linkage is crucial to the understanding of superoxide chemistry in general.

Moreover, extensive studies on the biological role of superoxide have focused attention on the possible occurrence of the reaction outlined in Equation 25.

$$O_2^{\overline{\cdot}} + ROOH \longrightarrow O_2 + RO\cdot + HO^- \qquad (25)$$

The essence of this reaction, dubbed the Haber-Weiss reaction[73] when R = H, is an electron transfer from $O_2^{\overline{\cdot}}$ to the peroxide linkage. Such a process would induce scission of the oxygen-oxygen bond, generating hydroxide ion and an alkoxy (or hydroxy, when R = H) radical.

1. Hydrogen Peroxide

In the case of H_2O_2, the generation of hydroxy radicals by the Haber-Weiss reaction could explain the crucial role $O_2^{\overline{\cdot}}$ plays in a variety of biological hydroxylations as well as the reason for its toxicity.[74] However, a series of kinetic studies[69,74-78] indicate that the Haber-Weiss reaction (Equation 25, R = H) is much too slow ($k < 1.3 \pm 0.7 \times 10^{-1}$)[75] to compete with the rapid disproportionation of $O_2^{\overline{\cdot}}$ (Equation 3) induced by protic solvent and/or the slightly acidic hydroperoxide itself. Already in 1947, George[79] had observed that solid KO_2 dispersed in 96.5% H_2O_2 yielded precisely the amount of oxygen gas predicted by a simple acid-catalyzed disproportionation (Equation 3, B = HOO), i.e., 1 mol of oxygen for every 2 mol of $O_2^{\overline{\cdot}}$ reacted, rather than the one-to-one correspondence required by electron transfer Reaction 25. A theoretical justification for the absence of an electron transfer mode has been suggested by Koppenol.[78]

2. Alkyl Hydroperoxides

The wave of interest in the Haber-Weiss reaction and its possible role in biological processes brought in its wake a series of studies on the reactions of alkyl hydroperoxides with $O_2^{\overline{\cdot}}$. Le Berre and Berguer[38,39] studied the reaction of triphenylmethyl hydroperoxide with dispersed sodium superoxide, NaO_2. They reported that while only simple disproportionation was observed in benzene (with the concomitant formation of a peroxy anion), a 15% yield of reductive cleavage was observed after 24 hr when tetrahydrofuran (THF) served as solvent. Since the same yield resulted even when the hydroperoxide was first deprotonated by potassium *t*-butoxide, the authors concluded that reductive cleavage occurs from the anion (Equations 26 and 27).

$$\phi_3COOH + 2NaO_2 \longrightarrow \phi_3COONa + HOONa + O_2 \qquad (26)$$

$$\phi_3COONa + 2NaO_2 \longrightarrow \phi_3CONa + Na_2O + 2O_2 \qquad (27)$$

This interpretation is problematic for two reasons. Firstly, Equation 27 requires electron transfer to a species already negatively charged in order to generate a poorly stabilized O^{-2} anion in aprotic weakly polar media — an unlikely sequence. More important, reductive cleavage occurs only in THF but not in benzene. The authors themselves note that the oxygen yields in THF are far below that predicted by the mechanism, suggesting that the solvent is slowly chewed up by the $O_2^{\overline{\cdot}}$. It is not clear what other species are produced in this latter process, but they may well be responsible for the reductive cleavage observed.

Recently, Peters and Foote[9] reported that tetramethylammonium superoxide reacts with either *t*-butyl or *t*-amyl hydroperoxide in acetonitrile, yielding the corresponding alcohols as the major product along with a small amount of acetone. These authors suggested that a Haber-Weiss type process is at play (Equations 28 and 29) and noted that the formation of acetone is easily explained by a β-cleavage of the intermediate alkoxy radical (Equation 30).

$$R{-}\underset{CH_3}{\overset{CH_3}{\underset{|}{\overset{|}{C}}}}{-}OOH + O_2^{\overline{\cdot}} \longrightarrow R{-}\underset{CH_3}{\overset{CH_3}{\underset{|}{\overset{|}{C}}}}{-}O\cdot + HO^- + O_2 \qquad (28)$$

$$R{-}\underset{CH_3}{\overset{CH_3}{\underset{|}{\overset{|}{C}}}}{-}O\cdot + R'H \longrightarrow R{-}\underset{CH_3}{\overset{CH_3}{\underset{|}{\overset{|}{C}}}}{-}OH + R\cdot' \qquad (29)$$

$$R{-}\underset{CH_3}{\overset{CH_3}{\underset{|}{\overset{|}{C}}}}{-}O\cdot \longrightarrow CH_3{-}\overset{O}{\overset{\|}{C}}{-}CH_3 + R\cdot \qquad (30)$$

In support of Foote's results, Pryor and co-workers[80,81] also found that *t*-butyl and linoleic hydroperoxides were reductively cleaved by KO_2/18-crown-6 polyether complex in acetonitrile-hexane or acetonitrile-DMSO solvent mixtures.

Gibian and Ungerman[82,83] and Stanley,[84] however, found that the peroxy linkage of *t*-butyl hydroperoxide was inert to $O_2^{\overline{\cdot}}$ when the reaction was carried out in pyridine, benzene, or toluene. In these cases, only acid-catalyzed disproportionation of $O_2^{\overline{\cdot}}$ was observed with concomitant formation of a stable peroxy anion. Furthermore, they report[83,84] that the results observed by Foote[9] and Pryor[80,81] can also be obtained in acetonitrile when KOH replaces KO_2. More important, the isolation of acetamide as a product of the reaction indicates that the solvent plays a crucial role in the course of the reaction. It is probable, therefore, that the mechanism of the acetonitrile reactions involves initial deprotonation of the hydroperoxides by $O_2^{\overline{\cdot}}$ as observed in toluene, benzene, and pyridine. The resulting peroxy anion reacts with nitrile solvent to form a perimidic ester which decomposes to produce acetamide, alcohol, and β-cleavage products. Overall this reaction is simply the well-precedented[85-90] free-radical chain decomposition of hydroperoxides in the presence of nitriles and base. Gibian and Ungerman[83] conclude, therefore, that no Haber-Weiss type reaction occurs with alkyl hydroperoxides in aprotic solvents.

Is the same true for aqueous solutions? Pryor[80,81] reports that enzymatically (xanthine-xanthine oxidase) generated $O_2^{\overline{\cdot}}$ does indeed convert hydroperoxides to alcohols. Bors and co-workers[91] have, however, generated $O_2^{\overline{\cdot}}$ by radiolytic techniques and conclude that superoxide anions do not react with hydroperoxides. They argue that, in contradistinction to their pulse radiolysis technique wherein $O_2^{\overline{\cdot}}$ is generated cleanly, the enzymatic system utilized by Pryor does not produce $O_2^{\overline{\cdot}}$ exclusively. It is presumably these other species which are causing the effects observed.

Thus, the data suggest that superoxide reacts with the protic peroxides primarily as a base; other potential reactions (e.g., the Haber-Weiss reaction) do not or cannot compete effectively.

3. Dialkyl Peroxides

As shown in Table 1, dialkyl peroxides show varying reactivity with $O_2^{\overline{\cdot}}$. In general

Table 1
REACTION OF $O_2^{\cdot -}$ WITH DIALKYL PEROXIDES

Entry	Peroxide	Type	Products	Ref.
1	Di-*t*-butyl peroxide	3°	N.R.	24,82,92
2	Biadamantylidene dioxetane	3°	N.R.	93
3	9,10-Diphenylanthracene-9,10-peroxide	3°	Meso-9,10-dihydroxy-9,10-diphenylanthracene	94
4	Di-*sec*-butyl peroxide	2°	*Sec*-butanol (50%); 2-butanone (30%)	92
5	Di-*n*-butyl peroxide	1°	*N*-butanol, propanal	36
6	Di-*n*-hexyl peroxide	1°	*N*-hexanol (65%)	92
7	1,4-Dihydro-2,3-benzodioxin	1°	Complex mixture	94
8	1,4-Dihydro-2,3-naphtho(2,3)dioxin	1°	Complex mixture	94

tertiary dialkyl peroxides such as di-*t*-butyl peroxide[24,82,92] and biadamantylidene dioxetane[93] are essentially inert (Table 1, entries 1 and 2). Nevertheless, Lee-Ruff et al.[94] report that the endoperoxide of 9,10-diphenylanthracene upon reaction with KO_2 in DMF for 5 days yields the corresponding diol in high yield (Table 1, entry 3). Primary and secondary peroxides[36,92,94] react slowly producing alcohols as the major isolable product. In several cases (see for example entries 4 and 5) carbonyl product is also isolated.

The question of mechanism is clearly of interest and has been discussed in some detail by San Filippo and co-workers.[92] One possibility is oxygen-oxygen scission via an electron transfer from $O_2^{\cdot -}$ to the peroxide linkage in a Haber-Weiss type of reaction (Equation 31).

$$ROOR + O_2^{\cdot -} \longrightarrow RO\cdot + RO^- + O_2 \qquad (31)$$

Such a mechanism, however, seems unlikely in light of the fact that tertiary dialkylperoxides appear to be generally inert.

A more likely possibility invokes the effective basicity of $O_2^{\cdot -}$ and suggests that we are observing another example of the Kornblum-DeLaMare reaction[95-99] (Equation 32). In this base-catalyzed process, a peroxide bearing an α-proton is decomposed directly to an alcohol and a carbonyl compound.

$$\underset{\displaystyle H \longleftarrow :B}{R_2C{-}O{-}O{-}CR_3} \longrightarrow R_2C{=}O + {}^-O{-}CR_3 + \overset{+}{B}H \qquad (32)$$

Such a mechanism[36,83,92] nicely explains the formation of both alcohol as well as carbonyl compounds as primary products in the reaction of $O_2^{\cdot -}$ with primary and secondary dialkyl peroxides (see entries 4 and 5). The observed low yield of carbonyl-containing product undoubtedly results from its further base catalyzed autoxidation to acid (see sections IV.B.4 and IV.E.1). Furthermore, it would explain the general inertness of tertiary peroxides as well as the greater than six-fold increase in the reaction half-life of di-*n*-hexyl-1,1,1′,1′-d_4 peroxide as compared to the tetraprotio analog.[92]

Nevertheless, the fact that alcohol is produced in excess of 50% yield in the case of di-*n*-hexyl peroxide (entry 6) requires that an additional pathway be partially operative. Such a secondary pathway is also required by the observation that the tertiary peroxide 9,10-diphenylanthracene-9,10-peroxide (entry 3) is cleaved to diol after being stirred for 5 days with KO_2 in DMF. What may be involved in these cases is a nucleophilic displacement at the peroxide oxygen.[96] Indeed when di-*n*-hexyl peroxide is reacted with

KOH/18-crown-6 in benzene, a 60% yield of 1-hexanol is obtained.[92] Similarly nucleophilic displacement on oxygen has been invoked to explain the production of diol in the reaction of trimethyldioxetane with alkali.[100] In any case, the role of trace metals[101] in catalyzing the homolytic scission of the peroxide linkage should not be overlooked.[97,102]

For the purpose of comparison, it should be noted that electrogenerated $O_2^{\cdot -}$ is reported[103] to react with dialkyl and diaryl disulfides catalyzing an interchange of substituents as shown in Equation 33. Sulfur oxidation products (e.g., RSO_3^- or RSO_2^-) were not observed, although oxidation products have been isolated when $O_2^{\cdot -}$ is generated from KO_2/18-crown-6.[104]

$$RSSR + R'SSR' \xrightarrow{O_2^{\cdot -}} RSSR' \qquad (33)$$

Interestingly, the interchange reaction occurs not only with disulfides having a higher half-wave potential than $O_2^{\cdot -}$, but also with those having a lower one. The mechanistic details still need to be elucidated.

B. Reaction of $O_2^{\cdot -}$ with Substrates Bearing Labile Hydrogens

In Section III we noted that, while the effective basicity of $O_2^{\cdot -}$ is quite high, its radical character is minimal. Hence, we would expect that, upon reaction with labile hydrogens, $O_2^{\cdot -}$ would act preferably as a Bronstead base rather than as a hydrogen atom acceptor. Let us turn now to the experimental data and see how valid our predictions are.

1. O—H Bond

We have already shown above that the major, if not sole, reaction of *hydroperoxides* with $O_2^{\cdot -}$ is acid-catalyzed disproportionation of the latter with the accompanying formation of a stable peroxy anion. *Alcohols*, too, react in a similar fashion yielding the corresponding alkoxides.[28,30] Stanley[94] reports that while methanol and ethanol apparently cause the instantaneous disproportionation of $O_2^{\cdot -}$, isopropanol requires several minutes for complete reaction and *tert*-butanol reacts at appreciable rates only at relatively high concentrations.

The reaction of $O_2^{\cdot -}$ with aryl hydroxylic substrates such as phenols,[105] *o*- and *p*-dihydroxyarenes,[22,36,54,106-109] and α-tocopherol and related 6-hydroxychroman compounds[106,110-113] have been studied and substrate oxidation products result. The observation of semiquinones in the case of dihydroxyarenes and chromanoxyl radicals in the case of 6-hydroxychroman compounds suggests that the initial steps of these oxidations involve net hydrogen atom transfer from ArOH to $O_2^{\cdot -}$ as outlined in Equation 34. Semiquinone formation merely entails subsequent proton transfer to base.

$$O_2^{\cdot -} + ROH \longrightarrow HO_2^- + RO\cdot \qquad (34)$$

Such a net hydrogen-atom transfer mechanism may be envisaged either as a concerted hydrogen atom abstraction by $O_2^{\cdot -}$ (Equation 35) or as a two-stepped process involving initial proton abstraction (Equation 36) followed by electron transfer from the hydroperoxyl group to the substrate anion (Equation 37).

$$O_2^{\cdot -} + HOR \longrightarrow (O_2^{\cdot -} \cdots H \cdots OR)^{\dagger} \longrightarrow HO_2^- + RO\cdot \qquad (35)$$

$$O_2^{\cdot -} + ROH \longrightarrow HO_2^{\cdot} + RO^- \qquad (36)$$

$$HO_2^{\cdot} + RO^- \longrightarrow HO_2^- + RO\cdot \qquad (37)$$

Sawyer has calculated that such a net H-atom transfer is not favored thermodynamically for either 3,5-di-*tert*-butylcatechol or α-tocopherol.[106]

Afanas'ev and Polozova[108,109] have studied the reaction of $O_2^{\overline{\cdot}}$ with catechol, trimethylhydroquinone, and dopamine. They noted the immediate formation of semiquinone radical-ions and the absence of substantial amounts of oxygen in the reaction mixture. Indeed, a molecular oxygen content of less than 5% (based on initial $O_2^{\overline{\cdot}}$ concentration) was observed. These authors argue that if the first step were proton abstraction, substantial amounts of O_2 should have been produced by the acid catalyzed dismutation of $O_2^{\overline{\cdot}}$ (Equations 1, 2, and 3). This evidence leads them to suggest *a single step mechanism* for the formation of quinones in which hydrogen peroxide is formed directly, as shown in Equation 38.

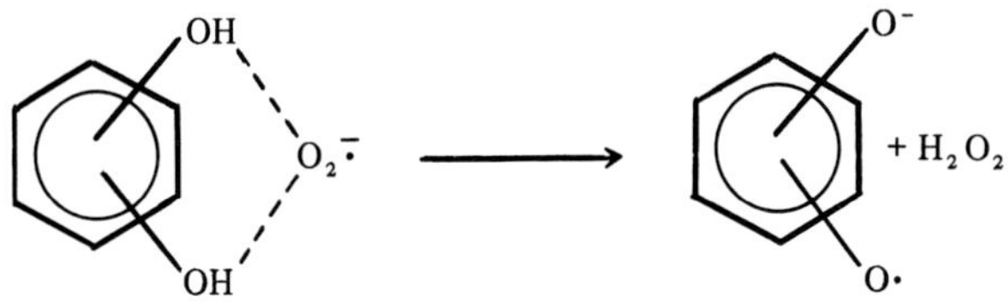
(38)

However, Sawyer and co-workers[106] have strong evidence for the transient formation of O_2 in a *reversible equilibrium* in the cases of 3,5-di-*tert*-butylcatechol and α-tocopherol. Based on this and other experimental electrochemical evidence, Sawyer et al. suggest that the oxidation of acidic substrates by superoxide ion involves an initial rate determining proton transfer from the substrate to $O_2^{\overline{\cdot}}$ (Equation 39). This is followed by rapid dismutation to give peroxide and molecular oxygen (Equation 40), with the latter oxidizing the substrate anion (Equation 41).

$$O_2^{\overline{\cdot}} + ROH \longrightarrow HO_2\cdot + RO^- \quad (39)$$

$$HO_2\cdot + O_2^{\overline{\cdot}} \longrightarrow HO_2^- + O_2 \quad (40)$$

$$O_2 + RO^- \longrightarrow RO\cdot + O_2^{\overline{\cdot}} \quad (41)$$

Following the formation of RO·, a variety of free radical oxidative processes can take over. In the case of *o*-dihydroxy arenes,[54] *o*-quinones and dicarboxylic acids (or their cyclization products) are formed perhaps by the mechanism shown in Scheme 1.

In the case of α-tocopherol model compounds (see Scheme 2), chroman **2a** (R = CO_2H) reacted with enzymatically generated $O_2^{\overline{\cdot}}$ in aqueous media to yield quinone **3**[110] via 8a-hydroxychroman-6-one **4**, while chroman **2b** (R = CH_3) yielded 5-hydroxychroman-6-one **5** when reacted with KO_2 suspended in THF.[111] When chroman **2b** (R = CH_3) was reacted with KO_2/18-crown-6 in acetonitrile, diepoxide **6** resulted.[113] The probable mechanisms for these reactions are shown in Scheme 2.

That the nature of the products change with reaction conditions is quite consistent with the role of $O_2^{\overline{\cdot}}$ as base. Very selective reactions, depending on solvent and temperature, have been reported in the base-catalyzed autoxidation of phenols.[114]

The oxidation of ascorbic acid[106,115,116] has also been studied in some detail. Here, however, the transient formation of O_2 in a reversible equilibrium is not observed.[106] This suggests two possibilities. One is that a net hydrogen transfer sequence (Equation 34) may be operative either by a concerted (Equation 35) or two-stepped process (Equations 36 and 37). Indeed, Sawyer's calculations suggest that net hydrogen transfer in the case of ascorbic acid is mildly exothermic ($E°_{rxn}$ = 0.05 V).[106]

Another more fascinating possibility[106] is that $O_2^{\overline{\cdot}}$ reacts by a concerted two-hydrogen atom (two-electron) transfer from ascorbic acid (Equation 42).

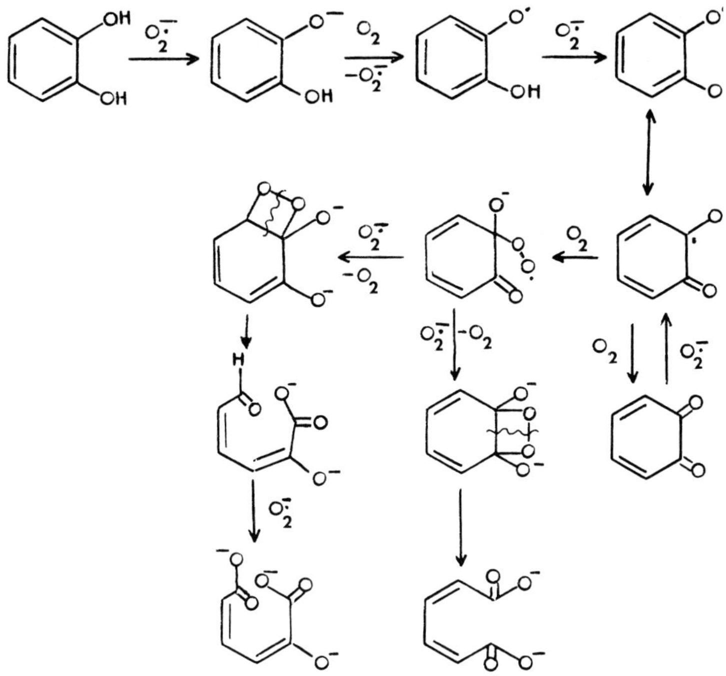

Scheme 1. Possible mechanism for the $O_2^{\cdot-}$-induced autoxidation of *o*-hydroquinone.

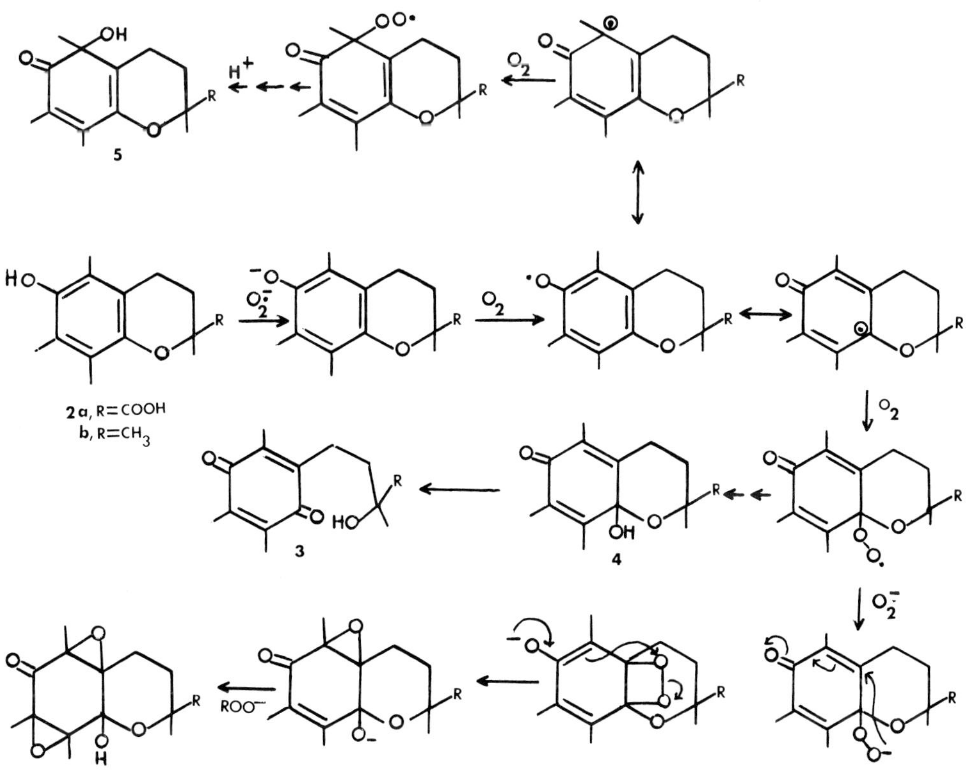

Scheme 2. Various reaction pathways for the reaction of $O_2^{\cdot-}$ with α-tocopherol model compounds.

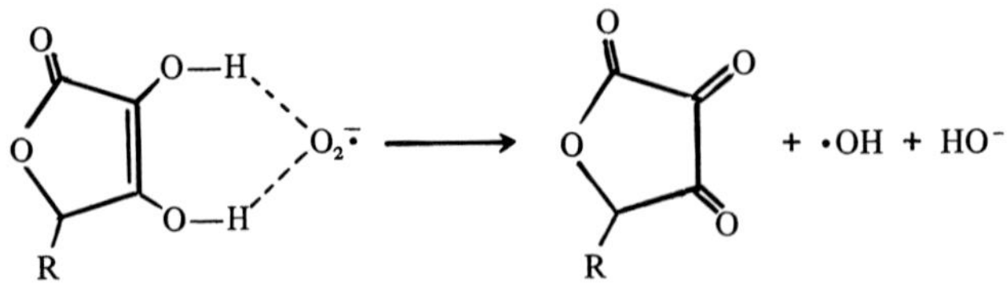

(42)

Sawyer calculates such a process to be strongly favored thermodynamically for ascorbic acid. Furthermore, such two-electron reductions of $O_2^{\cdot -}$ by substrates such as ascorbic acid may represent a serious biological hazard because the reduction product, ·OH, is a potent cytogenic oxidant.[106]

However, an additional possibility should be considered and that is a two-stepped process whose net result is equivalent to Equation 42. In the first step a hydrogen atom and proton would be concertedly transferred to $O_2^{\cdot -}$ (Equation 43), as suggested by Afanas'ev and Polozova[108,109] for dihydroxyarenes. The semidione would then be oxidized by hydrogen peroxide to dehydroascorbic acid (Equation 44).

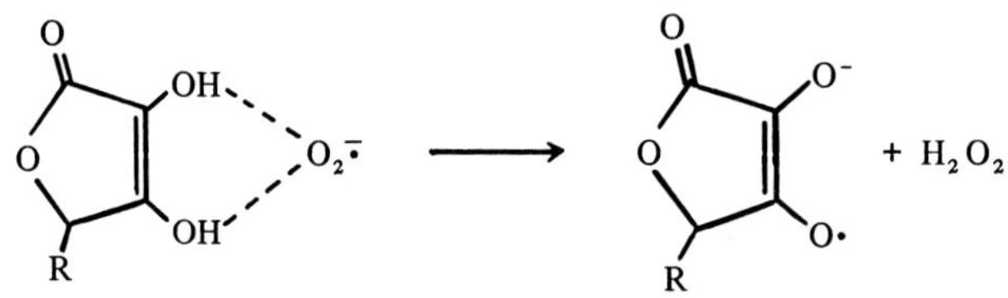

(43)

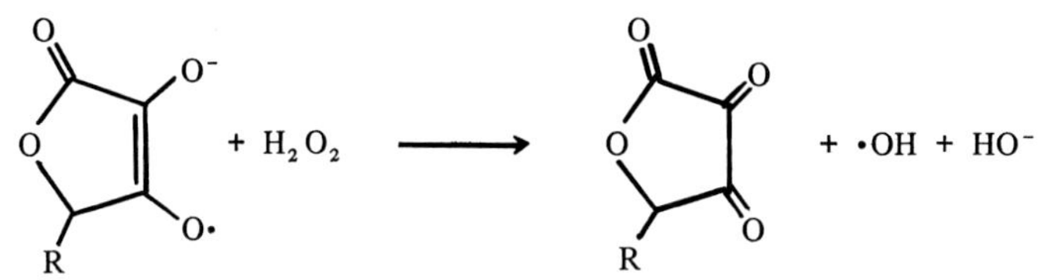

(44)

Frimer and Gilinsky have studied the reaction of $O_2^{\cdot -}$ with a variety of enols,[117,118] in particular 2-hydroxycylohexa-2,5-dien-1-ones (**9**) (Equation 45). The latter were in turn synthesized by reacting 4,4 or 5,5-disubstituted cyclohex-2-en-1-ones (**7** and **8**, respectively) with $O_2^{\cdot -}$ (KO_2/18-crown-6 in benzene),[118] a reaction we will discuss in Section IV.B.4. Interestingly, these enols (**9**) were oxidized to the corresponding lactols (**10**).

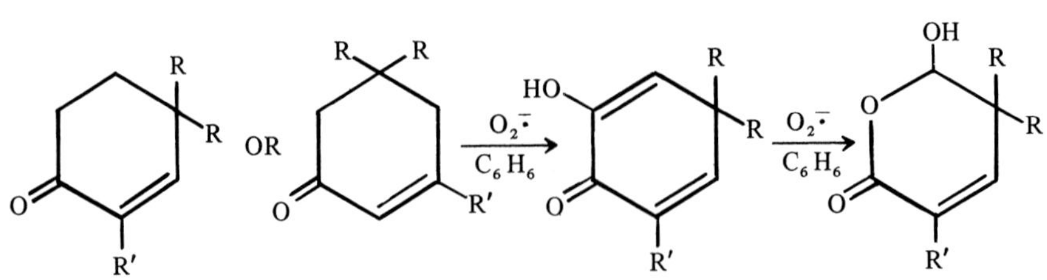

(45)

Considering that the reaction proceeded even more rapidly with *t*-butoxide and oxygen, these authors consider the $O_2^{\cdot -}$ process to be a base-catalyzed autoxidation. A possible mechanism[119a] is outlined in Scheme 3 and involves initial proton removal as suggested by Sawyer.[106]

Hosokawa et al.[119b] have recently reported that $O_2^{\cdot -}$ produced from KO_2 and Pd (II) catalysts, reacts with ketoximes and CH_2Cl_2 to produce methylene dioximes (Equation 46).

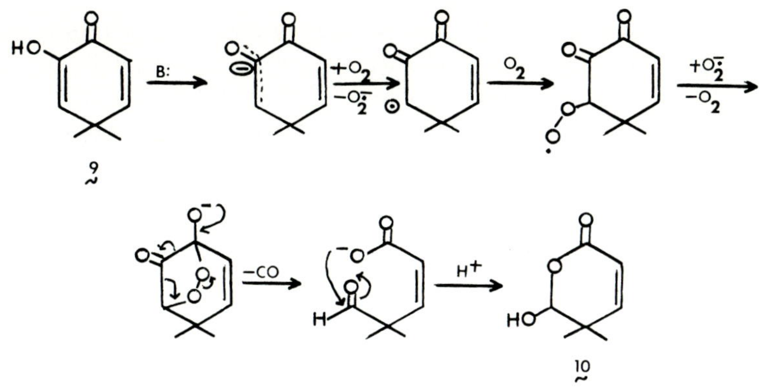

Scheme 3. Mechanism for the $O_2^{\overline{\cdot}}$-induced oxidation of 4,4-disubstituted-2-hydroxy-cylohexa-2,5-dien-1-ones.

$$(R_1)(R_2)C{=}N{-}OH + CH_2Cl_2 \xrightarrow[\text{Pd(II)}]{KO_2} [(R_1)(R_2)C{=}N{-}O_2]_2CH_2 \qquad (46)$$

2. S–H Bond

A variety of thiols[120-122] have been reacted with $O_2^{\overline{\cdot}}$ and generally disulfides result. Considering the greater acidity of S–H bonds as compared to the corresponding O–H bonds, the mechanism of these reactions is probably as outlined below (Equation 47).

$$RSH \xrightarrow{O_2^{\overline{\cdot}}} RS^- \xrightarrow{O_2} RS\cdot \longrightarrow RS{-}SR \qquad (47)$$

3. N–H Bond

A perusal of standard pK_a tables reveals that as a general rule the pK_a values of N–H bonds are 15 to 20 units higher than those of the corresponding O–H analogs. Hence, proton-catalyzed dismutation of $O_2^{\overline{\cdot}}$ is not expected to be a general phenomenon with nitrogen compounds. On the other hand, the average bond energy of an N–H bond (93 kcal) is substantially less than that of O–H (111 kcal) or a C–H bond (99 kcal).[48] This suggests that for N–H bonds we might expect H-atom abstraction by $O_2^{\overline{\cdot}}$ to be more prevalent. Indeed Sawyer and Nanni[123] have studied the electrogenerated $O_2^{\overline{\cdot}}$ oxidation of hydrophenazines, reduced flavins, hydroxylamine, and related substrates where the labile hydrogens are not acidic but are readily abstractable via free-radical processes. In these cases, electrochemical evidence strongly suggests that oxidation by $O_2^{\overline{\cdot}}$ occurs via initial hydrogen atom transfer. Similarly Hussey et al.[124] have shown that proton abstraction is not involved in the oxidation of *o*-phenylenediamine by electrogenerated $O_2^{\overline{\cdot}}$.

Poupko and Rosenthal[125] report that based on ESR data it is clear that secondary dialkyl amines are instantaneously oxidized to dialkylnitroxides by KO_2 in DMSO, even in the absence of atmospheric oxygen (Equation 48).

$$R_2NH + O_2^{\overline{\cdot}} \longrightarrow R_2N{-}O\cdot + HO^- \qquad (48)$$

Tertiary amines such as triphenylamine seem inert. Frimer et al.[126] have observed however, no reaction between $O_2^{\overline{\cdot}}$ (KO_2/18-crown-6) and diethyl or triethylamine.

The reaction of $O_2^{\overline{\cdot}}$ with aromatic amines has been studied by a variety of groups[120,124-127] and the major product is a symmetrical azobenzene. Since the same product yields are obtained if one begins with the corresponding 1,2-disubstituted hydrazines,[127,128] it is likely that the hydrazine, formed by the coupling of two anilinyl radicals, is a reactive intermediate in this process (Equation 49).

$$\text{Ar-NH}_2 \xrightarrow{O_2^{\bar{\cdot}}} \text{Ar-NH}\cdot \longrightarrow \text{Ar-NH-NH-Ar} \xrightarrow{O_2^{\bar{\cdot}}} \text{Ar-N=N-Ar} \qquad (49)$$

Further evidence for the intermediacy of the anilinyl radical comes from the work of Frimer, Ziv, and Aljadeff[126] on the reaction of aniline with KO_2/crown ether. These researchers found, in addition to azobenzene (26% yield) and nitrobenzene (8% yield), a 24% yield of *N*-nitrophenyl aniline. The latter undoubtedly results from the trapping of the anilinyl radical by nitrobenzene.

Despite the evidence for a free radical process, a base-catalyzed autoxidative process is also possible. Indeed Balogh-Hergovich et al.[127] report that azobenzenes can be obtained (although at a slower rate) when aryl amines are oxidized with molecular oxygen catalyzed by "naked" hydroxide ion (KOH/18-crown-6).

Superoxide anion radical reacts with hydrazo compounds and related substances in a variety of ways.[128,129] As mentioned previously, 1,2-diarylhydrazines are converted to the corresponding azo compounds (Equation 50), though 1,2-dialkylhydrazines seem to be unreactive. Monosubstituted alkyl and aryl hydrazines are oxidized in a reaction which appears to involve free alkyl or aryl radicals (Equation 51). 1,1-Disubstituted hydrazines are oxidized to *N*-nitroso amines (Equation 52) while the hydrazines of aryl ketones yield the corresponding azine in high yield (Equation 53). However, the hydrazine of the alkyl ketone cyclohexanone was unreactive.

$$\text{ArHNNHAr} \xrightarrow{O_2^{\bar{\cdot}}} \text{Ar-N=N-Ar} \qquad (50)$$

$$\text{R-NHNH}_2 \xrightarrow{O_2^{\bar{\cdot}}} \text{R-H} \qquad (51)$$

$$\text{R}_2\text{NNH}_2 \xrightarrow{O_2^{\bar{\cdot}}} \text{R}_2\text{N-NO} \qquad (52)$$

$$\text{Ar(R)C=N-NH}_2 \xrightarrow{O_2^{\bar{\cdot}}} \text{Ar(R)C=N-N=C(R)Ar} \qquad (53)$$

The mechanisms for these processes are unknown. Chern and San Filippo[128] have noted that the failure to observe any appreciable reaction between 1,2-dialkylhydrazines or dialkylhydrazones and superoxide may reflect the fact that the pK_a of alkyhydrazines and -hydrazones is significantly higher. This suggests of course that the ability of $O_2^{\bar{\cdot}}$ to affect the oxidation of these substrates is related to their acidity and that the initial step involves proton abstraction by $O_2^{\bar{\cdot}}$. Sawyer[106] agrees with this suggestion, noting that the fact that the oxidation of hydrazines is catalyzed by protons or metal ions is further support for this conclusion.

Nevertheless, it is as of yet unclear why compounds like phenylenediamine, hydroxylamine, and hydrophenazine should react with $O_2^{\bar{\cdot}}$ via initial hydrogen atom abstraction while phenylhydrazine should react by initial proton abstraction. The pK_a of the hydroxylamine is 12 to 13 due to the hydroxyl anions and should be a superb proton donor.[130] On the other hand, hydrazines show only a weak acidity and can be deprotonated only by exceptionally strong bases like sodium or sodium amide.[131] In any case, the pK_a of hydrazines and hydrazones should not be substantially different from that of phenylenediamine or hydrophenazine. Hence, it is not at all clear why there should be such a radical difference in reaction pathway.

Nitroaromatic amines such as *o-* and *p-*nitroaniline with acidic N—H bonds (pK_a = 17.9 and 18.4, respectively) have been shown by Hussey[132] to react with $O_2^{\cdot -}$ by proton transfer. The failure of *m-*nitroaniline to react with $O_2^{\cdot -}$ is not surprising since it should be much less acidic than the *o-* or *p-*isomers.

Purines, such as purine, adenine, and *N-*benzyladenine, also bear an acidic proton at the 7 position. It is not surprising, therefore, that upon reaction with $O_2^{\cdot -}$ (Equation 54), the corresponding salt is formed which readily regenerates the starting material in acid.

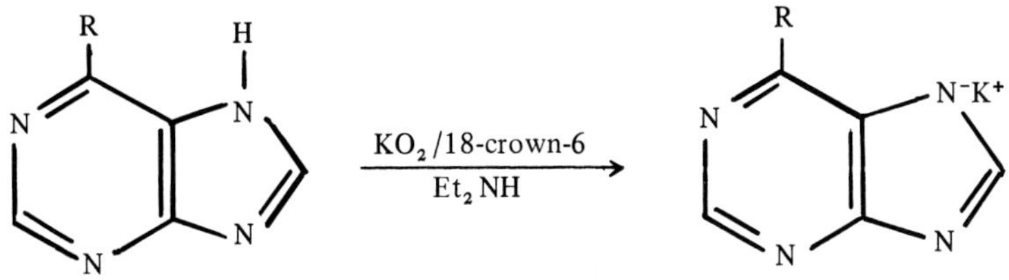

(54)

Superoxide does not react with diethyl- or triethylamine which are convenient solvents for this reaction.[126]

4. C—H Bond

C—H linkages with low pK_a values undoubtedly react with $O_2^{\cdot -}$ via initial proton transfer. Thus, Stanley[133] reports that cyclopentadiene (pK_a = 16)[134] and diethyl malonate (pK_a = 13)[134] induce instantaneous disproportionation of $O_2^{\cdot -}$. Dibenzoylmethane and 1,3-cyclohexadione are rapidly deprotonated by $O_2^{\cdot -}$, though the resulting anions are stable to O_2 and $O_2^{\cdot -}$.[118] Similarly the oxidation of benzoylacetonitrile,[5] malonitrile[135,136] (pK_a = 11.2),[134] benzyl cyanide,[135,136] and carbonyl compounds[57-59] (pK_a = 20)[134] at the α-position is expected to proceed via a base catalyzed autoxidative process outlined in Equations 55 to 61.

$$R-H + O_2^{\cdot -} \longrightarrow R^- + HOO\cdot \quad (55)$$

$$HOO\cdot + O_2^{\cdot -} \longrightarrow HOO^- + O_2 \quad (56)$$

$$R^- + O_2 \rightleftharpoons R\cdot + O_2^{\cdot -} \quad (57)$$

$$R\cdot + O_2 \longrightarrow RO_2\cdot \quad (58)$$

$$RO_2\cdot + O_2^{\cdot -} \longrightarrow RO_2^- + O_2 \quad (59)$$

$$RO_2^- + RH \longrightarrow ROOH + R^- \quad (60)$$

$$ROOH \longrightarrow R{=}O,\ ROH,\ RCOOH,\ \text{etc.} \quad (61)$$

There have been a large number of reports that upon reaction with $O_2^{\cdot -}$, ketones yield α-diketones and acids,[119,137-139] and that α-hydroxy ketones yield acids.[140,141] In the case of enones both the α′ and γ hydrogens may be removed (see also Section IV.B.1). Frimer and Gilinsky[117,118] have found that upon reaction with $O_2^{\cdot -}$, 4,4- and 5,5-disubstituted-cyclohex-2-en-1-ones (**7** + **8** respectively) yielded 2-hydroxycyclohexa-2,5-dien-1-ones (**9**). In the case of the former, α′-proton abstraction is invoked, while for the latter, γ-proton removal seems to be involved (Scheme 4). In the case of 6,6-disubstituted cyclohex-2-en-1-ones (**11**), epoxides and dimers result by a mechanism which most likely involves initial γ-proton removal as outlined in Scheme 5. Condensation of the resulting anion with starting material yields dimer, while oxidation of the anion yields peroxy anion. Either the latter or HOO^- (formed by the dismutation of $O_2^{\cdot -}$) can epoxidize the substrate.

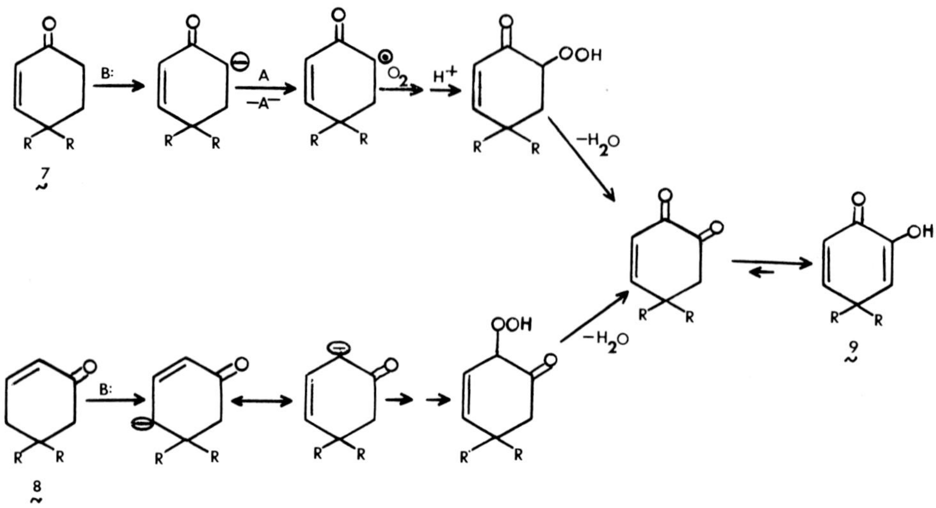

Scheme 4. Reaction mechanism for the oxidation of 4,4- and 5,5-disubstituted-2-cyclohexen-1-ones.

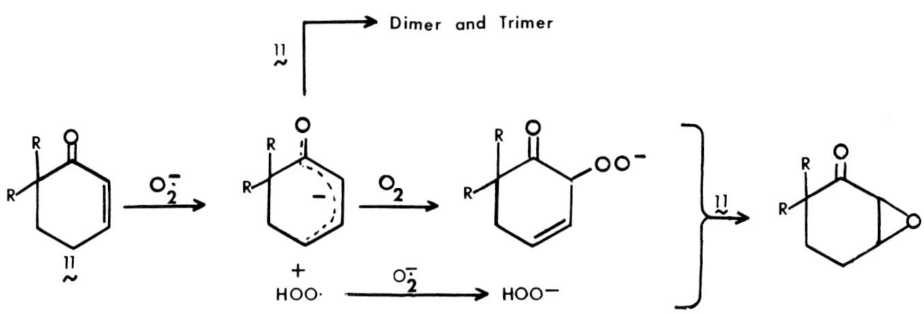

Scheme 5. $O_2^{\overline{\cdot}}$ -induced oxidation of 6,6-disubstituted-2-cyclohexen-1-ones.

Further proof for the base-catalyzed oxidative nature of this reaction is that the same products are obtained with "naked" hydroxide and *t*-butoxide[117,118] with the order of decreasing rates *t*-butoxide > superoxide > hydroxide. Additional evidence comes from the observation that the rate of the $O_2^{\overline{\cdot}}$ reaction is essentially the same whether carried out in air or under argon. As seen from Equation 3, for every molecule of substrate deprotonated, a molecule of O_2 is formed. Hence, the autoxidation may proceed even in the absence of an oxygen atmosphere.

In light of all the above, quite surprising is the recent report of Sawyer and coworkers[5] that a survey (by cyclic voltametric generation of $O_2^{\overline{\cdot}}$) indicates no loss of $O_2^{\overline{\cdot}}$ in the presence of cyclohexanone, acetone, or 2-butanone. These authors suggest that KO_2-derived oxidants (perhaps HOO^-) are responsible for the reactions observed. Hence,the exact mechanistic details have yet to be clarified.

In the case of substrates with pK_a values greater than 20, the question of mechanism is not always clear. Sagae et al.[142,143] have reported that *o*- and *p*- (but not *m*-) nitrotoluenes are oxidized by electrogenerated $O_2^{\overline{\cdot}}$ to the corresponding benzoic acids (Equation 62). These authors suggest an initial H-atom abstraction by $O_2^{\overline{\cdot}}$. However, nitrotoluene also undergoes base-catalyzed autoxidation.[57,58]

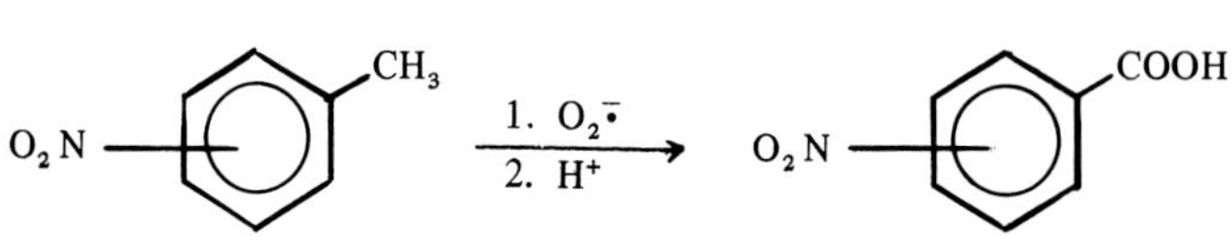

(62)

9,10-Dihydroanthracene[144,145] reacts instantaneously with $O_2^{\cdot-}$ in a nitrogen atmosphere to give a 9:1 ratio of anthracene and anthraquinone (Equation 63).

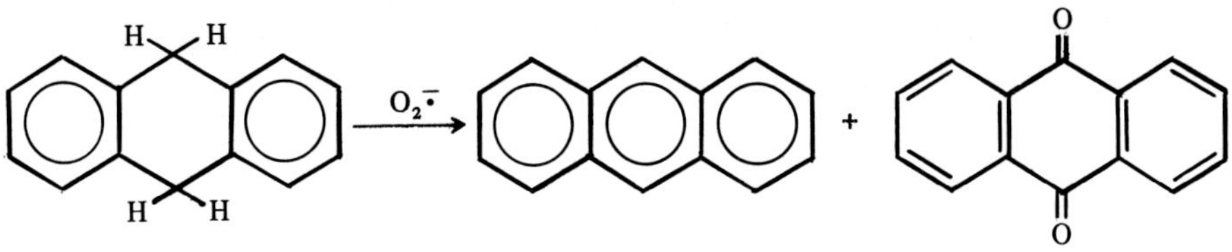

(63)

In the presence of oxygen, this ratio drops to 2:1. Similarly 1,3- and 1,4-hexadiene[144] yield benzene as the sole product. It is presumed here as well that the reaction involves initial H-atom abstraction, although conversion of dihydroanthracene to anthracene and anthraquinone also occurs under base-catalyzed conditions.[146] Nevertheless, in the latter case the conditions are much more severe (55° for 2 hr) and the anthraquinone is the major product even at low oxygen levels. Hence, it seems fair to presume that the reaction mechanisms differ in the two cases.

In addition to dihydroanthracene and 1,4- and 1,3-cyclohexadiene, Moro-oka et al.[144] report that anthrone, fluorene, xanthine, and diphenylmethane are oxidized by $O_2^{\cdot-}$ to anthraquinone, fluorenone, xanthone, and benzophenone, respectively. However, 9,10-dihydrophenanthrene, acenaphthene, cyclohexene, 2,3-dimethyl-2-butene, tetraline, and cumene are not. This indicates[144] that $O_2^{\cdot-}$ abstracts hydrogen from the substrate only when the radical formed can be highly stabilized by conjugating π-electrons. Obviously then, H-atom abstraction by $O_2^{\cdot-}$ depends on the C–H bond dissociation energy of the substrate. Moro-oka's[144] suggestion that the borderline seems to be roughly 80 kcal/mol at room temperature is in line with our prediction of Section III.C.

Lee-Ruff and Timms[147] have studied the superoxide mediated autoxidation of fluorene ($pK_a = 23$),[134] diphenylmethane ($pK_a = 34$),[134] and distyrylmethane, each of which is converted to the corresponding ketone. Fluorine is of particular interest since its pk_a value[134] of 23 lies at the upper limit of the effective basicity of $O_2^{\cdot-}$ as predicted by Sawyer[3,22] (See Section III.A.) Furthermore, the facile base-catalyzed autoxidation of fluorene and diphenylmethane has been well investigated.[57-59,148,149] Nevertheless, these researchers argue in favor of an initial hydrogen atom abstraction (Equation 64, path a) based on the following evidence. First, they argue that although hydroxide indeed catalyzed the autoxidation of fluorene, it did so only at rates 20 to 100 times slower than $O_2^{\cdot-}$. However, we note that in light of Sawyer's[3,22] suggestion that the effective basicity of $O_2^{\cdot-}$ is eight orders of magnitude greater than that of HO^-, this observation by no means excludes a base-catalyzed autoxidative process (Equation 64, path b).

$$O_2^{\cdot-} + Ar_2CH_2 \xrightarrow{a} Ar_2CH\cdot \xrightarrow{O_2} Ar_2CHO_2\cdot \longrightarrow Ar_2CHOOH \xrightarrow{-H_2O} Ar_2C{=}O \qquad (64)$$

$$O_2^{\cdot-} + Ar_2CH_2 \xrightarrow{b} Ar_2CH^- \xrightarrow{+O_2,\ -O_2^{\cdot-}} Ar_2CH\cdot$$

Secondly, they point out that the fluorene autoxidation is about 30 times faster than diphenylmethane, which is consistent with the relative order expected for H-abstraction based on bond dissociation energies. However, this is also the relative order of pK_a's of the two substrates.

The most convincing evidence for the radical nature of fluorene autoxidation mediated by $O_2^{\cdot-}$ is the presence of an induction period typical for free radical processes. However, base-catalyzed autoxidations in many cases also show induction periods such as in the case of benzhydrol[59] and nitropropane.[57] Although Russel does mention[57] that the base-catalyzed autoxidation of fluorene has no induction period, his experiments were carried out in protic solvents. The situation in aprotic solvents may be substantially different. Hence, while the Lee Ruff-Timms evidence is consistent with a hydrogen-atom abstraction as the first step, it is by no means conclusive.

Top and co-workers[149] report that under a N_2 atmosphere neither diphenylmethane nor fluorene were autoxidized even in the presence of a large excess of $O_2^{\overline{\cdot}}$ at 90°C. Nevertheless, the $Cr(CO)_3$ complexes of these arylmethanes are highly reactive. This suggests that on complexation by $Cr(CO)_3$, the effective acidity of these substrates has increased by several pK_a units, bringing it within the superoxide's proton removal capacity. As we've seen, this, in turn, will catalyze the dismutation of $O_2^{\overline{\cdot}}$ and generate molecular oxygen via Equations 1 to 3.

Frimer et al.[126,150,151] have studied the reaction of $O_2^{\overline{\cdot}}$ with benzyl amines and report that the major product is the corresponding amide (Equation 65). The reaction is presumed to be a free radical process in which the benzyl hydrogen is abstracted.

$$\text{Ar}-\text{CH}_2-\text{NH}-\text{R} \xrightarrow{O_2^{\overline{\cdot}}} \text{Ar}-\overset{\overset{\text{O}}{\|}}{\text{C}}-\text{NHR} \qquad (65)$$

It is further suggested that such a mechanism helps to explain the role cytokinin plays in inhibiting senescence in plants. It should be pointed out, however, that a similar oxidation occurs with naked *t*-butoxide, although only a trace of products is observed with hydroxide. Furthermore, the reaction proceeds even under an argon atmosphere, suggesting that oxygen is produced in the course of the reaction via Equations 1 to 3.[126]

Methylpyridines and methylpyridine-*N*-oxides are oxidized by electrogenerated $O_2^{\overline{\cdot}}$ to the corresponding carboxylic acids in low yields. Here, too, a H-atom abstraction is assumed to be the first step.[152]

$O_2^{\overline{\cdot}}$ can also induce E_2 eliminations producing olefins from the corresponding quaternary ammonium salts[153] and halides,[36,38,39,92,138,154,155] imines from *N*-chloramines,[156] and diphenylacetylene from dibromostilbene.[157]

C. Reaction of $O_2^{\overline{\cdot}}$ with Alkyl Halides and Sulfonates

We have already noted the supernucleophilicity of $O_2^{\overline{\cdot}}$, and indeed aliphalic halides and sulfonates undergo rapid nucleophilic substitution with $O_2^{\overline{\cdot}}$ to produce peroxides, hydroperoxides, alcohols, aldehydes, or acids, depending on reaction conditions and work-up.[24,25,33,36,38,39,82,92,94,138,140,155,158-164] The reaction bears all the characteristics of an S_N2 mechanism. The usual reactivity order is present: 1° > 2° > 3° is the order of reactivity in substrate while I > Br > OTs > Cl is the order in leaving group. What's more, the reaction is highly stereoselective and Walden inversion is observed. While substitution predominates with primary substrates, substantial elimination occurred with secondary and tertiary systems. Remembering of course that $O_2^{\overline{\cdot}}$ is a basic nucleophile, this is not unexpected. E_2 eliminations are discussed briefly at the end of the previous section (IV.B.4).

D. Aryl Systems

Aromatic hydrocarbons per se are inert to $O_2^{\overline{\cdot}}$. Thus benzene, toluene, pyridine, 2,5-diphenylfuran, naphthalene, anthracene, and rubrene are unreactive,[35] as are nitrobenzene, chloro-, and bromobenzene.[165] Even the highly reactive and readily oxidized 1,3-diphenylisobenzofuran is reported by several groups to be unreactive to $O_2^{\overline{\cdot}}$.[64,65,166-168] However, rings activated by one or two electron-withdrawing groups readily undergo nucleophilic aromatic substitutions.[10,169,170] Thus 1-halo-2,4-dinitro-

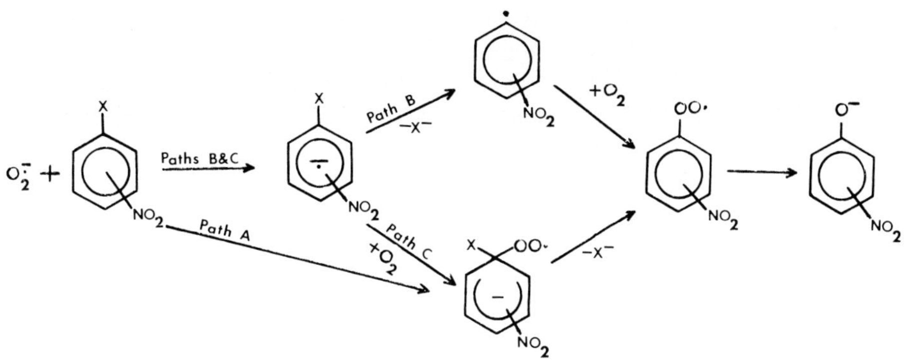

Scheme 6. Possible reaction mechanisms for the reaction of $O_2^{\cdot-}$ with nitrobenzenes (x = halogen or nitro group).

benzenes react rapidly (F > B ∿ I > Cl) to yield the corresponding 2,4-dinitrophenols. *o*- and *p*-Bromonitrobenzene as well as *o*-, *m*-, and *p*-dinitrobenzene react with $O_2^{\cdot-}$ to yield the corresponding nitrophenols.

Three mechanisms are possible for the above reactions and are outlined in Scheme 6. Path A involves direct addition of $O_2^{\cdot-}$ to the aromatic ring according to a typical nucleophilic aromatic substitution mechanism.[171] Paths B and C involve initial electron transfer from $O_2^{\cdot-}$ to the substituted benzene yielding an anion radical. In Path B, bond scission between the nucleofugic substituent and the aryl ring occurs, a process typical of nucleophilic radical aromatic substitution.[172,173] The resulting aryl radical is scavenged by oxygen. In path C, oxygen scavenges the radical anion directly and only subsequently does the nucleofugic group depart.

The fact that upon reaction with $O_2^{\cdot-}$ well-resolved EPR spectra were obtained for the radical anions of 1,3- and 1,4-dinitrobenzenes as well as for several other nitrobenzenes tends to affirm the intermediacy of an electron transfer step. Additional evidence against path A was provided by the use of $K^{18}O_2$,[10,11,165] which indicated that only a small amount of label entered the product. Path B seems to be ruled out because no coupling product was observed, as would have been expected if phenyl radicals had been formed.[174] Hence, path C seems to be the likely mechanism.

A related reaction has been observed in the case of halogenoazines by Yamaguchi and van der Plas.[175] These researchers observed the conversion of 3-bromoquinoline, 2-chloroquinoline, and 1-bromoisoquinoline to the corresponding hydroxy analogs, upon reaction with $O_2^{\cdot-}$ (Equation 66). Again no coupling product was observed and a "path C" type mechanism is involved.

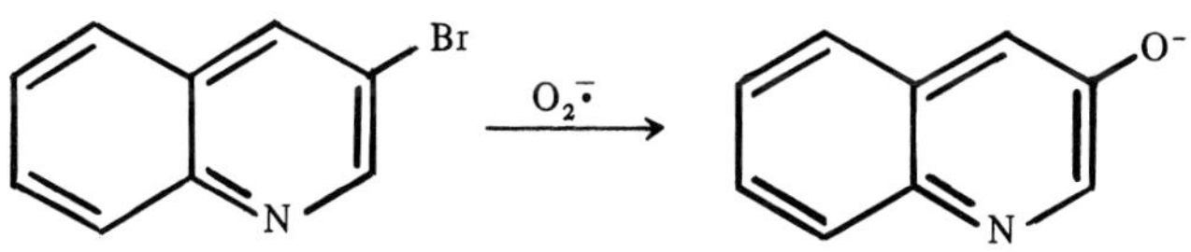

(66)

E. Addition of $O_2^{\cdot-}$ to Double Bond

1. Carbonyl (C=O) Compounds

a. Aldehydes and Ketones

There have been a vast number of reports indicating that aldehydes are oxidized by $O_2^{\cdot-}$ to carboxylic acids.[7,38,39,137,154,176-178] Gibian et al.[5] have investigated the reaction of electrogenerated $O_2^{\cdot-}$ with benzaldehyde in some detail. When impure aldehyde is used, a rapid decomposition of $O_2^{\cdot-}$ results. However, when the aldehyde is freshly and carefully purified, there is no decomposition of the $O_2^{-\cdot}$ beyond that observed in the absence of substrate. Storage of the purified benzaldehyde under argon for even 1 day results in increased rates of $O_2^{-\cdot}$ decomposition. Thus, it would seem that not $O_2^{-\cdot}$ but some derived oxidant(s) is responsibile for the conversion of benzaldehyde to benzoic acid.

This research group also reports that when benzaldehyde is reacted with $O_2^{-\cdot}$ (generated from KO_2/18-crown-6), benzyl alcohol is formed at a rate one half that of the disappearance rate of substrate. The benzyl alcohol is in turn oxidized to benzoic acid.

These results suggest that the oxidation of benzaldehyde involves some base B^- which is not $O_2^{-\cdot}$. The reaction occurs via a Cannizzaro-type process yielding benzyl alcohol and an oxidized benzaldehyde species.

$$B^- + \phi\text{-}\overset{\overset{\displaystyle O}{\|}}{C}H \rightleftharpoons \phi\text{-}\underset{\underset{\displaystyle B}{|}}{\overset{\overset{\displaystyle O^-}{|}}{C}}\text{-}H \xrightarrow{\phi\overset{\overset{O}{\|}}{C}\text{-}H} \phi\overset{\overset{\displaystyle O}{\|}}{C}B + \phi CH_2O^- \longrightarrow \phi\overset{\overset{\displaystyle O}{\|}}{C}O \qquad (67)$$

The nature of the base B^- is not clear but may well be HOO^- formed as described in Equations 1 to 3.

Ketones lacking acidic α-hydrogens such as benzophenone[5,137] are unreactive toward $O_2^{-\cdot}$. Those bearing acidic protons are oxidized by base-catalyzed autooxidative processes described in Section IV.B.4.

Diketones and keto-acids are cleaved by $O_2^{-\cdot}$ to carboxylic acids.[140,179] Three mechanisms are possible (Scheme 7). One involves nucleophilic attack on the carbonyl group (path A). The second and third alternatives (paths B and C) involve initial ketyl formation. This radical ion may then be scavenged by oxygen (path B) or $O_2^{-\cdot}$ (path C). Path A may be ruled out because the electrochemical studies of Boujlel and Simonet[179] suggest that the oxidative cleavage occurs in good yield only when the fixed potential of the cathode allows for the reduction of the diketone, i.e., when $O_2^{-\cdot}$ and (diketone) are simultaneously present at the interface (and later in solution). These authors therefore prefer path C. However, we have noted that radical coupling with $O_2^{-\cdot}$ is not observed (Section III.C), and this should be all the more true in this case where the coupling would have to be between two negatively charged radicals. Thus, path B seems the most likely mechanism.

α-Hydroxy and α-halo ketones, esters, and carboxylic acids are also oxidatively cleaved to carboxylic acids,[140] presumably via the corresponding α-diketones and α-ketoacids.

The simple enone moiety per se is unreactive to $O_2^{-\cdot}$, be it via electron transfer or Michael addition. Thus 4,4,6,6-tetrasubstituted cyclohex-2-en-1-ones proved totally inert to $O_2^{-\cdot}$ even after being in contact for several days.[118] However, as discussed previously, this moiety does labilize the adjacent hydrogens towards proton abstraction (see Section IV.B.4). If the π system is extended, then electron transfer from $O_2^{-\cdot}$ to substrates may be observed. Thus, several research groups report that anion radicals can be detected in the reaction of $O_2^{-\cdot}$ with *N*-methylacridone[180] and various benzoquinones.[125,181]

Frimer and Rosenthal[7,10,35] have studied the oxidative cleavage of tetracyclone and chalcones by the action of $O_2^{-\cdot}$. The mechanism suggested for this process involves initial electron transfer (Scheme 8). A Michael type addition to the enone system was excluded on the basis of $K^{18}O_2$ experiments.

Saito et al.[182] have reported that 2,3-dimethyl-1,4-napthoquinone and similar compounds related to vitamin K_1 react with KO_2/18-crown-6 to give the corresponding 2,3-oxides and phthalic acid. The latter seems to be a secondary oxidation product of the oxirane. The mechanism proposed is outlined in Scheme 9.

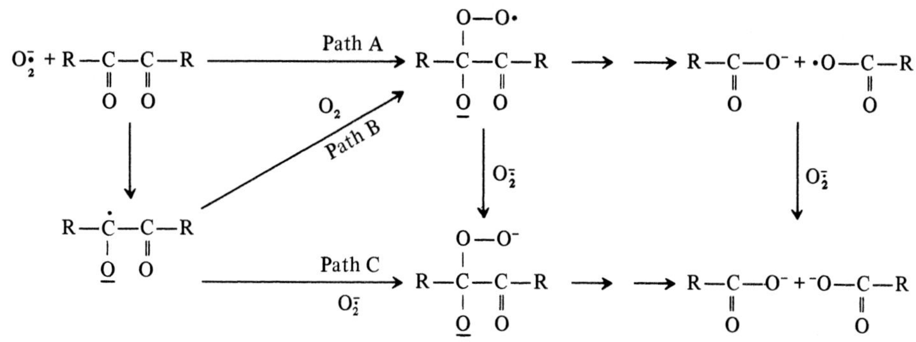

Scheme 7. Mechanistic possibilities for the reaction of $O_2^{\cdot-}$ with α-diketones.

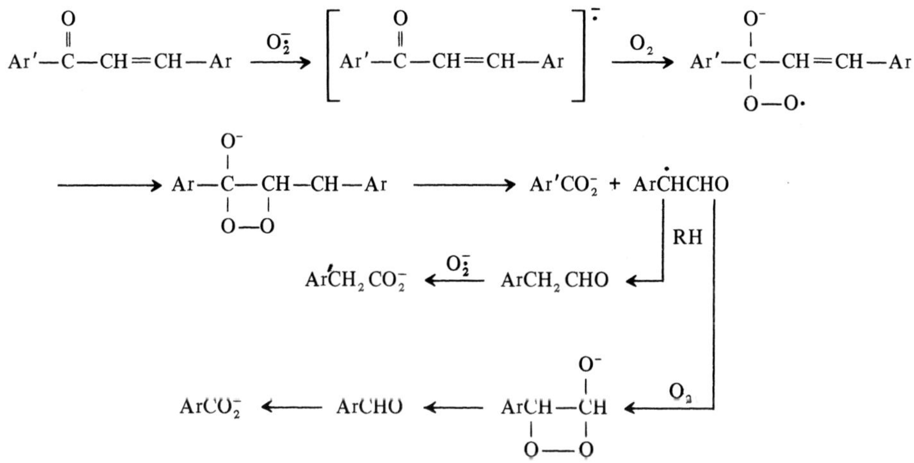

Scheme 8. Mechanism proposed for the oxidative cleavage of chalcones by $O_2^{\cdot-}$.

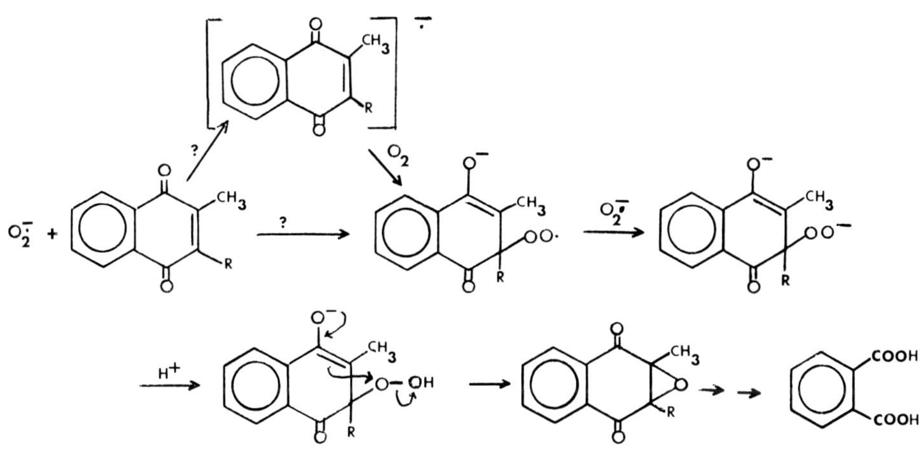

Scheme 9. Mechanism suggested for the reaction of $O_2^{\cdot-}$ with vitamin K.

These authors suggest that either an initial electron transfer or direct nucleophilic attack is involved. However, the ESR data on other quinones[125,180,181] as well as the $K^{18}O_2$ experiments on chalcone[7,10] would seem to affirm the first of these two alternatives. Nevertheless, here a third alternative exists. The authors, in their mechanism,

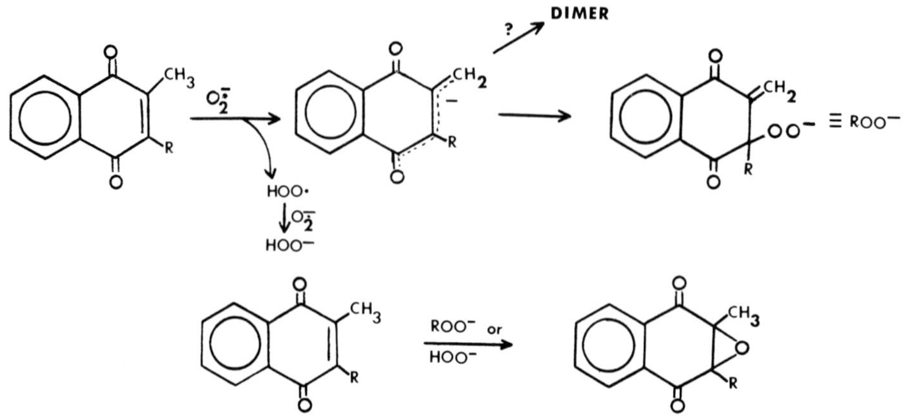

Scheme 10. Alternative mechanism for the epoxidation of vitamin K by $O_2^{\cdot-}$.

require a proton source to protonate the peroxy anion on the way to oxirane formation. If such a proton source exists, it should catalyze the dismutation of $O_2^{\cdot-}$ to HOO^-, the latter being a known enone epoxidizing agent. Furthermore, the 2,3-epoxide and phthalic acid represent only 25 to 35% of the total product yield with the remaining products unidentified. Based on the work of Frimer and Gilinsky with cyclohexanones,[118] an alternative mechanism (Scheme 10) might involve deprotonation of the 2-methyl group producing HOO^- and an enolate. The anion may dimerize or be oxidized. The resulting peroxy anion or HOO^- can epoxidize the substrate. Thus the mechanistic details of this reaction are not at all clear.

b. Carboxylic Acid Derivatives

Acid chlorides[39,154,183] and *anhydrides*[39,84,154] react rapidly with $O_2^{\cdot-}$ in aprotic solvents producing diacylperoxides (50 to 74% yield)[183] and small amounts of peracid.[84,154] Approximately 1 mol of oxygen gas is evolved for every 2 mol of $O_2^{\cdot-}$ consumed, and in the case of anhydrides the corresponding acid is formed as a primary product. The mechanism suggested for this reaction is outlined in Equations 68 to 70 below.[154]

$$R-\overset{O}{\overset{\|}{C}}-X + O_2^{\cdot-} \longrightarrow R-\overset{O}{\overset{\|}{C}}-OO\cdot + X^- \tag{68}$$

$$R-\overset{O}{\overset{\|}{C}}-OO\cdot + O_2^{\cdot-} \longrightarrow R-\overset{O}{\overset{\|}{C}}-OO^- + O_2 \tag{69}$$

$$R-\overset{O}{\overset{\|}{C}}-OO^- + R-\overset{O}{\overset{\|}{C}}-X \longrightarrow R-\overset{O}{\overset{\|}{C}}-OO-\overset{O}{\overset{\|}{C}}-R + X^- \tag{70}$$

Reaction times of 1 to 3 hr are usually sufficient for these reactions, even when slurries of KO_2 in benzene (without crown ether) are used as the $O_2^{\cdot-}$ source. Excess $O_2^{\cdot-}$ is not only not required to drive the reaction to completion, but is deleterious, since the diacylperoxide is efficiently converted to 2 mol of carboxylic acid (vide infra).[178,183]

By comparison, the reaction of *carboxylic esters* with $O_2^{\cdot-}$ is in general[184] quite slow (1 to 3 days, excess KO_2/crown ether). Nevertheless, good to excellent yields of the corresponding acid and alcohol are obtained subsequent to aqueous work-up.[38,39,154,177,178,184] The mechanistic details of this reaction have been investigated to some extent.[5,177,178] Considering first of all the substantial nucleophilic character of $O_2^{\cdot-}$, one mechanistic possibility[177] would be $O_2^{\cdot-}$ attack at the alcohol carbon resulting in displacement of a carboxylate anion (Equation 71).

$$O_2^{\cdot-} \quad R\text{—}O_2CR \longrightarrow ROO\cdot + RCO_2^- \tag{71}$$

Such a mechanism requires Walden inversion at R; however, 99% net retention of configuration at the chiral carbon has been observed in the case of the acetate ester of (−) − (R)-2-octanol.[178]

An alternative mechanism[5,178] might involve nucleophilic attack at the carbonyl carbon in an addition-elimination process (Equations 72 and 73). For such a situation k_{73} will be highly dependent upon the stability of the leaving group (i.e., pK_a of ROH).

$$R-\overset{\overset{\large O}{\|}}{C}-OR' + O_2^{\cdot-} \underset{k_{-72}}{\overset{k_{72}}{\rightleftharpoons}} R-\underset{\underset{OO\cdot}{|}}{\overset{\overset{O^-}{|}}{C}}-OR' \tag{72}$$

$$R-\underset{\underset{OO\cdot}{|}}{\overset{\overset{O^-}{|}}{C}}-OR' \underset{k_{-73}}{\overset{k_{73}}{\rightleftharpoons}} R-\overset{\overset{\large O}{\|}}{C}-OO\cdot + R'O^- \tag{73}$$

The overall rate, on the other hand, depends upon the competition between the loss of $O_2^{\cdot-}$ (k_{-72}) and the loss of $R'O^-$ (k_{73}). Such a mechanism is therefore consistent with the observation that the structure of the leaving group has a strong effect on the overall rate of reaction. In particular it has been observed[5,177,178] that the relative rate of reaction decreases in the order R′ = acyl > *p*-chlorophenyl > phenyl > primary > secondary > tertiary.

Once the peracid radical is formed (Equation 73), it is suggested[3-5,154,177,178] that reactions such as Equations 69 and 70 follow, with the ultimate cleavage of the *diacyl peroxide* by excess $O_2^{\cdot-}$ resulting in the observed carboxylic acid. The mechanistic details of this last transformation are by no means simple, and it is to this reaction that we now turn our attention.

Until recently it was assumed that the conversion of *alkoyl and aroyl peroxides* to carboxylic acids proceeds by a simple two-stepped electron transfer from $O_2^{\cdot-}$ to the peroxy linkage[3-5,154,178] (Equations 74 and 75).

$$O_2^{\cdot-} + R\overset{\overset{\large O}{\|}}{C}-O-O-\overset{\overset{\large O}{\|}}{C}R \longrightarrow R\overset{\overset{\large O}{\|}}{C}-O\cdot + R\overset{\overset{\large O}{\|}}{C}-O^- + O_2 \tag{74}$$

$$R\overset{\overset{\large O}{\|}}{C}-O\cdot + O_2^{\cdot-} \longrightarrow R\overset{\overset{\large O}{\|}}{C}-O^- + O_2 \tag{75}$$

Several pieces of evidence gainsay this mechanism. As discussed previously (Section IV.A), electron transfer to the oxygen-oxygen bond has yet to be observed for any other peroxide. More important, Stanley[84] has observed that an epoxidizing species is formed in the reaction of $O_2^{\cdot-}$ with diacyl peroxides and anhydrides. For example, when acetic anhydride or lauroyl peroxide are reacted with $O_2^{\cdot-}$ in the presence of tetramethylethylene, a 35% yield of the corresponding epoxide is formed. Similarly Oae and Takata[185] report that the action of $O_2^{\cdot-}$ on benzoyl chloride leads to a species which can efficiently (40% yield) epoxidize chalcones and stilbenes. The exact nature of the epoxidizing species is not known, although it may be

$$R''\overset{O}{\overset{\|}{C}}OO\cdot \qquad \text{(Ref. 84)}$$

This would lead one to suggest that the initial step in the conversion of diacylperoxides to carboxylic acids is essentially the same as proposed for acyl chlorides, anhydrides, and esters, i.e., nucleophilic attack at the carbonyl carbon[84] (Equations 76 and 77).

$$O_2^{\overline{\cdot}} + R\overset{O}{\overset{\|}{C}}O O\overset{O}{\overset{\|}{C}}R \longrightarrow R\overset{O}{\overset{\|}{C}}OO\cdot + R\overset{O}{\overset{\|}{C}}OO^- \qquad (76)$$

$$R\overset{O}{\overset{\|}{C}}OO\cdot + O_2^{\overline{\cdot}} \longrightarrow R\overset{O}{\overset{\|}{C}}OO^- + O_2 \qquad (77)$$

Decomposition of the peroxycarboxylate anion[96] to the corresponding carboxylate requires hydrogen ion catalysis which may occur to some extent during work-up. Alternatively, a nucleophilic attack on the peroxy oxygen may be involved (Equations 78 to 80) in which case

$$R''\overset{O}{\overset{\|}{C}}OOO\cdot$$

may well be the epoxidizing species. In any case, the question of mechanism still remains a matter of speculation.

$$O_2^{\overline{\cdot}} + R\overset{O}{\overset{\|}{C}}O O\overset{O}{\overset{\|}{C}}R \longrightarrow R\overset{O}{\overset{\|}{C}}OOO\cdot + R\overset{O}{\overset{\|}{C}}O^- \qquad (78)$$

$$R\overset{O}{\overset{\|}{C}}OOO\cdot + O_2^{\overline{\cdot}} \longrightarrow R\overset{O}{\overset{\|}{C}}OOO^- + O_2 \qquad (79)$$

$$R\overset{O}{\overset{\|}{C}}OOO^- \longrightarrow R\overset{O}{\overset{\|}{C}}O^- + O_2 \qquad (80)$$

Amides[178] and *nitriles*[5,178] seem to be essentially inert to $O_2^{\overline{\cdot}}$. There have been several reports[83,184] in which nitriles were converted to amides in superoxide reactions. However, these reactions involve not $O_2^{\overline{\cdot}}$ but rather ROO^- generated *in situ.*[85-95]

2. Carbon Nitrogen Double Bonds (C=N)

As discussed above (Section IV.B.3) the reaction of $O_2^{\overline{\cdot}}$ with hydrazones[128] (Equation 81) and with ketoximes and methylene chloride[119b] (Equation 82) yields hydrazones and methylene dioximes, respectively. The C=N functionality itself remained untouched.

$$R_2C{=}N{-}NH_2 \xrightarrow{O_2^{\overline{\cdot}}} R_2C{=}N{-}N{=}CR_2 \qquad (81)$$

$$R_2C{=}N{-}OH \xrightarrow[CH_2Cl_2]{O_2^{\overline{\cdot}},\ Pd(II)} R_2C{=}N{-}CH_2{-}N{=}CR_2 \qquad (82)$$

On the other hand, certain diazo compounds are oxidatively cleaved.[128] Thus, diazofluorene yields an 86% yield of fluorenone, while diphenyldiazomethane was essen-

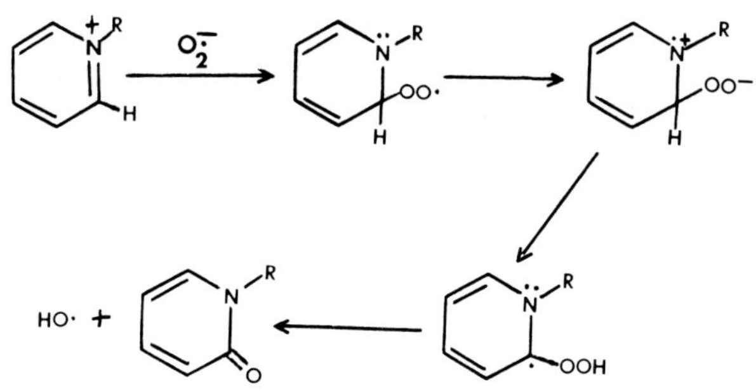

Scheme 11. Proposed mechanism for the reaction of $O_2^{\cdot-}$ with immonium compounds **12** and **13**.

tially unreactive. The mechanistic details of this reaction are not clear, though an electron transfer mechanism is likely.[185]

When the nitrogen of the C=N bond bears a positive charge, nucleophilic attack at carbon becomes a facile process. Thus, several nitrones,[187-192] notably 5,5-dimethyl-1-pyrroline-1-oxide (11) (Equation 83), have been used for spin trapping superoxide anion radical.

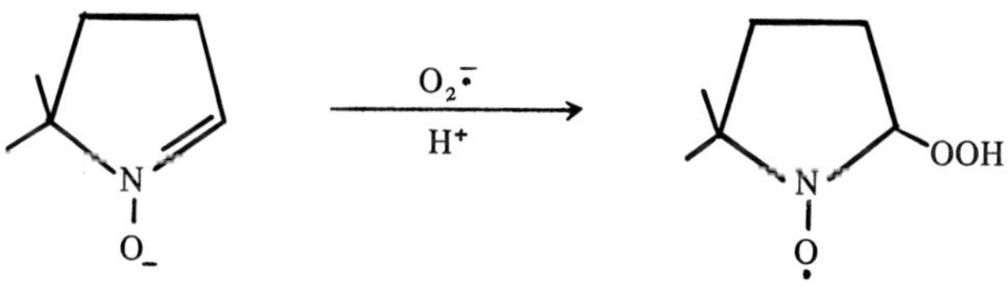

11

(83)

In fact, however, since the spin trapping is carried out in aqueous media, it is not really clear whether it is $O_2^{\cdot-}$ or its conjugate acid HOO· that is being scavenged.[188]

Picot et al.[193] report that immonium fluorosulfonates **12a** and **13a** react with $O_2^{\cdot-}$ to give a quantitative yield of the corresponding amides **14a** and **15a**, respectively (Equations 84 and 85).

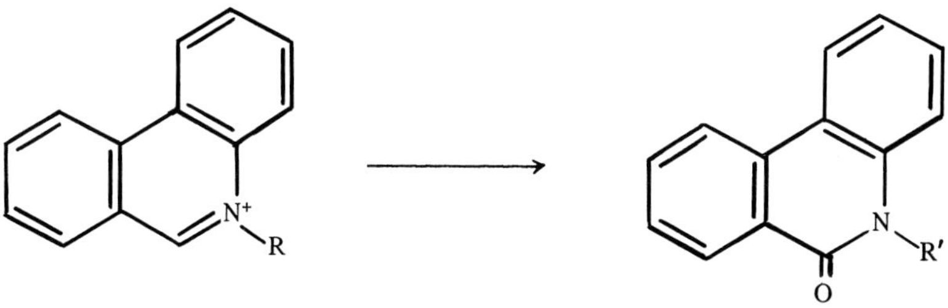

12: a, R=CH_3
b, R=O^-

14: a, R′=CH_3
b, R′=OH

(84)

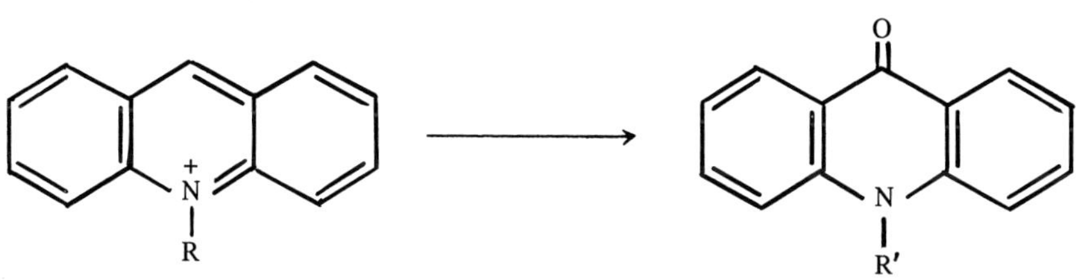

13: a, R=CH_3
b, R=O^-

15: a, R′=CH_3
b, R′=OH

(85)

Similarly nitrones **12b** and **13b** were oxidized to the hydroxamic acids **14b** and **15b**, respectively, in a 75% yield. The mechanism suggested by these authors is outlined in Scheme 11.

3. Nitroso Compounds (N=O)

Aurich[194a] has studied the reaction of $O_2^{\overline{\cdot}}$ (KO_2/18-crown-6) with the spin trap 2-methyl-2-nitrosopropane in aprotic solvents (ether, benzene, and toluene). The results are quite similar to those he observed for the reaction of 2-methyl-2-nitrosopropane anion radical with oxygen,[194b] suggesting that the initial product is the same in both cases (Equation 86). The products obtained (as determined by ESR) are 2-methyl-2-nitropropane anion radical, di-*tert*-butylnitroxide, and solvent addition products.

$$\text{t-Bu–N=O} + O_2^{\overline{\cdot}} \longrightarrow \text{t-Bu–}\underset{\substack{|\\ O^-}}{\text{N}}\text{–O–O}\cdot \longleftarrow \text{t-Bu–}\dot{\text{N}}\text{–O}^- + O_2 \qquad (86)$$

4. Olefins (C=C)

As noted in Section III.C, unlike the case of other free radicals, the direct addition of $O_2^{\overline{\cdot}}$ to olefinic double bonds is not expected to be a thermodynamically favored process, nor is it experimentally observed. Thus, simple olefins such as tetramethyl- and trimethylethylene, 2-methyl-2-pentene, cyclohexene and its 1,2-dimethyl analog, and cholesterol are totally unreactive to $O_2^{\overline{\cdot}}$.[35,36,195] Tetraphenylethylene[35] and styrene[38,39] yielded only traces of polar compounds. Feroci and Roffia[40] have studied the reaction of electrogenerated $O_2^{\overline{\cdot}}$ with styrene in detail and have isolated only small amounts of ethylbenzene and 1,4-diphenylbutane. The absence of oxygenated compounds supports the hypothesis that the only reaction between $O_2^{\overline{\cdot}}$ and styrene is electron transfer. Protonation of the styryl radical ion yields ethylbenzene, while dimerization produces 1,4-diphenylbutane.

Interestingly, electron poor nitro and cyano olefins are oxidatively cleaved by $O_2^{\overline{\cdot}}$.[10,137] When $^{18}O_2^{\overline{\cdot}}$ was used, only very little of the label was actually incorporated into the product. Hence, electron transfer would seem to be the reaction pathway here as well. A likely mechanism is outlined in Scheme 12. Cinnamonitrile, ØCH=CHCN, is reported[5] to be stable to KO_2/18-crown-6. It would seem than that one nitro or two cyano groups are required to obtain proper electron transfer conditions.

Dietz et al.[36] have reported that benzylidenefluorene is converted to fluorenone by electrogenerated $O_2^{\overline{\cdot}}$. The authors suggest a mechanism (Scheme 13) involving initial nucleophilic attack of $O_2^{\overline{\cdot}}$ upon the substrate (path A). However, in light of the work on styrene and electron-poor olefins discussed above, it would seem that a more plausible mechanism would involve initial electron transfer to the extended π system. Molecular oxygen then scavenges the resulting radical ion (path B).

The reaction of $O_2^{\overline{\cdot}}$ with enones has been discussed above (Section IV.E.1.a and IV.B.4).

F. Miscellaneous Sulfur Compounds

Several sulfides have been reported to be inert to $O_2^{\overline{\cdot}}$. Included in this list are diphenylsulfide,[68] di-t-butylsulfide,[185,196] di-*t*-butyldisulfide,[104] and thianthrene.[68] Presumably other sulfides are oxidized. Thus Takata and co-workers[104] report that $O_2^{\overline{\cdot}}$ (KO_2/18-crown-6) converts aryl and alkyl disulfides, thiosulfinates, thiosulfonates, and sodium thiolates to the corresponding sulfinic and sulfonic acids.

$$\text{RSSR, RS(O)SR}', \text{RS(O)}_2\text{SR, or RS}^-\text{Na}^+ \xrightarrow[\text{pyridine}]{O_2^{\overline{\cdot}}} \text{RSO}_2^- + \text{RSO}_3^- \qquad (87)$$

In the case of thiosulfinates high yields (35 to 45%) of the symmetrical disulfide

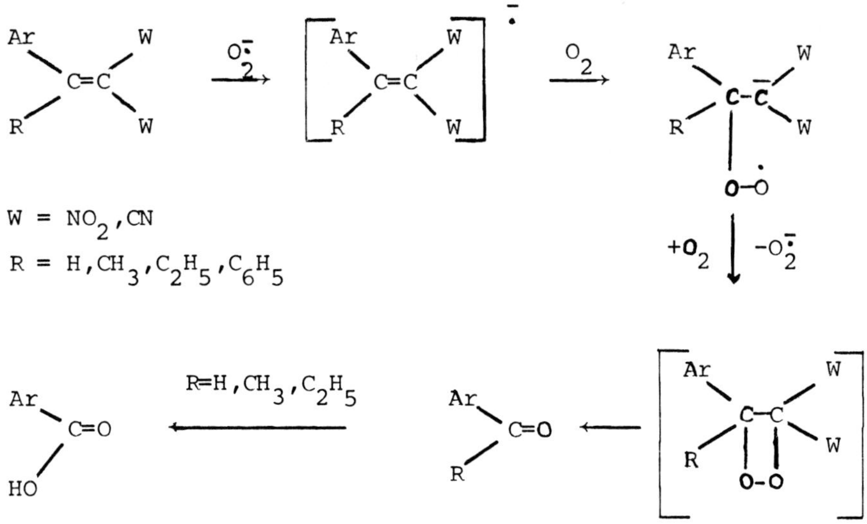

Scheme 12. Mechanism for the oxidative cleavage of nitro- and cyanonoolefins by $O_2^{\overline{\cdot}}$.

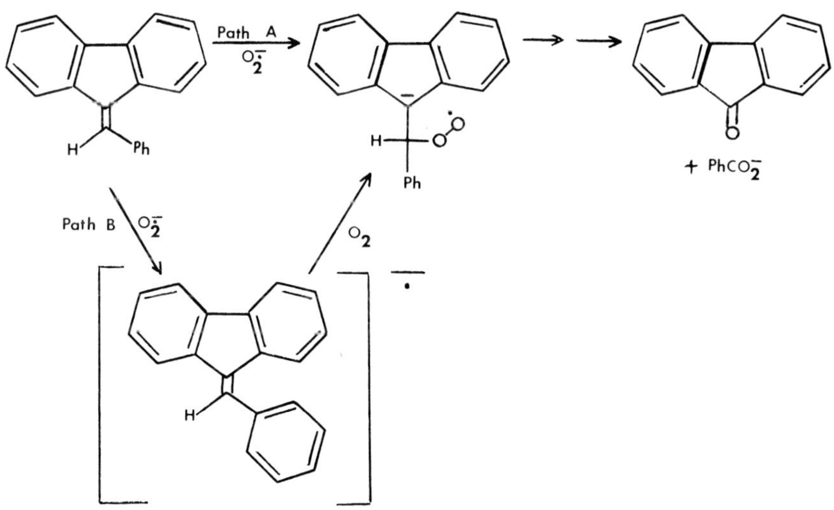

Scheme 13. Possible mechanistic pathways for the reaction of $O_2^{\overline{\cdot}}$ with benzylidenefluorene.

R′-SS-R′ was also isolated. The authors posit that it results from the coupling of two thiyl radicals formed in the nucleophilic attack of $O_2^{\overline{\cdot}}$ on the sulfinyl sulfur of the thiosulfinate. This fact is supported by the exclusive formation of the acids from the sulfinylside.

In the reaction of thiosulfonates a small amount of symmetrical disulfide was also detected, though the main products were the acids. In this case, attack of $O_2^{\overline{\cdot}}$ occurs at both the sulfenyl and sulfonyl sulfur atoms of the thiosulfonates.

As noted previously (Section IV.A.3), Nagano et al. have reported[102] that a mixture of symmetrical disulfides react with electrogenerated $O_2^{\overline{\cdot}}$ to yield disulfides in which there has been an interchange of substituents. Oxidation products were not observed in this instance.

Sulfonyl chloride, sulfinyl chloride, and thiosulfonate react with $O_2^{\overline{\cdot}}$ generating *in situ* peroxysulfenate and peroxysulfinate.[185] The latter epoxidize chalcones, stilbene, and acenaphthylene, presumably via nucleophilic attack on the double bond.

The reaction of $O_2^{\overline{\cdot}}$ with thiols has been discussed above (Section IV.B.2).

G. Reaction of $O_2^{\overline{\cdot}}$ with Cations and Cation Radicals

The nucleophilic attack of $O_2^{\overline{\cdot}}$ at positively charged centers has been reported by several research groups. Thus, the reaction of tropylium ion with $O_2^{\overline{\cdot}}$ (and O_2^{-2}) leads to benzaldehyde, benzene, cycloheptatriene, and carbon monoxide in a 2:2:2:1 ratio.[197] The likely intermediate is tropyl peroxide. Pyrillium ion[198]

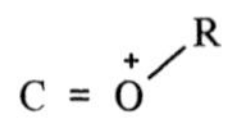

and immonium ion[193]

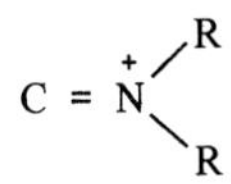

as well as nitrones[187-193]

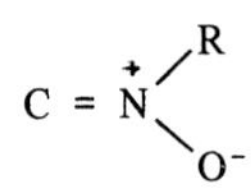

react as expected at the carbon end of the double bond.

Although radical coupling is not generally observed with $O_2^{\overline{\cdot}}$, reaction between cation radicals and $O_2^{\overline{\cdot}}$ has been suggested as the mechanism of photo-induced electron transfer oxidations. For example, Foote et al.[199-203] report that the electron-deficient sensitizer 9,10-dicyanoanthracene (DCA) and 9-cyanoanthracene (CNA) sensitize the photooxidation of 1,1-diphenylethylene, *trans*- and *cis*-stilbene, and tetrophenylethylene in oxygen-saturated polar solvents. The photoproducts in CH_3CN include the cleavage products $Ph_2C{=}O$ and PhCHO, epoxides, and in the case of the stilbenes, *cis-trans* isomerized olefin. All the experimental results are consistent with an electron-transfer mechanism (Equation 88) involving a donor radical cation ($D^{+\cdot}$)and a sensitizer radical anion (sens$^{\overline{\cdot}}$) which subsequently reduces oxygen to superoxide. The coupling of the donor radical cation and $O_2^{\overline{\cdot}}$ lead to oxidation products.

$$\text{Sens} \longrightarrow {}^1\text{Sens}^* \xrightarrow{D} \text{Sens}^{\overline{\cdot}} + D^{\dot{+}} \xrightarrow{{}^3O_2} \text{Sens} + D^{\dot{+}} + O_2^{\overline{\cdot}} \longrightarrow DO_2 \qquad (88)$$

Presumably the coupling product is a peroxy biradical which can close to dioxetane, cleavage of which leads to the carbonyl compounds observed. Alternatively, the biradical can dimerize to tetroxide. Expulsion of O_2 and closure of the resulting biradical leads to epoxide (Scheme 14).

Stable dioxetanes were isolated by Schaap and co-workers[204] in the electron-transfer photooxidation of 2,3-disubstituted-1,4-dioxenes and adamantylidene-adamantane. Furthermore, sulfides yield sulfoxides and sulfones, while 1,3-cyclohexadienes give endoperoxides[200] (Equation 89). Diphenylacetylene yields benzil and benzoic acid, presumably via a dioxetene[205,206] (Equation 90).

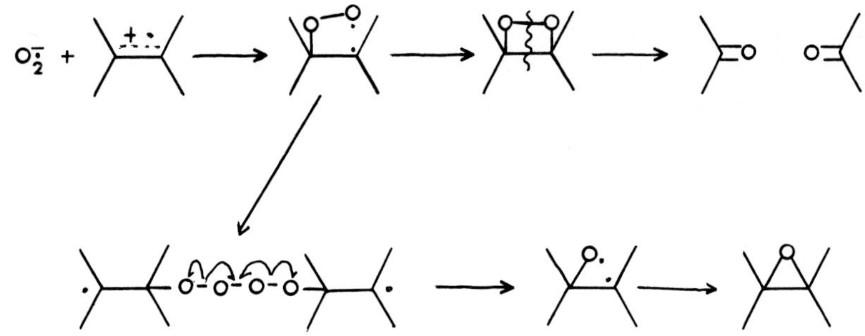

Scheme 14. Proposed mechanism for the formation of oxidation products in photoinduced electron transfer oxidations of olefins.

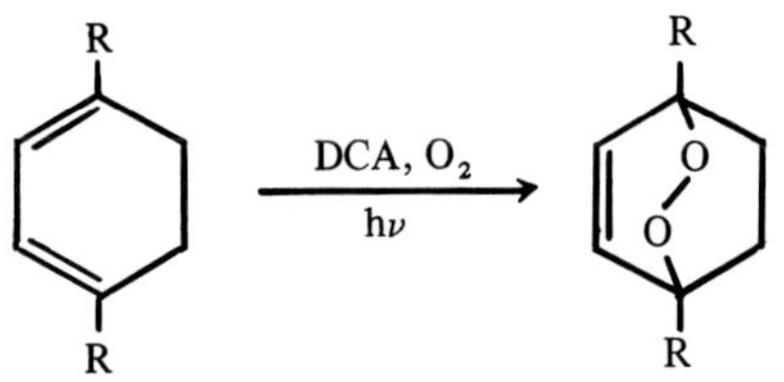

(89)

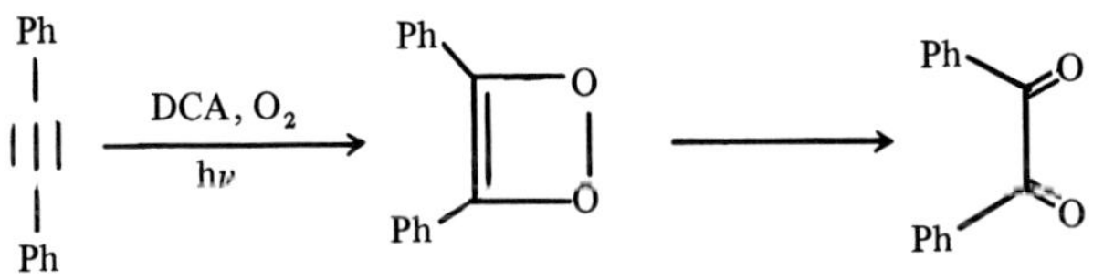

(90)

DCA-sensitized photooxidation of methyl 1,2-diphenylcyclopropene-3-carboxylate led primarily to epoxide rearrangement products[207] (Equation 91).

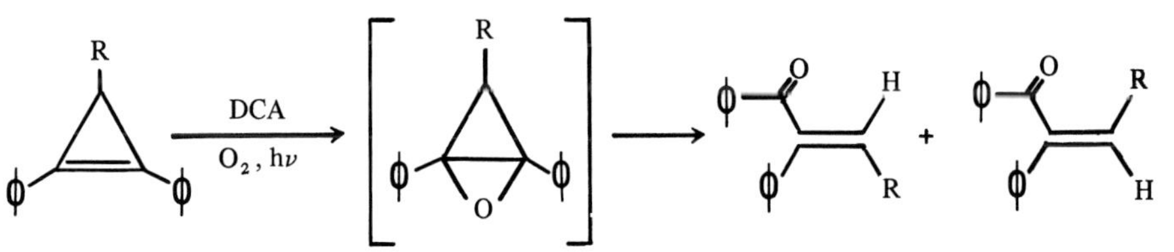

(91)

Several recent reports indicate that photoinduced electron transfer oxidations can be sensitized by metal complexes[208] and semiconductors[209] as well. Thus, the irradiation of TiO_2 or CdS suspended in organic solvents containing 1,1-diphenylethylene and its derivatives under oxygen afforded the corresponding epoxides and benzophenone.

It should be pointed out, however, that there is no conclusive evidence for the intermediacy of $O_2^{\cdot-}$. Foote[199] favors an $O_2^{\cdot-}$ mechanism since electron transfer from DCA^- or CNA^- to O_2 is calculated to be exothermic and thus should occur at diffusion-controlled rates. However, several researchers[207,210,211] have suggested an alternative mechanism (Equation 92) involving attack by $^3O_2$ on the substrate radical cation followed by back-transfer of the electron at a later stage.

$$(\mathrm{Sens}^{\cdot-} + \mathrm{D}^{\cdot+}) \xrightarrow{{}^3\mathrm{O}_2} [\mathrm{Sens}^{\cdot-} + (\mathrm{DO}_2)^{\cdot+}] \longrightarrow \mathrm{DO}_2 + \mathrm{Sens} \quad (92)$$

Such a mechanism would not implicate $O_2^{\cdot-}$. Indeed Saito et al.[210] have studied the DCA-sensitized oxidation of the alkyl side chain of alkylbenzenes. They report the $O_2^{\cdot-}$ generated from KO_2/18-crown-6 did not react with the substrate under the photooxidation conditions in the presence of DCA. More important, the DCA was rapidly

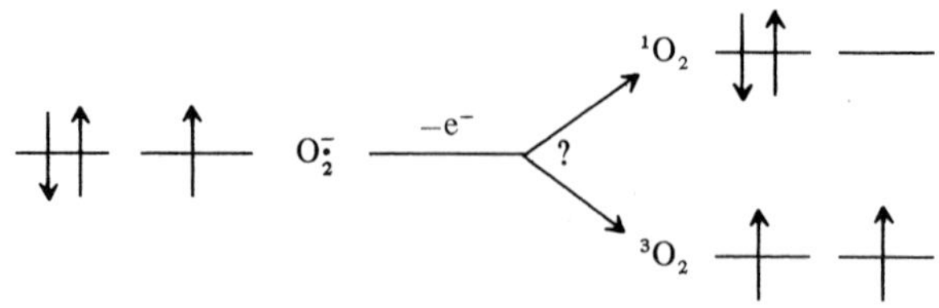

Scheme 15. Schematic presentation of the electron occupancy[212] in the two degenerate $2\pi_g^*$ orbitals of $O_2^{\overline{\cdot}}$, $^1O_2$, and $^3O_2$.

destroyed by KO_2/18-crown-6. Attempts to trap $O_2^{\overline{\cdot}}$ with phenyl *n*-caprate were unsuccessful; the expected phenol and *n*-caprylic acid were not detected. These results strongly suggest that $O_2^{\overline{\cdot}}$ is not generated in appreciable amounts during the DCA catalyzed photooxygenation of alkylbenzenes and places into question the role of $O_2^{\overline{\cdot}}$ in all such oxidations.

Similarly, Mayeda and Bard[66] report that the reaction between the electrogenerated cation radicals of either 1,3-diphenylisobenzofuran or ferrocene with electrogenerated $O_2^{\overline{\cdot}}$ results not in coupling but electron transfer with concomitant formation of molecular oxygen, which these researchers indicate is of singlet multiplicity (Equation 93).

$$D^{\dot{+}} + O_2^{\overline{\cdot}} \longrightarrow D + {}^1O_2 \tag{93}$$

Nishinaga et al.[64] also report that stable 2,5-dialkyl-1,4-dimethoxybenzene radical cations oxidize KO_2 to $^1O_2$ in 82 to 84% yield.

On the other hand, several groups report coupling between $O_2^{\overline{\cdot}}$ and a cation radical. Thus, Ando and co-workers[68] report that while electron transfer is indeed observed in the reaction of thianthrene cation radical perchlorate and KO_2 in acetonitrile, coupling also occurs leading to sulfinyl oxides. Similarly, Sawyer and Nanni[67] report that when DMF solutions of reduced methyl viologen (MV^+) and of $O_2^{\overline{\cdot}}$ are combined in one-to-one stoichiometry, the reversible formation of a dioxygen adduct (MVO_2) results. The reaction chemistry of (MVO_2) conforms with what one would expect for a dioxetane.

Thus the role of $O_2^{\overline{\cdot}}$ in photoinduced electron-transfer oxidations is moot.

H. Electron Transfer; Singlet Oxygen ($^1O_2$) from $O_2^{\overline{\cdot}}$

We have seen on several occasions throughout this review an electron transfer from $O_2^{\overline{\cdot}}$ to a reducible substrate. For example, the reaction of $O_2^{\overline{\cdot}}$ with chalcones,[7,10] tetracyclone,[35] *n*-methylacridone,[180] quinones,[125,181] styrene,[40] cation radicals,[64,66,68] phenoxy[64] and peroxy radicals,[14] cyano- and nitroethylenes,[10,137] as well as with halogenoazines,[175] nitrohalo- and dinitrobenzenes,[10,169,170] all presumably proceed via an initial reduction of the substrate by $O_2^{\overline{\cdot}}$. The question that arises is whether the multiplicity of the oxygen generated is singlet or triplet[212] (Scheme 15).

Because of its biological importance a great deal of effort has been invested studying the acid-catalyzed dismutation of $O_2^{\overline{\cdot}}$ (Equations .1 to 3). Theoretical studies[72,78,213-215] suggest that indeed $^1O_2$ may be formed in the crucial second electron transfer step of this process (Equation 94; cf. Equation 2).

$$HOO\cdot + O_2^{\overline{\cdot}} + H^+ \longrightarrow HOOH + {}^1O_2 \tag{94}$$

Nevertheless, there has been no real experimental substantiation for its occurrence. Indeed, the fact that the well-known $^1O_2$ scavengers 2,3-methyl-2-butene,[125] 2-methyl-2-pentene,[125] cyclohexene[125] and its 1,2-dimethyl analog,[216] 1,3-diphenylisobenzofuran

(DPBF),[168] and cholesterol[63] are unaffected by the dismutating $O_2^{\cdot -}$ suggests that $^1O_2$ is in fact not formed.

Kahn,[215] one of the first to suggest a role for $^1O_2$ in $O_2^{\cdot -}$ dismutation,[217] argues that since $O_2^{\cdot -}$ is also an efficient quencher of $^1O_2$[218] only very low yields of $^1O_2$ are to be expected. However, Foote et al.[63] have recently demonstrated that if $^1O_2$ results from $O_2^{\cdot -}$ dismutation, it is produced in amounts of no more than 0.2%, and this value includes correction for trapping efficiency and $O_2^{\cdot -}$ quenching.

In light of all this evidence, the report of Mayeda and Bard[167] that $^1O_2$ can indeed be trapped by DPBF from the dismutation of electrochemically generated $O_2^{\cdot -}$ is somewhat surprising. It has been argued[69] that these results are suspect because DPBF is quite susceptible to free radical autoxidation.[219] However, several research groups have found DPBF resistent to $O_2^{\cdot -}$ oxidation.[64,65,166-168] Clearly this experiment deserves reexamination.

No $^1O_2$ could be detected in the electron transfer reaction between $O_2^{\cdot -}$ and phenoxy radical[64] or nitrohalobenzenes.[157] Nevertheless, there are several superoxide reactions in which $^1O_2$ is clearly generated. We have seen, for example, that the cation radicals of DPBF,[66] ferrocene,[66] 2,5-dialkyl-1,4-dimethoxybenzene,[64] and thianthrene[68] react with $O_2^{\cdot -}$ producing $^1O_2$ in substantial yields. The production of $^1O_2$ was verified by DPBF scavenging,[64,66] and spectroscopically in the case of thianthrene.[68] The generation of $^1O_2$ has also been observed in the reaction of $O_2^{\cdot -}$ with diacyl peroxides.[216] The presence of $^1O_2$ in this system was verified by DPBF, 1,2-dimethylcyclohexene, and tetramethylethylene scavenging. Furthermore, the $^1O_2$ was quenched by β-carotene and $O_2^{\cdot -}$ as expected.

These results suggest that in some very specific cases the deleterious effects on metabolic systems which have been attributed to $O_2^{\cdot -}$ or $\cdot OH$ may actually be due to singlet oxygen produced by oxidation of $O_2^{\cdot -}$ with biological oxidizing agents. However, it is also clear that $^1O_2$ production from $O_2^{\cdot -}$ is not a general process.

V. CONCLUSION

Despite the pivotal role of free radical processes in nature, free radical damage presents a serious and constant threat to living organisms.[220-222] One of the clearest sources of radicals in the body is superoxide anion radical $O_2^{\cdot -}$, which is formed in a large number of reactions of biological importance in both enzymic and nonenzymic processes.[223] Fluxes of $O_2^{\cdot -}$, generated enzymatically or photochemically, have been shown to inactivate virus, induce lipid peroxidation (a suspected source of senescence and carcinogenesis[222]), damage membranes, and kill cells.[223] There are indications that in many cases $O_2^{\cdot -}$ was not necessarily the species that actually caused these effects; rather, it was the precursor of a more potent oxidant, perhaps $\cdot OH$, whose generation depended on the simultaneous presence of H_2O_2.[223]

The cell, as would be expected, has several endogenous mechanisms for protection against oxidative or free radical damage. For example, in the human body, the antioxidants vitamins E and C scavenge free radicals in the lipid and aqueous phases of the cell. Peroxides, a common by-product of oxidative reactions and a troublesome source of free radicals, are decomposed by catalases and peroxidases. Superoxide radicals are disproportionated to oxygen and hydrogen peroxide by superoxide dismutase. When these mechanisms are lacking, major damage can result leading perhaps to disease, aging, or cancer.[220-222]

Because of the importance of $O_2^{\cdot -}$ in biological processes, it is clearly of value to understand its organic chemistry. As Fridovich[224] has poignantly noted: "If we are going to know how it does its dirty work, we have to know what it is capable of doing." We trust this review has displayed the manifold ways in which $O_2^{-\cdot}$ can react, and enlightened those who still wonder why this petite little molecule has become the center of so much scientific attention.

REFERENCES

1. **Cassebaum, H. and Schufle, J. A.,** Scheele's priority for the discovery of oxygen, *J. Chem. Ed.*, 52, 442, 1975.
2. **Fridovich, I.,** Oxygen radicals, hydrogen peroxide and oxygen toxicity, in *Free Radicals in Biology*, Vol. 1, Pryor, W. A., Ed., Academic Press, New York, 1976,chap. 6.
3. **Sawyer, D. T. and Gibian, M. J.,** The chemistry of superoxide ion, *Tetrahedron*, 35, 1471, 1979.
4. **Wilshire, J. and Sawyer, D. T.,** Redox chemistry of dioxygen species, *Acc. Chem. Res.*, 12, 105, 1979.
5. **Gibian, M. J., Sawyer, D. T., Ungermann, T., Tangpoonpholvivat, R., and Morrison, M. M.,** Reactivity of superoxide ion with carbonyl compounds in aprotic solvents, *J. Am. Chem. Soc.*, 101, 640, 1979.
6. **Valentine, J. S. and Curtis, A. B.,** A convenient preparation of solutions of superoxide anion and the reaction of superoxide anion with a copper (II) complex, *J. Am. Chem. Soc.*, 97, 224, 1975.
7. **Rosenthal, I. and Frimer, A.,** Superoxide radical anion cleavage of chalcones, *Tetrahedron Lett.*, p. 2805, 1976.
8. **McElroy, A. D. and Hashman, J. S.,** Synthesis of tetramethylammonium superoxide, *Inorg. Chem.*, 40, 1798, 1964.
9. **Peters, J. W. and Foote, C. S.,** Chemistry of superoxide ion. II. Reaction of hydroperoxides, *J. Am. Chem. Soc.*, 98, 873, 1976.
10. **Frimer, A. A. and Rosenthal, I.,** Chemical reactions of superoxide anion radical in aprotic solvents, *Photochem. Photobiol.*, 28, 711, 1978.
11. **Rosenthal, I.,** The synthesis of ^{18}O-enriched KO_2, *J. Labelled Comp. Radiopharm.*, 12, 317, 1976.
12. **Valentine, J. S.,** The chemical reactivity of superoxide anion in aprotic versus protic media: a review, in *Biochemical and Clinical Aspects of Oxygen*, Caughey, W. S., Ed., Academic Press, New York, 1979, 659.
13. **Behar, D., Czapaki, G., Rabani, J., Dorfman, L. M., and Schwarz, H. A.,** The acid dissociation constant and decay kinetics of the perhydroxyl radical, *J. Phys. Chem.*, 74, 3209, 1970.
14. **Bielski, B. H. J. and Allen, A. O.,** Mechanism of the disproportionation of superoxide radicals, *J. Phys. Chem.*, 81, 1048, 1977.
15. **Yamadagni, R., Payzant, J. D., and Kebarle, P.,** Solvation of Cl^- and O_2^- with H_2O, CH_3OH and CH_3CN in the gas phase, *Can. J. Chem.*, 51, 2507, 1973.
16. **Parker, A. J.,** Solvation of ions. II. Dipolar aprotic solvents as media for nucleophilic substitution reactions at saturated carbon atom, *J. Chem. Soc.*, 1328, 1961.
17. **Parker, A. J.,** Rates of bimolecular substitution reactions in protic and dipolar aprotic solvents, *Adv. Phys. Org. Chem.*, 5, 173, 1967.
18. **Parker, A. J.,** Protic-dipolar aprotic solvent effects on rates of bimolecular reactions, *Chem. Rev.*, 69, 1, 1969.
19. **Sheppard, W. A. and Sharts, C. M.,** *Organic Fluorine Chemistry*, W. A. Benjamin, New York, 1969, 14.
20. **Liotta, C. L. and Harris, H. P.,** The chemistry of "naked" anions. I. Reactions of the 18-crown-6 complex of potassium fluoride with organic substrates in aprotic organic solvents, *J. Am. Chem. Soc.*, 96, 2250, 1974.

21a. **Dolman, D. and Stewart, R.,** Strongly basic systems. VIII. The H_--function for DMSO-H_2O-tetramethylammonium hydroxide, *Can. J. Chem.*, 45, 911, 1967.

21b. **Dzidic, I., Carrol, D. I., Stillwell, R. N., and Horning, E. C.,** Gas phase reactions. Ionization by proton transfer to superoxide anions, *J. Am. Chem. Soc.*, 96, 5258, 1974.

22. **Sawyer, D. T., Gibian, M. J., Morrison, M. M., and Seo, E. T.,** On the chemical reactivity of superoxide ion, *J. Am. Chem. Soc.*, 100, 627, 1978.
23. **Goolsby, A. D. and Sawyer, D. T.,** The electrochemical reduction of superoxide ion and the oxidation of hydroxide ion in dimethylsulfoxide, *Anal. Chem.*, 40, 83, 1968.
24. **Danen, W. C. and Warner, R. J.,** The remarkable nucleophilicity of superoxide anion radical. Rate constants for reaction of superoxide ion with aliphatic bromides, *Tetrahedron Lett.*, p. 989, 1977.

25. **Danen, W. C., Warner, R. J., and Arudi, R. L.,** Nucleophilic reaction of superoxide anion radical, in *Organic Free Radicals,* A.C.S. Symp. Ser., Vol. 69, Pryor, W. A., Ed., American Chemical Society, Washington, D.C., 1978, 244.
26. **Edwards, J. O.,** Bimolecular nucleophilic displacement reactions, *J. Chem. Ed.,* 45, 386, 1968.
27. **Edwards, J. O. and Pearson, G. O.,** The factors determining nucleophilic reactivities, *J. Am. Chem. Soc.,* 84, 16, 1962.
28. **McIssac, J. E., Jr., Subbaraman, L. R., Subbaraman, J., Mulhausen, H. A., and Behrman, E. J.,** The nucleophilic reactivity of peroxy anions, *J. Org. Chem.,* 37, 1037, 1972.
29. **Liotta, C. L., Grisdale, E. E., and Hopkins, H. P., Jr.,** Relative nucleophilicities of naked anions, *Tetrahedron Lett.,* p. 4205, 1975.
30. **Young, L. B., Lee-Ruff, E., and Bohme, D. K.,** Gas phase nucleophilicities of anions: H^-, F^-, OH^-, and NH_2^-, *Chem. Commun.,* 35, 1973.
31. **Liotta, C. L., Harris, H. P., McDermott, M., Gonzales, T., and Smith, K.,** Chemistry of naked anions. II. Reactions of the 18-crown-6 complex of potassium acetate with organic substrates in aprotic organic solvents, *Tetrahedron Lett.,* p. 2417, 1974.
32. **Cook, F. L., Bowers, C. W., and Liotta, C. L.,** Chemistry of naked anions. III. Reactions of the 18-crown-6 complex of potassium cyanide with organic substrates in aprotic organic solvents, *J. Org. Chem.,* 39, 3416, 1974.
33. **Johnson, R. A., Nidy, E. G., and Merrit, M. V.,** Superoxide chemistry. Reaction of superoxide with alkyl halides and alkyl sulfonate esters, *J. Am. Chem. Soc.,* 100, 7960, 1978.
34. **Curci, R. and DiFuria, F.,** The influence of solvent and crown polyethers on the nucleophilic reactivity of potassium tert-butylperoxide, potassium tert-butoxide, and some other oxygen bases, *Int. J. Chem. Kinet.,* 7, 341, 1975.
35. **Rosenthal, I. and Frimer, A.,** The reaction between tetracyclone and superoxide anion radical, *Tetrahedron Lett.,* p. 3731, 1975.
36. **Dietz, R., Forno, A. E. J., Larcombe, B. E., and Peover, M. E.,** Nucleophilic reactions of electrogenereated superoxide ion, *J. Chem. Soc.,* B, 816, 1970.
37. **Bors, W., Saran, M., Michel, C., Elstner, E. F., Grosch, W., and Laskaway, G.,** unpublished data, 1979.
38. **Le Berre, A. and Berguer, Y.,** Reactions des superoxydes alcalins avec des composés organiques. I. Reactions radicalaires, *Bull. Soc. Chim. Fr.,* 2363, 1966.
39. **Le Berre, A. and Berguer, Y.,** Sur quelque reactions des superoxydes alcalin avec divers composés organiques, *C. R. Acad. Sci. (Paris),* 260, 1995, 1965.
40. **Feroci, G. and Roffia, S.,** Study of the reaction between styrene and electrogenerated superoxide ion in dimethylformamide, *J. Electroanal. Chem.,* 81, 387, 1977.
41. **Richter, A. F.,** Dissociation and bond energies of the peroxide bond energies, *Casopis Lekary Ceskych,* 104, 1076, 1965.
42. **Zemke, W. T., Das, G., and Wahl, A. C.,** Theoretical determination of the electron affinity of O_2 molecule from the bonding energy of O_2 , *Chem. Phys. Lett.,* 14, 310, 1972.
43. **Tiernan, T. O. and Wu, R. L. C.,** Thermochemical data for molecular negative ions from collisional-dissociation thresholds, *Adv. Mass. Spec.,* 7A, 136, 1978.
44. **Nangia, P. S. and Benson, S. W.,** Thermochemistry and kinetics of ozonation reactions, *J. Am. Chem. Soc.,* 102, 3105, 1980.
45a. **Benson, S. W. and Nangia, P. S.,** Electron affinity of HO_2 and HO_x radicals, *J. Am. Chem. Soc.,* 102, 2843, 1980.
45b. **Bierhaum, V. M., Schmitt, R. J., De Puy, C. H., Mead, R. D., Schulz, P. A., and Lineberger, W. C.,** Experimental measurement of the electron affinity of the hydroperoxy radical, *J. Am. Chem. Soc.,* 103, 6262, 1981.
46. **Benson, S. W. and Shaw, R.,** Thermochemistry of organic peroxides, hydroperoxides, polyoxides and their radicals, in *Organic Peroxides,* Vol. 1, Swern, D., Ed., Wiley-Interscience, New York, 1970, 105.
47. **Benson, S. W.,** Bond energies, *J. Chem Ed.,* 42, 502, 1965.
48. **Roberts, J. D. and Caserio, M. J.,** *Basic Principles of Organic Chemistry,* 2nd ed., W. A. Benjamin, Menlo Park, California, 1977, 77.
49. **Knox, B. E. and Palmer, H. B.,** Bond dissociation energies in small hydrocarbon molecules, *Chem. Rev.,* 61, 247, 1961.
50. **Golden, D. M. and Benson, S. W.,** Free radical and molecule thermochemistry from studies of gas phase iodine atom reactions, *Chem. Rev.,* 69, 125, 1969.
51. **Valentine, J. S.,** personal communication regarding Reference 12, 1981.
52. **Garst, J. F.,** Electron transfer reactions in organic chemistry, in *Free Radicals,* Vol. 1, Kochi, J. K., Ed., Wiley-Interscience, New York, 1973, 503.
53. **Alder, R. W., Baker, R., and Brown, J. M.,** *Mechanism in Organic Chemistry,* Wiley-Interscience, Great Britain, 1971, 158.

54. **Moro-oka, Y. and Foote, C. S.**, Chemistry of superoxide ion. I. Oxidation of 3,5-di-tert-butylcatechol with KO_2, *J. Am. Chem. Soc.*, 98, 1510, 1976.
55. **Moro-oka, Y., Chung, P. J., Arakawa, H., and Ikawa, T.**, Chemistry of superoxide ion with substrates having labile hydrogens, *Chem. Lett.*, p. 1293, 1976.
56. **Haynes, R. K. and Musso, H.**, Uber den hachweis freier Radikale bei der Autoxidation des Orcins, *Chem. Ber.*, 107, 3723, 1974.
57. **Russel, G. A., Janzen, E. G., Bemis, A. G., Geels, E. J., Moye, A. J., Mak, S., and Strom, E. T.**, Oxidation of hydrocarbons in basic solution, *Adv. Chem. Ser.*, 51, 112, 1965.
58. **Russel, G. A.**, Detection of paramagnetic intermediates in the oxidation of carbanions, *Pure Appl. Chem.*, 15, 185, 1967.
59. **Russel, G. A., Bemis, A. G., Geels, E. J., Janzen, E. G., and Moye, A. J.**, Oxidation of carbanions. Oxidation of diarylmethanes and diarycarbinols in basic solution, *Adv. Chem. Ser.*, 75, 174, 1968.
60. **Boujlel, K. and Simonet, J.**, On the electrochemical reduction of α-diketones in the presence of oxygen, *Tetrahedron Lett.*, p. 1063, 1979.
61. **Merenyi, G. and Lind, J. S.**, Role of a peroxide intermediate in the chemiluminescence of luminol. A mechanistic study, *J. Am. Chem. Soc.*, 102, 5830, 1980.
62. **Shakashiri, B. Z. and Williams, L. G.**,Singlet oxygen in aqueous solution: a lecture demonstration, *J. Chem. Ed.*, 53, 358, 1976.
63. **Foote, C. S., Shook, F. C., and Abakerli, R. A.**, Chemistry of superoxide ion. IV. Singlet oxygen is not a major product of dismutation, *J. Am. Chem. Soc.*, 102, 2503, 1980.
64. **Nishinaga, A., Shimizu, T., and Matsuura, T.**, Reaction of potassium superoxide with phenoxy radicals. On the mechanism of base catalyzed oxygenation of phenols, *Chem. Lett.*, p. 547, 1977.
65. **Nishinaga, A., Itahara, T., Shimizu, T., Tomita, H., Nishizawa, K., and Matsuura, T.**, The role of dioxygen species in the base- and cobalt-catalyzed oxygenation of hindered phenols, *Photochem. Photobiol.*, 28, 687, 1978.
66. **Mayeda, E. A. and Bard, E. J.**, The production of singlet oxygen in electrogenerated radical ion electron transfer reactions, *J. Am. Chem. Soc.*, 95, 6223, 1973.
67. **Nanni, E. J., Jr., Angelis, C. T., Dickson, J., and Sawyer, D. T.**, Oxygen activation by radical coupling between superoxide ion and reduced methyl viologen, *J. Am. Chem. Soc.*, 103, 4626, 1981.
68. **Ando, W., Kabe, Y., Kobayashi, S., Takyu, C., Yamagishi, A., and Inaba, H.**, Formation of sulfinyl oxide and singlet oxygen in the reaction of thianthrene cation radical and superoxide ion, *J. Am. Chem. Soc.*, 102, 4526, 1980.
69. **Fee, J. A. and Valentine, J. S.**, Chemical and physical properties of superoxide, in *Superoxide and Superoxide Dismutase,* Michelson, A. M., McCord, J. M., and Fridovich, I., Eds., Academic Press, New York, 1977, 19.
70. **Chevalet, J., Rouelle, F., Gierst, L., and Lambert, J. P.**, Electrogeneration and some properties of superoxide ion in aqueous solutions, *J. Electroanal. Chem.*, 39, 201, 1972.
71. **Dietz, R., Peover, M. E., and Rothbaum, P.**, Electrode reactions of organic compounds with electrochemically generated O_2^- ion, *Chem. Ing. Tech.*, 42, 185, 1970.
72. **Koppenol, W. H.**, Reactions involving singlet oxygen and the superoxide anion, *Nature,* 262, 420, 1976.
73. **Haber, F. and Weiss, J.**, The catalytic decomposition of hydrogen peroxide by iron salts, *Proc. R. Soc. London,* A147, 332, 1934.
74. **Ferradini, C., Foos, J., Houee, C., and Pucheault, J.**, The reaction between superoxide anion and hydrogen peroxide, *Photochem. Photobiol.*, 28, 697, 1978.
75. **Weinstien, J. and Bielski, B. H. J.**, Kinetics of the interactions of HO_2 and $O_2^{\overline{\cdot}}$ radicals with hydrogen peroxide. The Haber-Weiss reaction, *J. Am. Chem. Soc.*, 101, 58, 1979.
76. **Melhuish, W. H. and Sulton, H. C.**, Study of the Haber-Weiss reaction using a sensitive method of detection of HO· radicals, *Chem. Commun.*, p. 970, 1978.
77. **Czapski, G. and Ilan, Y. A.**, On the generation of the hydroxylation agent from superoxide radical. Can the Haber-Weiss reaction be the source of HO· radicals?, *Photochem. Photobiol.*, 28, 651, 1978.
78. **Koppenol, W. H., Butler, J., and van Leeuwen, J. W.**, The Haber-Weiss cycle, *Photochem. Photobiol.*, 28, 655, 1978.
79. **George, P.**, Some experiments on the reactions of potassium superoxide in aqueous solution, *Discuss. Farad. Soc.*, 2, 196, 1947.
80. **Thomas, M. J., Mehl. K. S., and Pryor, W. A.**, The role of superoxide anion in the xanthine-oxidase-induced autoxidation of linoleic acid, *Biochem. Biophys. Res. Commun.*, 83, 927, 1978.
81. **Pryor, W. A.**, The formation of free radicals and the consequences of their reactions in vivo, *Photochem. Photobiol.*, 28, 787, 1978.
82. **Gibian, M. J. and Ungerman, T.**, Reaction of tert-butyl hydroperoxide anion with dimethyl sulfoxide. On the pathway of superoxide-alkyl halide reaction, *J. Org. Chem.*, 41, 2500, 1976.

83. **Gibian, M. J. and Ungerman, T.**, The unlikelihood of an electron transfer (Haber-Weiss) reaction between superoxide and peroxides, *J. Am. Chem. Soc.*, 100, 1291, 1978.
84. **Stanley, J. P.**, Reactions of superoxides and peroxides, *J. Org. Chem.*, 45, 1413, 1980.
85. **Kharasch, M. S., Fono, A., Nudenberg, W., and Bischof, B.**, The chemistry of hydroperoxides. XI. Hydroperoxides as oxidizing and reducing agents, *J. Org. Chem.*, 17, 207, 1952.
86. **Denney, D. B. and Rosen, J. D.**, Concerning the mechanism of the induced decomposition of t-alkyl and aralkyl hydroperoxides, *Tetrahedron*, 20, 271, 1964.
87. **Berger, H.**, Decomposition of organic peroxides. V. Decomposition of tert-butyl hydroperoxide catalyzed by nitriles and base, *Trans. Farad. Soc.*, 58, 1137, 1962.
88. **Wiberg, K. B.**, The mechanism of hydrogen peroxide reactions. I. The conversion of benzonitrile to benzamide, *J. Am Chem. Soc.*, 75, 3961, 1953.
89. **Payne, G. B., Deming, P. H., and Williams, P. H.**, Reactions of hydroperoxides. VII. Alkali catalyzed epoxidation and oxidation using a nitrile as coreactant, *J. Org. Chem.*, 26, 659, 1961.
90. **Wiberg, K.**, The mechanism of hydrogen peroxide reactions. II. A comparison of the reactivity of hydroxyl ion and hydroperoxide ion toward benzonitrile, *J. Am. Chem. Soc.*, 77, 2519, 1955.
91. **Bors, W., Michel, C., and Saran, M.**, Superoxide anions do not react with hydroperoxides, *F.E.B.S. Lett.*, 107, 403, 1979.
92. **Chern, C.-I., Di Cosimo, R., De Jesus, R., and San Filippo, J., Jr.**, A study of superoxide reactivity. Reaction of potassim superoxide with alkyl halides, *J. Am. Chem. Soc.*, 100, 7317, 1978.
93. **Jefford, C. W. and Boschurg, A. F.**, The reaction of biadamantylidene with singlet oxygen in the presence of dyes, *Helv. Chim. Acta*, 60, 2673, 1977.
94. **Dao, L. H., Hopkinson, A. C., Lee-Ruff, E., and Rigaudy, J.**, Reaction of bishalomethylarene with superoxide, *Can. J. Chem.*, 55, 3791, 1977.
95. **Kornblum, N. and DeLaMare, H. E.**, The base catalyzed decomposition of dialkyl peroxides, *J. Am. Chem. Soc.*, 73, 880, 1951.
96. **Curci, R. and Edwards, J. O.**, Peroxide reaction mechanisms-polar, in *Organic Peroxides*, Vol. 1, Swern, D., Ed., Wiley-Interscience, New York, 1970, chap. 4.
97. **Hiatt, R.**, Hydroperoxides, in *Organic Peroxides*, Vol. 2, Swern, D., Ed., Wiley-Interscience, New York, 1971, chap. 1.
98. **Hiatt, R.**, Alkyl peroxides, in *Organic Peroxides*, Vol. 3, Swern, D., Ed., Wiley-Interscience, New York, 1972, chap. 1.
99. **Ohloff, G.**, Singlet oxygen: a reagent in organic synthesis, *Pure Appl. Chem.*, 43, 481, 1975.
100. **Mumford, C.**, Something new — the 1,2-dioxetanes, *Chem. Br.*, 11, 402, 1975.
101. **Bielski, B. H. J. and Allen, A. O.**, Mechanism of the disproportionation of superoxide radicals, *J. Phys. Chem.*, 81, 1048, 1977.
102. **Sosnovsky, G. and Rawlinson, D. J.**, Metal ion catalyzed reactions of symmetric peroxides, in *Organic Peroxides*, Vol. 1, Swern, D., Ed., Wiley-Interscience, New York, 1970, chap. 9.
103. **Nagano, T., Arakane, K., and Hirobe, M.**, Disulfide interchange reactions by superoxide, *Tetrahedron*, 21, 5021, 1980.
104. **Takata, T., Kin, Y. H., and Oae, S.**, Reaction of organic sulfur compounds with superoxide anion: oxidation of disulfides, thiosulfinates and thiosulfonates to their sulfinic and sulfonic acids, *Tetrahedron Lett.*, p. 821, 1979.
105. **James, H. J. and Broman, R. F.**, Proton transfer in the two step electrochemical reduction of oxygen in N,N-dimethylformamide, *J. Phys. Chem.*, 75, 4019, 1971.
106. **Nanni, E. J., Stallings, M. D., and Sawyer, D. T.**, Does superoxide ion oxidize catechol, α-tocopherol and ascorbic acid by direct electron transfer?, *J. Am. Chem. Soc.*, 102, 4481, 1980.
107. **Lee-Ruff, E., Lever, A. B. P., and Rigaudy, J.**, The reaction of catechol and derivatives with potassium superoxide, *Can. J. Chem.*, 54, 1837, 1976.
108. **Afanas'ev, I. B. and Polozova, N. I.**, Formation of semiquinones in the reaction between O_2^- ion and hydroquiones, *J. Org. Chem. U.S.S.R.*, 12, 1799, 1976.
109. **Afanas'ev I. B. and Polozova, N. I.**, One electron oxidation of p- and o-dihydrobenzenes by the oxygen radical anion in an aprotic medium, *J. Org. Chem. U.S.S.R.*, 14, 947, 1978.
110. **Nishikimi, M. and Machlin, L. J.**, Oxidation of α-tocopherol model compound by superoxide anion, *Arch. Biochem. Biophys.*, 170, 684, 1975.
111. **Matsumoto, S. and Matsuo, M.**, The reaction of α-tocopherol model compound with KO_2; a new oxidation product of 6-hydroxy-2,2,5,7,8-pentamethylchroman, *Tetrahedron Lett.*, p. 1999, 1977.
112. **Ozawa, T., Hanaki, A., Matsumoto, S., and Matsuo, M.**, Electron spin resonance of radicals obtained by the reaction of α-tocopherol and its model compound with superoxide anion, *Biochim. Biophys. Acta*, 531, 72, 1978.
113. **Matsuo, M., Matsumoto, S., Titaka, Y., Hanaki, A., and Ozawa, T.**, Epoxidation of an α-tocopherol model compound, 2,2,5,7,8-pentamethylchroman-6-ol with potassium superoxide; X-ray crystal structure of 4a,5; 7,8-diepoxy-4a,7,8,8a-tetrahydro-8a-hydroxy-2,2,5,7,8-pentamethylchroman-6(5H)-one, *Chem. Commun.*, p. 105, 1979.

114. **Matsuura, T.,** Bio-mimetic oxygenation, *Tetrahedron,* 33, 2869, 1977.
115. **Nishikimi, M.,** Oxidation of ascorbic acid with superoxide anion generated by the xanthine-xanthine oxidase system, *Biochem. Biophys. Res. Commun.,* 63, 463, 1975.
116. **Elstner, E. F. and Kramer, R.,** Role of the superoxide free radical ion in the photosynthetic ascorbate oxidation and ascorbate mediated photophosphorylation, *Biochim. Biophys. Acta,* 314, 340, 1973.
117. **Frimer, A. A. and Gilinsky, P.,** Oxasteroid synthesis with superoxide anion radical-base catalyzed autoxidation of enones, in *Oxygen and Oxy-Radicals in Chemistry and Biology,* Powers, E. L. and Rodgers, M. A. J., Eds., Academic Press, New York, 1981, 636.
118. **Frimer, A. A. and Gilinsky, P.,** The reaction of superoxide anion radical (O_2^-) with cyclohex-2-en-1-ones, *Tetrahedron Lett.,* p. 4331, 1979.
119a. **Utaka, M., Matsushita, S., Yamasaki, H., and Takula, A.,** Micellar catalysis in the base catalyzed autoxidation of diosphenol, *Tetrahedron Lett.,* 21, 1063, 1980.
119b. **Hosokawa, T., Ohta, T., Okamoto, Y., and Murahashi, S.-I.,** Pd (II)-catalyzed condensation reaction of ketoximes and dichloromethane using potassium superoxide, *Tetrahedron Lett.,* 21, 1259, 1980.
120. **Crank, G. and Makin, M. I. H.,** Organic chemistry of superoxide. I. Oxidation of aromatic compounds, *Tetrahedron Lett.,* p. 2169, 1979.
121. **Assada, K. and Kanematsu, S.,** Reactivity of thiols with superoxide radicals, *Agr. Biol. Chem.,* 40, 1891, 1976.
122. **Armstrong, D. A. and Buchanan, J. D.,** Reactions of O_2^-, H_2O_2 and other oxidants with sulfhydryl enzymes, *Photochem. Photobiol.,* 28, 743, 1978.
123. **Nanni, E. J. and Sawyer, D. T.,** Superoxide-ion oxidation of hydrophenazines, reduced flavins, hydroxylamine, and related substrates via hydrogen-atom transfer, *J. Am. Chem. Soc.,* 102, 7591, 1980.
124. **Hussey, C. L., Laher, T. M., and Achord, J. M.,** Observations on the mechanism of the reaction of electrogenerated superoxide ion with 2-phenlenediamine, *J. Electrochem. Soc.,* 127, 1865, 1980.
125. **Poupko, R. and Rosenthal, I.,** Electron transfer interactions between superoxide ion and organic compounds, *J. Phys. Chem.,* 77, 1722, 1973.
126. **Frimer, A. A., Ziv, Y., and Aljadeff, G.,** manuscript in preparation.
127. **Balogh-Hergovich, E., Speier, G., and Winkelmann, E.,** The reaction of *o*-phenylenediamine and some aromatic amines with superoxide anion radical and molecular oxygen, *Tetrahedron Lett.,* p. 3541, 1979.
128. **Chern, C. I. and San Filippo, J., Jr.,** The reaction of superoxide with hydrazines, hydrazones and related compounds, *J. Org. Chem.,* 42, 178, 1977.
129. **Misra, H. P. and Fridovich, I.,** The oxidation of phenylhydrazine: superoxide and mechanism, *Biochemistry,* 15, 681, 1976.
130. **Smith, P. A. S.,** *The Chemistry of Open-Chain Organic Nitrogen Compounds,* Vol. 2, W. A. Benjamin, New York, 1966, 121.
131. **Smith, P. A. S.,** *The Chemistry of Open-Chain Organic Nitrogen Compounds,* Vol. 2, W. A. Benjamin, New York, 1966, 3.
132. **Hussey, C. L., Laher, T. M., and Achord, J. M.,** The reaction of electrogenerated superoxide ion with nitroaromatic amines, *J. Electrochem. Soc.,* 127, 1484, 1980.
133. **Stanley, J. P.,** Reactions of superoxides and peroxides, *J. Org. Chem.,* 45, 1413, 1980 (footnote 20).
134. **Streitweiser, A., Jr. and Heathcock, C. H.,** *Introduction to Organic Chemistry,* Macmillan, New York, 1976, 1191.
135. **Tezuka, M., Hamada, H., Ohkatsu, Y., and Osa, T.,** Reactivity of electrogenerated superoxide ion. II. Correlation between stability and reactivity of superoxide ion, *Denki Kagaku,* 44, 17, 1976.
136. **DiBiase, S. A., Wolak, R. P., Jr., Dishong, D. M., and Gokel, G. W.,** Crown cation complex effects. X. Potassium tert-butoxide mediated penultimate oxidative hydrolysis of nitriles, *J. Org. Chem.,* 45, 3630, 1980.
137. **Frimer, A. A., Rosenthal, I., and Hoz, S.,** The reaction of superoxide anion radical with electron poor olefins, *Tetrahedron Lett.,* p. 4631, 1977.
138. **Corey, E. J., Nicolaou, K. C., Shibasaki, M., Machida, Y., and Shiner, C. S.,** Superoxide anion as a synthetically useful oxygen nucleophile, *Tetrahedron Lett.,* p. 3183, 1975.
139. **Lissel, M. and Dehmlow, E. V.,** Oxidative Spaltung von Ketones mit Kalumsuperoxid (Anwendungen de Phasentransfer-katalyse 8), *Tetrahedron Lett.,* p. 3689, 1978.
140. **San Filippo, J., Jr., Chern, C.-I., and Valentine, J. S.,** Oxidative cleavage of α-keto, α-hydroxy and α-halo ketones, esters and carboxylic acids by superoxide, *J. Org. Chem.,* 41, 1077, 1976.
141. **Caton, M. P. L., Darnbrough, G., and Parker, T.,** Prostaglandins. VI. Base-catalyzed autoxidation of α-hydroxycyclopentanones and the synthesis of 9,10-diketoprostanoic acid, *Tetrahedron Lett.,* 21, 1685, 1980.
142. **Sagae, H., Fujihira, M., Lund, H., and Osa, T.,** Oxidation of nitrotoluenes with electrogenerated superoxide ion, *Bull. Chem. Soc. Jpn.,* 53, 1537, 1980.

143. **Sagae, H., Fujihira, M., Osa, T., and Lund, H.,** Oxidation of nitroalkylbenzene with electrogenerated superoxide ion, *Chem. Lett.*, p. 793, 1977.
144. **Moro-oka, Y., Chung, P. J., Arakawa, H., and Ikawa, T.,** Chemistry of superoxide ion. Reaction of superoxide ion with substrates having labile hydrogens, *Chem. Lett.*, p. 1293, 1976.
145. **Tezuka, M., Ohkatsu, Y., and Osa, T.,** Reactivity of electrogenerated superoxide ion. I. Autoxidation of 9,10-dihydroanthracene, *Bull. Chem. Soc. Jpn.*, 48, 1471, 1975.
146. **Hawthorne, J. O., Schowalter, K. A., Simon, A. W., Wilt, M. H., and Morgan, M. S.,** Base catalyzed autoxidation of 9,10-dihydroanthracene and related compounds, *Adv. Chem. Ser.*, 75, 203, 1968.
147. **Lee-Ruff, E. and Timms, N.,** Superoxide mediated autoxidation of hydrocarbons, *Can. J. Chem.*, 58, 2138, 1980.
148. **DiBiase, S. A. and Gokel, G. W.,** Crown-cation complex effects. VIII. Reactions of crown ether activated tert-butoxide ion, *J. Org. Chem.*, 43, 447, 1978.
149. **Top, S., Jaoven, G., and McGlinckey, M.,** Reaction of superoxide ion with hydrocarbons activated by the tricarboxylchromium unit, *Chem. Commun.*, p. 643, 1980.
150. **Leshem, Y. Y., Liftman, Y., Grossman, S., and Frimer, A. A.,** Free radicals and pea foliage senescence: increase of ESR signals with progress of aging and cytokinin-induced changes, in *Oxygen and Oxy-Radicals in Chemistry and Biology,* Powers, E. L. and Rodgers, M. A. J., Eds., Academic Press, New York, 1981, 676.
151. **Leshem, Y., Grossman, S., Frimer, A., and Ziv, J.,** Endogenous lipoxygenase control and lipid associated free radical scavenging as modes of cytokinin action in plant senescence retardation, in *Biochemistry and Physiology of Plant Lipids,* Appelqvist, L. A. and Liljenberg, C., Eds., Elsevier, Amsterdam, 1979, 193.
152. **Sage, H., Fujihara, M., Lund, H., and Osa, T.,** Oxidation of methylpyridines and methylpyridine-N-oxides with electrogenerated superoxide ion, *Heterocycles,* 13, 321, 1979.
153. **Feroci, G. and Roffia, S.,** On the reduction of oxygen in dimethylformamide, *J. Electroanal. Chem.*, 71, 191, 1976.
154. **Le Berre, A. and Berguer, Y.,** Reactions des superoxydes alcalins avec des composes organiques. II. Reactions ionique, *Bull. Soc. Chim. Fr.*, 2368, 1966.
155. **San Filippo, J., Jr., Chern, C. I. , and Valentine, J. S.,** The reaction of superoxide with alkyl halides and tosylates, *J. Org. Chem.*, 40, 1678, 1975.
156. **Scully, F. E., Jr. and Davis, R. C.,** Superoxide in organic synthesis: a new mild method for the oxidation of amines to carboxyls via N-chloramines, *J. Org. Chem.*, 43, 1467, 1978.
157. **Frimer, A. A. and Rosenthal, I.,** unpublished results, 1979.
158. **Merrit, M. V. and Sawyer, D. T.,** Electrochemical studies of the reactivity of superoxide ion with several alkylhalides in dimethylsulfoxide, *J. Org. Chem.*, 35, 2157, 1970.
159. **Johnson, R. A. and Nidy, E. G.,** Superoxide chemistry. A convenient synthesis of dialkyl peroxides, *J. Org. Chem.*, 40, 1680, 1975.
160. **Corey, E. J., Nicolaou, K. C., and Shibasaki, M.,** Simple stereospecific routes to 9-epi-prostaglandin $F_{2\alpha}$ ($PGF_{2\beta}$) and 11-epi-prostaglandin $F_{2\alpha}$, *Chem. Commun.*, p. 658, 1975.
161. **Divisek, J. and Kastening, B.,** Electrochemical generation and reactivity of the superoxide ion in aqueous solutions, *J. Electroanal. Chem.*, 65, 603, 1975.
162. **Merrit, M. V. and Johnson, R. A.,** Spin trapping, alkylperoxy radicals and superoxide-alkyl halide reaction, *J. Am. Chem. Soc.*, 99, 3713, 1977.
163. **Osawa, Z., Nakano, H., Mitsui, E., and Nakano, M.,** Reaction of poly (vinyl chloride) with superoxide anion, *J. Polymer Sci. Polymer Chem. Ed.*, 17, 139, 1979.
164. **Perkle, W. H. and Rinaldi, P. L.,** Synthesis and enantiomeric purity determination of the optically active epoxide Disparlure, sex pheromone of the gypsy moth, *J. Org. Chem.*, 44, 1025, 1979.
165. **Frimer, A. A. and Rosenthal, I.,** Nucleophilic radical aromatic substitution with superoxide ion, *Tetrahedron Lett.*, 2809, 1976.
166. **Mayeda, E. A. and Bard, A. J.,** The production of singlet oxygen in electrogenerated radical ion electron transfer reaction, *J. Am. Chem. Soc.*, 95, 6223, 1973.
167. **Mayeda, E. A. and Bard, A. J.,** Singlet oxygen. The suppression of its production in dismutation of superoxide ions by superoxide dismutase, *J. Am. Chem. Soc.*, 96, 4023, 1974.
168. **Nilsson, R. and Kearns, D. R.,** Role of singlet oxygen in some chemiluminescence and enzyme oxidation reactions, *J. Phys. Chem,* 78, 1681, 1974.
169. **Levonwich, P. F., Tannenbaum, H. P., and Dougherty, R. C.,** Negative chemical ionization as a model for reactions in solution: new nucleophilic reactions of superoxide, *Chem. Commun.*, p. 597, 1975.
170. **Sagae, H., Fujihara, M., Komazawa, K., Lund, H., and Osa, T.,** Reaction of halogenonitrobenzenes with electrogenerated superoxide ion, *Bull. Chem. Soc. Jpn.*, 53, 2188, 1980.
171. **Ingold, C. K.,** *Structure and Mechanism in Organic Chemistry,* 2nd ed., G. Bell and Sons, London, 1969, 389.

172. **Kim, J. K. and Bunnet, J. F.**, Evidence for a radical mechanism of aromatic "nucleophilic" substitution, *J. Am. Chem. Soc.*, 92, 7463, 1970.
173. **Kim, J. K. and Bunnet, J. F.**, Alkali metal promoted aromatic "nucleophilic" substitution, *J. Am. Chem. Soc.*, 92, 7464, 1970.
174. **Russel, G. A. and Bridger, R. F.**, The reactivity of phenyl radicals toward molecular oxygen, *J. Am. Chem. Soc.*, 85, 3765, 1963.
175. **Yamaguchi, T. and van der Plas, H. C.**, Nucleophilic aromatic substitution with superoxide ion in halogenoazines, *Rec. Trav. Chim. Pays-Bas*, 96, 89, 1977.
176. **Lee-Ruff, E.**, The organic chemistry of superoxide, *Chem. Soc. Rev.*, 6, 195, 1977.
177. **Magno, F. and Bontempelli, G.**, On the reaction kinetics of electrogenerated superoxide ion with aryl benzoates, *J. Electroanal. Chem.*, 68, 337, 1976.
178. **San Filippo, J. Jr., Romano, L. J., Chern, C.-I, and Valentine, J. S.**, Cleavage of esters by superoxide, *J. Org. Chem.*, 41, 586, 1976.
179. **Boujlel, K. and Simonet, J.**, On the electrochemical reduction of α-diketones in the presence of oxygen, *Tetrahedron Lett.*, p. 1063, 1979.
180. **Rosenthal, I. and Bercovici, T.**, Chemiluminescence of N-methylacridone induced by potassium superoxide, *Chem. Commun.*, p. 200, 1973.
181. **Palet, K. B. and Wilson, R. L.**, Semiquinone free radicals and oxygen, *J. Chem. Soc. Faraday Trans. I*, 69, 814, 1973.
182. **Saito, I., Otsuki, T., and Matsuura, T.**, The reaction of superoxide ion with vitamin K, and its related compounds, *Tetrahedron Lett.*, p. 1693, 1979.
183. **Johnson, R. A.**, Superoxide chemistry. A new convenient synthesis of diacyl peroxides, *Tetrahedron Lett.*, p. 331, 1976.
184. **Kornblum, N. and Singaram, S.**, Conversion of nitriles to amides and esters to acids by the action of sodium superoxide, *J. Org. Chem.*, 44, 4727, 1979.
185. **Oae, S. and Takata, T.**, New epoxidation of olefins with superoxide anion in the presence of organic sulfur compounds, *Tetrahedron Lett.*, 21, 3689, 1980.
186. **Kauffmann, T. and Hage, S. M.**, Stable free radicals formed by the action of sodium on to diazomethane derivatives or phenyl azide, *Angew. Chem. Int. Ed.*, 2, 156, 1963.
187. **Evans, C. A.**, Spin trapping, *Aldrichimica Acta*, 12, 23, 1979.
188. **Finkelstein, E., Rosen, G. M., and Rauchmann, E. J.**, Spin trapping. Kinetics of the reaction of superoxide and hydroxyl radicals with nitrones, *J. Am. Chem. Soc.*, 102, 4994, 1980.
189. **Sanderson, D. G. and Chedekel, M. R.**, Spin trapping of the superoxide radical by 4-(N-methylpyridinium) t-butyl nitrone, *Photochem. Photobiol.*, 32, 573, 1980.
190. **Beuttner, G. R., Oberley, L. W., and Chan Leuthauser, S. W. H.**, The effect of iron on the distribution of superoxide and hydroxyl radicals as seen by spin trapping and on the superoxide dismutase assay, *Photochem. Photobiol.*, 28, 693, 1978.
191. **Kalyanaraman, B., Perez-Reyes, R., and Mason, R. P.**, Spin trapping and direct ESR investigations of the redox metabolism of quinone anticancer drugs, *Biochim. Biophys. Acta*, 630, 119, 1980.
192. **Van Ginkel, G. and Raison, J. K.**, Light induced formation of $O_2^{\cdot -}$. Oxygen radicals in systems containing chlorophyl, *Photochem. Photobiol.*, 32, 793, 1980.
193. **Picot, A., Milliet, P., Cherest, M., and Lusinchi, X.**, Action de l'anion superoxide sur les fluosulfonates de N-methyl benzo-pyridinium et les N-oxydes de benzo-pyridine, *Tetrahedron Lett.*, p. 3811, 1977.
194a. **Aurich, H. G.**, personal communication, 1980.
194b. **Aurich, H. G. and Dersch, W.**, Formation of radicals by treatment of 2-methyl-2-nitrosopropane with potassium tert-butoxide, *Z. Naturforsch.*, 28, 525, 1973.
195. **Smith, L. L., Kulig, M. J., and Teng, J. I.**, Sterol metabolism. XL. On failure of superoxide radical anion to react with cholesterol, *Chem. Phys. Lipids*, 20, 211, 1977.
196. **Bartlett, P. D.**, Free radical aspects of photooxidation, in *Organic Free Radicals*, Pryor, W. A., Ed., A.C.S. Symp. Ser. #69, American Chemical Society, Washington, D.C., 1978, 15.
197a. **Kobayashi, S., Tezuka, T., and Ando, W.**, A novel conversion of tropylium ion to benzenoid compounds in the reaction with peroxide ion, *Tetrahedron Lett.*, p. 261, 1979.
197b. **Kobayashi, S. Tezuka, T., and Ando, W.**, Nucleophilic and electron transfer oxidations of troponoid compounds by superoxide ion, *Chem. Commun.*, 508, 1979.
198. **Kobayashi, S. and Ando, W.**, Superoxide ion chemistry. Reaction of superoxide ion with 2,4,6-trisubstituted pyrylium tetrafluoroborates, *Chem. Lett.*, p. 1159, 1978.
199. **Eriksen, J. and Foote, C. S.**, Electron transfer photooxidation. V. Oxidation of phenyl-substituted alkenes sensitized by cyanoanthracenes, *J. Am. Chem. Soc.*, 102, 6083, 1980.
200. **Eriksen, J., Foote, C. S., and Parker, T. L.**, Photosensitized oxygenations of alkenes and sulfides via a non-singlet-oxygen mechanism, *J. Am. Chem. Soc.*, 99, 6455, 1977.
201. **Spada, L. T. and Foote, C. S.**, Electron-transfer photooxidation. III. Detection of radical ion intermediates in the cyanoaromatic sensitized photooxidation of *trans* and *cis*-stilbene, *J. Am. Chem. Soc.*, 102, 391, 1980.

202. Mauring, L. E., Eriksen, J., and Foote, C. S., Electron transfer photooxidation. IV. Photooxygenation of *trans*-stilbene sensitized by methylene blue, *J. Am. Chem. Soc.*, 102, 4275, 1980.
203. Eriksen, J. and Foote, C.S., Electron-transfer fluorescence quenching and exciplexes of cyano-substituted anthracene, *J. Phys. Chem.*, 82, 2659, 1978.
204. Schaap, A. P., Zaklika, K. A., Kashar, B., and Fung, L. W.-M, Mechanisms of photooxgenation. II. Formation of 1,2-dioxetanes via 9,10-dicyanoanthracene-sensitized electron-transfer processes, *J. Am. Chem. Soc.*, 102, 289, 1980.
205. Maltes, S. L. and Farid, S., Photooxygenation via electron transfer and its susceptibility to catalysis, *Chem. Commun.*, p. 456, 1980.
206. Berenjian, N., de Mayo, P., Phoenix, F. H., and Weedon, A. C., Photosensitized oxidation of acetylenes, *Tetrahedron Lett.*, 4179, 1979.
207. Brown-Wensley, K. A., Maltes, S. L., and Farid, S., Photochemical electron-transfer and triplet reactions of 1,2-diphenylcyclopropene-3-carboxylate, *J. Am. Chem. Soc.*, 100, 4162, 1978.
208. Whilten, D. G., Photoinduced electron transfer reactions of metal complexes in solution, *Acc. Chem. Res.*, 13, 83, 1980.
209. Kanno, T., Ogachi, T., Sakuragi, H., and Tokumaru, K., Semiconductor-catalyzed photooxygenation of aromatic olefins, *Tetrahedron Lett.*, 21, 467, 1980.
210. Saito, I., Tamoto, K., and Matsuura, T., Photosensitized oxygenation initiated by electron transfer. A new type of photooxygenation of alkylbenzenes, *Tetrahedron Lett.*, p. 2889, 1979.
211. Tang, R., Yue, H. J., Wolf, J. F., and Mares, F., Oxygenation of cyclic dienes to endoperoxides, *J. Am Chem. Soc.*, 100, 5248, 1978.
212. Cade, P. E., Bader, R. F. W., and Pelletier, J., Molecular charge distributions and chemical bonding. V. Molecular excitation, ionization and electron attachment, *J. Chem. Phys.*, 54, 3517, 1971.
213. Kuhn, A. V., Theory of electron transfer generation and quenching of singlet oxygen [$^1\Sigma^+_g$ and $^1\Delta_g$] by superoxide anion. The role of water in the dismutation of O_2^-, *J. Am. Chem. Soc.*, 99, 370, 1977.
214. Koppenol, W. H. and Butler, J., Mechanism of reactions involving singlet oxygen and the superoxide anion, *F.E.B.S. Lett.*, 83, 1, 1977.
215. Kahn, A. U., Activated oxygen: singlet molecular oxygen and superoxide anion, *Photochem. Photobiol.*, 28, 615, 1978.
216. Danen, W. C. and Arudi, R. L., Generation of singlet oxygen in the reaction of superoxide anion radical with diacyl peroxides, *J. Am. Chem. Soc.*, 100, 3944, 1978.
217. Khan, A. U., Singlet molecular oxygen from superoxide anion and sensitized fluorescence of organic molecules, *Science*, 168, 476, 1970.
218. Guiraud, H. J. and Foote, C. S., Chemistry of superoxide ion. III. Quenching of singlet oxygen, *J. Am. Chem. Soc.*, 98, 1984, 1976.
219. Howard, J. A. and Mendenhall, G. D., Autoxidation and photooxidation of 1,3-diphenylisobenzofuran: a kinetic and product study, *Can. J. Chem.*, 53, 2199, 1975.
220. Pryor, W. A., The formation of free radicals and the consequences of their reactions in vivo, *Photochem. Photobiol.*, 28, 787, 1978.
221. Parker, L. and Walton, J., Antioxidants vs. aging, *Chem. Tech.*, 7, 278, 1977.
222. Bland, J., Biochemical consequences of lipid peroxidation, *J. Chem. Ed.*, 55, 151, 1978.
223. Fridovich, I., The biology of oxygen radicals, *Science*, 201, 875, 1978.
224. Spikes, J. D. and Swartz, H. M., International conference on singlet oxygen and related species in chemistry and biology — review and general discussion, *Photochem. Photobiol.*, 28, 921, 1978.

Chapter 6

SUPEROXIDE DISMUTASE AND CANCER

Larry W. Oberley

TABLE OF CONTENTS

I. INTRODUCTION

Cancer is a disease that has resisted man's efforts at understanding since the beginning of its study. This lack of understanding is obviously because a vital link in the biochemistry of the cancer cell has been missing. Many groups around the world are now examining whether superoxide dismutase (SOD) is that missing link. As we have read in Volume I, Chapter 1, the enzyme itself was not discovered until 1968. The first studies of superoxide dismutase in cancer cells were not made until 1974. Thus, if superoxide dismutase is important in cancer, it is not difficult to understand why this disease has remained so enigmatic until the present.

In this chapter, we will examine what has been ascertained to date about superoxide dismutase in the cancer cell. It has been found that copper-zinc containing SOD (Cu-Zn SOD) activity is generally but not always lowered in the tumor cell, whereas manganese containing SOD (Mn SOD) activity is always lowered. The consequences of this enzymatic deficiency will be discussed.

Keep in mind when reading this chapter that it difficult to compare SOD activities from study to study. This is because different assay systems are used with different sensitivities. The reader should only compare the tumor tissue to a relevant normal tissue within the same study. It should be emphasized that the term "SOD level or amount'' may sometimes be used for simplicity, but in all cases SOD activity is what is being measured. No studies of SOD in tumors have measured the amount of SOD protein rather than activity.

II. IN VITRO MEASUREMENT OF SUPEROXIDE DISMUTASE ACTIVITY IN MALIGNANT CELLS

The first study on SOD in tumor cells has remained the only one attempted on in vitro tissue cell cultures. It is also the only definitive study on virally induced tumors. Yamanaka and Deamer studied SOD activity in WI-38 human lung cell fibroblasts in vitro.[1] Using gel electrophoresis, they showed that this normal cell strain had both Cu-Zn and Mn SOD activity. When these cells were transformed with simian virus 40 (SV 40), the Mn SOD activity band was diminished or absent, but the total SOD activity was somewhat higher than in the normal WI-38 cells. Trypsinization of confluent normal WI-38 cell cultures caused total SOD activity to decrease to half the original level, while trypsinization of SV-40 transformed cells had no effect on SOD activity.

III. IN VIVO MEASUREMENTS OF SUPEROXIDE DISMUTASE ACTIVITY IN MALIGNANT CELLS

Mn SOD activity has been found to be lowered in in vitro, in vivo, human, rat, mouse, virally induced, chemically induced, spontaneous, and transplanted tumors. The great majority of this work has been done on in vivo tumors. In this section we will examine SOD activities in various types of in vivo tumors.

A. Transplanted Tumors

The first in vivo observations of SOD activities in malignant neoplastic tissues were made by Dionisi et al.[2] They studied isolated mitochondria from bovine heart, rat Morris hepatoma 3924A, and mouse Ehrlich ascites tumor (Lettré mutant). Rat heart mitochondria possessed high levels of SOD activity, while Ehrlich ascites tumor mitochondria contained only a small amount of SOD activity, and hepatoma mitochondria did not contain SOD activity. The measured SOD activity was presumably due to Mn SOD since mitochondrial SOD is mostly Mn SOD.

Before this report was published, our group at the University of Iowa had also studied SOD activity in Ehrlich ascites tumor cells.[3] Using gel electrophoresis, we showed that normal mouse liver mitochondria contained both Cu-Zn and Mn SOD activity while Ehrlich ascites tumor mitochondria contained only Cu-Zn SOD activity. Thus, the small amount of SOD activity reported by Dionisi et al.[2] in Ehrlich ascites tumor mitochondria was probably due to some residual Cu-Zn SOD present in the intermembrane space of the mitochondrial membranes.

Our group at the University of Iowa has also measured SOD activity in various types of mouse liver.[4] We have shown by both direct assay and gel electrophoresis that normal mouse liver contains both Cu-Zn SOD (122 units/mg) and Mn SOD (35 units/mg) activity. Regenerating liver, a rapidly dividing normal cell system, also contained both forms of the enzyme. Moreover, the Cu-Zn SOD and Mn SOD activities per milligram protein changed with time after partial hepatectomy. Cu-Zn SOD activity per milligram protein was found to be at a minimum at 4 days postsurgery (46 units/mg protein), while Mn SOD activity per milligram protein peaked at 2.5 days (65 units/mg protein) and then fell back to normal levels by 4 days. In regenerating mouse liver, DNA synthesis starts at about 2 days after surgery, and the first wave of nearly synchronous cell division occurs at 4 days postsurgery. On the other hand, we have found that H6 hepatoma tumor cells, which are also a rapidly dividing cell system, contain Cu-Zn SOD (40 units/mg protein) but no Mn SOD. Isolated mitochondria from normal liver contained both Cu-Zn and Mn SOD, while isolated mitochondria from H6 hepatoma contained neither Cu-Zn nor Mn SOD. From these studies, it can be concluded that normal mouse liver, whether quiescent or dividing, posesses both forms of SOD activity. On the other hand, hepatoma tumor cells contain Cu-Zn SOD, but no detectable Mn SOD activity. Thus, it appears that in the mouse liver system the loss of Mn SOD activity is characteristic of tumor cells and not of normal cells.

We have also measured SOD activity in whole crude homogenates from several other mouse tumor systems. Normal thymus lymphocytes were found to have 15.5 units of total SOD activity per milligram protein, with 5.5 units/mg protein being Mn SOD.[5] In contrast, L1210 leukemia cells contained 26.3 units of total SOD activity per milligram protein with none of the activity due to Mn SOD. Yamanaka et al. have also reported that L1210 cells have no Mn SOD, using polyacrylamide gel electrophoresis to indicate the presence of the enzyme.[6] We have shown that the C3H mouse mammary carcinoma tumor has 15 units total SOD activity per milligram protein.[5] None of this activity is due to Mn SOD.

We have also measured SOD activity in two in vivo tumors derived from tissue of neural crest origin. Mouse S91 melanoma crude homogenate had 67 units of total SOD activity per milligram protein, again with no activity due to Mn SOD.[5] In contrast, neuroblastoma tumor cells contained a small amount of Mn SOD.[7] However, the level of enzyme activity in the tumor cell was much smaller than the corresponding normal cell. We found that normal mouse brain extract had total SOD activity of 9.0 units/mg protein, with 4.5 units/mg protein being Mn SOD. Prenatal brain extract had total SOD activity of 5.4 units/mg protein, with 2.5 units/mg protein due to Mn SOD. Extracts from neuroblastoma tissue had both lowered Cu-Zn and lowered Mn SOD, with 3.8 units total SOD activity per milligram protein and 1.3 units/mg protein due to Mn SOD. This tumor was the first reported to have nonzero amounts of Mn SOD activity. We will later see that this is not unusual in tumors. However, all tumors examined to date have lower activities of Mn SOD than their normal differentiated cell counterparts.

A number of early studies looked at total SOD and Cu-Zn SOD rather than Mn SOD activity. Peskin's group in Russia has developed a different theory about SOD activity in cancer cells.[8,9] They believe that both cytosol and mitochondrial SOD activities are lowered in the tumor cells of mice and rats. Thus, for hepatoma 27, Zajdela hepatoma, and Lewis lung carcinoma, they found that the ratio of cytosol SOD activities to those in homologous normal tissues was 0.1 to 0.2. They also found the total specific SOD activity in mitochondria from hepatoma 27 to be about 3.5-fold lower than in liver mitochondria. In the presence of 1 m*M* cyanide, both tumor and liver mitochondria displayed similar SOD activities of approximately 2 to 3 units/mg protein. They concluded that there is no loss of mitochondrial SOD and that the net decrease in the enzyme merely reflects the decline of the Cu-Zn SOD localized in the intermembrane space of the mitochondria. However, the author believes that this conclusion is unjustified for several reasons. First, it is very doubtful that they are working with isolated liver mitochondria because the SOD activity in the mitochondria that they have isolated is almost entirely Cu-Zn SOD (34 units/mg protein, of which only 3 units/mg protein is Mn SOD). By way of contrast, we have reported almost 90% Mn SOD in isolated liver mitochondria.[4] Thus, it appears that their mitochondria were contaminated with cytosol protein. Moreover, the reported enzyme activities are very low for isolated mitochondria.[4] Second, we have shown that at least 5 to 8 m*M* cyanide is necessary to inhibit the Cu-Zn SOD totally.[4] Tyler has also shown that 1 m*M* cyanide inhibits the Cu-Zn SOD by only 70%, while 2 m*M* inhibits by 86%, and 3 m*M* inhibits by 91%.[10] Hence, in the work reported by Peskin, if there is any cytosol enzyme in these mitochondrial preparations, its activity will not be totally inhibited by 1 m*M* cyanide. The small amount of residual activity that they measured in isolated mitochondria is thus most probably due to the Cu-Zn and not to Mn SOD. It is also possible that the activity measured is due not to Cu-Zn or Mn SOD but to some other compound with SOD activity.

There is, however, evidence to favor at least part of Peskin's hypothesis — that cytosol SOD is lowered in tumor cells. We found that H6 hepatoma had lower Cu-Zn SOD than does normal liver.[4] Lankin and Gurevich measured total SOD activity in mice given injections of Ehrlich's ascites carcinoma cells and found the activity at most times to be lower than normal.[11] The total activity was maximal during the period of intensive growth (6 to 9 days after injection); at this time, the SOD activity actually exceeded that found in the livers of normal mice. However, the latter observation does not agree with our studies. We found that, 8 days after injection, Ehrlich ascites tumor cells had about one half the total SOD activity of normal liver cells.[3] At the terminal stage of development of Ehrlich's carcinoma, Lankin and Gurevich found that the SOD activity in tumor cells was greatly lowered. In these studies, total SOD can be equated with cytosol SOD, inasmuch as the mitochondrial SOD is very low in these cells.

Bozzi et al. have measured cytosol SOD activity in normal liver and tumor tissue from mice and rats.[12] Liver cells from normal mice and Ehrlich ascites tumor-bearing mice had 1 μg of cytosol SOD per milligram of protein, while all tumor cells had lowered levels of cytosol SOD. Thus, in mice, Ehrlich ascites tumor cells had 0.25 μg SOD per milligram protein, while transplantable methylcholanthrene-induced rhabdomyosarcoma MCI-A cells had 0.50 μg/mg protein. In rats, Yoshida ascites tumor cells had 0.53 μg SOD per milligram protein, and Novikoff hepatoma ascites tumor cells had 0.50 μg SOD per milligram protein.

Many recent reports have strengthened the early observations about the activities of SOD in transplantable tumors. Bartoli et al. have measured the SOD activity of nuclei isolated from Ehrlich-Lettré ascites tumor cells.[13] A significant amount of SOD activity was measured in these nuclei, although the results indicated that the endogenous enzyme was low in activity and did not provide complete protection from superoxide radicals generated by nuclear membranes. It was not ascertained whether the SOD activity was due to Cu-Zn or Mn SOD.

VanBalgooy and Roberts have done an extensive survey of transplantable mouse and rat tumors.[14] Total SOD activity was in general lower in the malignant tissues studied than in normal tissues. Using gel electrophoresis, they found barely detectable Mn SOD levels in the following mouse tumors: L1210/6MP lymphoid leukemia, L1210/5FU lymphoid leukemia, P388 lymphocytic leukemia, B16 melanotic melanoma, Lewis lung carcinoma, L1210/0 lymphoid leukemia, EL-4 ascites, and Ehrlich ascites tumor cells. Cu-Zn patterns were essentially normal in these tumors. Normal Cu-Zn SOD electrophoretic patterns and barely detectable Mn SOD levels were also found in the following rat tumors: Novikoff hepatoma, mammary adenocarcinoma, Dunning leukemia, and Walker carcinoma.[256]

Docampo et al.[15] have studied SOD activity in sarcoma 180 ascites tumor cells. No catalase activity could be detected in sarcoma homogenates, but a low activity of total SOD was measured (37 μg bovine SOD per milligram protein). This is considerably lower than the 1 mg SOD per gram body weight found in normal mouse liver. Dinescu-Romalo and Mihai have measured SOD activity in Guérin T_8 ascites tumor cells.[16] They could not detect SOD activity either in homogenates or in mitochondria from these cells.

Our group at Iowa has recently completed an extensive study of SOD activity in the rat Morris hepatomas.[17] Total SOD and Mn SOD specific activities were measured in tissue homogenates and in isolated mitochondria from normal rat liver and three Morris hepatomas of different growth rates. These tumors are exceedingly interesting to study because they range in phenotype from fast growing, undifferentiated to slow growing, well-differentiated tumors. It was found that total SOD and Mn SOD specific activities were decreased in all tumor homogenates when compared to normal liver; the lowest activity was associated with the fastest growing tumor. These results are consistent with the hypothesis that total Mn SOD specific activity is decreased in all tumors. Even the slow-growing, well-differentiated Morris hepatomas have lower Mn SOD activity than does normal liver tissue. This is an impressive finding, because many other parameters have been thought to be characteristic of tumors, but when a slow-growing Morris hepatoma was measured, this parameter was equal to that of normal liver and, thus, not characteristic of tumors.

In isolated mitochondria, Mn SOD specific activity was similar to the total SOD specific activity indicating that mitochondrial SOD is almost entirely manganese containing. This activity was decreased in the fast- and medium-growth rate hepatomas but was slightly increased in the tumor with the slowest growth rate when compared to normal liver. The normal or higher than normal mitochondrial Mn SOD specific activity in the slow growth rate tumor indicates that decreased mitochondrial Mn SOD specific activity is not characterisitic of all tumors. It appeared from our studies that Mn SOD was found in another location besides mitochondria, and that Mn SOD in this location was diminished in all tumors.

Bartoli et al. have also measured superoxide dismutase content of Morris hepatomas.[18] The content of cytosolic superoxide dismutase was determined in Morris hepatoma 3924A (fast growing) and 44 (slow growing) and in ascites tumor cells (Novikoff hepatoma and Ehrlich Lettrê). SOD activity was decreased in all the tumors examined with the lowest amounts found in the tumors with the fastest growth rates. This result agrees closely with that of Bize et al.[17]

Tarakhovskii et al. have also examined SOD activities in transplantable tumors.[19] Schwetz's erythromyelosis, Plisse's lymphosarcoma, and Herren's adenocarcinoma had both low Cu-Zn and Mn SOD activity when compared to normal rat liver. In Schwetz's erythromyelosis, the ascitic form of the tumor had no Mn SOD activity, whereas the solid form had residual Mn SOD activity, suggesting a loss of Mn SOD upon conversion to the ascitic form.

B. Chemically Induced Tumors

The first study of the superoxide dismutase activity of a chemically induced tumor was performed by Petkau et al.[20] Mammary tumors were induced in female Sprague-Dawley rats by 7,12-dimethylbenz(a)anthracene. They found that at the center and margin of the carcinoma, the concentration of total SOD activity was 54 ± 10 and 117 ± 38 μg/g protein, respectively, while in the tumor as a whole it is 104 ± 32 μg/g. The latter value was not significantly different from the enzyme concentration in mammary tissue from lactating rats (113 ± 35 μg/g). Exposure of the tumor-bearing rats to hyperoxia did not increase the tumor protein but raised the SOD concentration at the center and margin of the carcinoma to 162 ± 73 and 286 ± 103 μg/g, respectively.

In the above article, no attempt was made to measure Mn SOD activity. For this reason, we examined SOD activity in another chemically induced tumor.[21] Large bowel tumors were induced by subcutaneous injection of 1,2-dimethylhydrazine (DMH). This model was chosen not only because of its close similarity to human cancer, but also because of its complexity. Intestinal tissue is composed of many cell types. We sought to answer whether those SOD changes that have been seen in systems with single cell types could also be measured in a system with many cell types. It was found that normal intestine had 58 ± 7 units (mean $\pm$ SE) of total SOD per milligram protein, with 38 ± 4 units (U) of Mn SOD per milligram protein and 20 ± 8 U of Cu-Zn SOD per milligram protein. In contrast, chemically induced large bowel adenocarcinoma had 258 ± 28 U of total SOD per milligram protein, with 27 ± 4 U of Mn SOD per milligram protein and 231 ± 28 U of Cu-Zn SOD per milligram protein. The levels of total SOD and Cu-Zn SOD were statistically different from normal intestine levels ($p < 0.01$). However, the Mn SOD activity was not statistically different in the tumor from the control. (Note in the original paper some of the *p* values were incorrect.)

The tissue (supposedly normal) adjacent to the tumor had the same total SOD activity as that of the normal intestine, but the Mn SOD activity was greatly reduced ($p < 0.01$) when compared to that of normal controls. The percentage of Mn SOD was the same in the adjacent tissue as in the tumor. In contrast, intestinal tissue distant from the tumor (∼4 cm from the tumor) had the same amount of total SOD as that of the control but slightly elevated levels of Mn SOD ($p < 0.1$). However, the percentage of Mn SOD was equal to that of the control. Thus, DMH seemed to affect SOD levels whether or not the tissue becomes malignant, because both tissues adjacent and distant to the tumor were altered; however, cancer has characteristic SOD levels not seen in normal tissue.

As mentioned earlier, the tumor data conflict with earlier reports of SOD activities in tumors. In these earlier studies, both Cu-Zn SOD and Mn SOD were greatly diminished in tumors when compared to that of normal tissue counterparts. Because of this discrepancy, we examined the effect of the many cell types present in normal large intestine on the measured SOD activities. Colon tissue is at least 50% muscle, and this tissue will have a great effect on the SOD activity measured. As a first approximation, we looked at large bowel muscle and mucosal scrapings. Muscle was found to have very low SOD activity, with 33 ± 1 U of total SOD activity per milligram protein, 27 ± 2 U of Mn SOD per milligram protein, and 6 ± 2 U Cu-Zn SOD per milligram protein. Although both total and Mn SOD activities were lower in the muscle than in the whole intestine, neither of these values was statistically significantly different.

Conversely, large bowel mucosa had SOD activities higher than those of normal intestine. Thus large bowel mucosal scrapings contained 112 ± 18 U of total SOD per milligram protein, 30 ± 10 U Mn SOD per milligram protein, and 82 ± 21 U Cu-Zn SOD per milligram protein. These data indicate that the low SOD activity seen in normal large intestine was mostly due to the low SOD activity of the muscle tissue. The total SOD activity was higher in mucosal scrapings than in normal intestine ($p < 0.1$), but lower than in large bowel adenocarcinoma. The Mn SOD activity was the same in mucosal scrapings as in normal intestine and large bowel adenocarcinoma. This finding was somewhat surprising because the Mn SOD level of the muscle was lower than that of normal intestine; thus, one might expect the mucosal Mn SOD levels to be higher than normal. This unexpected result was probably due to contamination of the scrapings by muscle and other tissue such as mucous tissue.

For an accurate comparison of SOD activities, the tumor cell should not be compared to the whole intestine or to muscle or mucosa, but it should be compared to the epithelial cell from which the adenocarcinoma arose. The most likely candidates for this epithelial cell are the crypt and surface epithelial cells of the mucosal lining. Most of the earlier comparisons have been done between tumor cells and normal, mature, nondividing cells as controls. For this reason, we first examined the surface epithelial cells, which are a well-differentiated cell type. Isolated surface epithelial cells had 140 ± 8 U of total SOD activity per milligram protein, with 72 ± 2 U Mn SOD activity per milligram protein and 68 ± 8 U of Cu-Zn SOD activity per milligram protein. Total SOD activity and Mn SOD activity were different from both whole intestine and tumor cells ($p < 0.01$). Thus, when compared to the mature epithelial cell, Mn SOD activity was vastly lowered in the tumor cell (about one third of normal). However, Cu-Zn

SOD activity was still significantly higher in the tumor cell than in the mature normal cell ($p < 0.01$); this observation was in contrast to most other observations on tumors in which Cu-Zn SOD was also low.

Crypt cells were also isolated; these cells are thought by most investigators to be stem cells of the intestinal epithelium. Isolated crypt cells were found to contain 217 ± 17 U of total SOD per milligram protein, with 65 ± 13 U of Mn SOD per milligram protein and 152 ± 21 U Cu-Zn SOD per milligram protein. Total SOD activity ($p <$ 0.01) and Mn SOD activity ($p = 0.1$) were different from that in the whole intestine. Total SOD activity in the crypt cells was not different from that in the tumor cells. Mn SOD activity was significantly lowered in the tumor cells compared to that in the crypt cells ($p < 0.05$). Tumor cells most resembled crypt cells in their SOD activity, though their Mn SOD activity was significantly lower.

C. Human Tumors

Many studies have been made on the activity of SOD in human tissues. It should be emphasized that when one examines human tumors the inducer of the tumor is unknown. Thus, the tumor could be spontaneous, virally induced, chemically induced, or radiation induced. Probably each of these inducing agents are represented when one examines a wide variety of human cancer.

Yamanaka was the first to measure SOD activities in human cells.[6,22] Total SOD activities were increased in myelocytic, monocytic, and lymphocytic leukemia cells as compared to mature blood cells of similar types. With polyacrylamide gel electrophoresis, it was shown that Mn SOD was diminished or absent, especially in immature type malignant cells.

Sykes et al. have investigated the cytosol SOD activity of 46 human tumors.[23] They used an extraction procedure coupled with the epinephrine assay, but did not measure any normal control tissue. They found in general that the SOD content of the tumor correlated with its radiation sensitivity.

Galeotti et al. have recently measured Cu-Zn SOD activity in human epidermis and squamous cell epithelioma.[24] It was found that the epithelioma had about one half the Cu-Zn SOD activity as normal human epidermis. Mn SOD activity was not determined.

Baur and Wendel have measured antioxidant enzyme levels in human colon carcinoma.[25] The parameters of the peroxide-metabolizing system were investigated in eight human colon carcinoma excisions and compared with apparently nonneoplastic colon tissue from the same patients. The mean malondialdehyde level (an index of lipid peroxidation) was 147 pmol/mg protein in carcinoma compared to 40 pmol/mg in healthy tissue. The following relations of specific activities in carcinoma and healthy tissue were found: total superoxide dismutase 3.9 μg/mg protein (carcinoma) to 2.4 μg/mg (healthy); glutathione peroxidase 23 mU/mg to 10 mU/mg; GSH-transferase 38 mU/mg to 25 mU/mg; catalase 2.3 mK/mg to 4.1 mK/mg. The mean selenium concentration in carcinoma was 36 ng/mg protein compared to 24 ng/mg in the control. The difference between carcinoma and normal tissue showed an overall significance of $p < 0.002$ for the whole set of data. No significant differences were found in glutathione (GSH) levels (12.6 nmol/mg protein in carcinoma to 9.4 nmol/mg in normal colon), glutathione disulfide (GSSG) levels (0.6 nmol/mg to 0.7 nmol/mg), and in the protein content (70 ± 23.5 mg/g wet tissue to 66 ± 20 mg/g). It was concluded that the enhanced lipid peroxidation in tumor tissue may be due to peroxisomally generated H_2O_2 in conjunction with a lowered catalase activity.

Marklund and Westman have measured both Cu-Zn and Mn SOD activity in a wide variety of human tissues and human malignant tumors.[26] They have devised a direct enzyme assay which is quite different from the indirect assays commonly used. They found both lowered Cu-Zn and Mn SOD activity in the tumor tissues when compared to similar normal tissues. This study is especially valuable because it used the direct assay to confirm the results obtained with indirect assays.

D. Summary

The general conclusion of this large amount of work is that tumor cells, regardless of the source, usually, but not always have lowered Cu-Zn SOD activity and always have lowered Mn SOD activity when compared to their differentiated, normal cell counterparts.

Lowered Cu-Zn SOD activity is not a universal characteristic of tumor cells because at least four exceptions have been found. First, Yamanaka and Deamer found that after SV 40 transformation, total SOD activity was higher than in normal lung fibroblast WI-38 cells.[1] Since no Mn SOD was present, this meant that Cu-Zn SOD activity increased after transformation. The same must be true of Petkau's study,[20] in which the total SOD activity was the same after tumor induction by a chemical carcinogen, and Yamanaka's study of human leukemia cells, in which total SOD was elevated, but little Mn SOD was seen.[6,22] More direct measurements of elevated Cu-Zn SOD activity have been made in the case of mouse leukemia cells[5] and chemically induced intestinal tumors in the rat.[21] In these cases, Cu-Zn SOD activity was increased in the malignant cell when compared to normal controls.

The cause of the loss of SOD activity is still unknown. The activity could be lost because of lack of synthesis, increased degradation, presence of an inhibitor, production of an inactive protein, or other causes. Many fundamental questions remain to be answered in this area.

E. Superoxide Dismutase Activity in Tumor-Bearing Animals

Two reports have appeared which show that SOD activity is lowered in the nontumor-containing tissues of animals bearing tumors. Tarakhovskii et al. have reported that the livers of tumor-bearing animals have decreased total and Cu-Zn SOD activity, while the Mn SOD activity remains unchanged.[19] Leuthauser et al. have reported just the opposite — Mn SOD activity was lowered with Cu-Zn activity variable.[27] In the latter study Mn SOD was lowered at some point after implantation in liver, spleen, and kidney of mice bearing intramuscular Ehrlich ascites carcinoma cells even though no tumor cells were present in these organs. These authors felt the Russians reached the opposite conclusion because they used homogenization in Triton® and centrifugation to prepare their tissue, while Leuthauser et al. used homogenization without Triton® or centrifugation. Loss of SOD activity, if true, is reminisecent of the effects of a tumor on liver catalase activity. Many studies have shown that tumors produce a substance (called toxohormone) which depresses liver and kidney catalase activity.[28]

One report reached different conclusions. Dinescu-Romalo and Mihai have reported $O_2^{\cdot -}$ production was increased in lymphocyte and liver homogenates and also in isolated mitochondria from liver of tumor-bearing rats.[16] SOD levels showed no change in this study. Obviously, in all three cases, the $O_2^{\cdot -}$/SOD ratio was changed in supposedly normal organs after introduction of a tumor. The cause of this change is still unknown. The significance of this enzyme loss remains to be elucidated, but one speculation is that the host animal may be less able to meet oxidative stress due to the decline in its organ SOD and catalase.

IV. SUPEROXIDE RADICAL IN CANCER CELLS

A. Introduction

If the rate of production of superoxide ion in tumor mitochondria is comparable to that found in the mitochondria from normal tissue, then the loss of Mn SOD activity would result in a net increase in the level of superoxide ion in the tumor cell. This could have vast metabolic consequences due to production of chemical species derived from superoxide. Indeed, Fridovich[29] has provided impressive evidence that SOD is needed to maintain life in all oxygen-metabolizing cells. On the other hand, if the production of superoxide in tumor cell mitochondria is greatly reduced compared to normal mitochondria, then the loss of Mn SOD should not lead to any such harmful effects. Mn SOD could be low in a tumor cell because it is not being induced by its substrate, the superoxide radical. In this case, research should focus on the loss of superoxide-producing ability and not on the loss of Mn SOD. From these considerations, it can be seen that, in order to understand the significance of the loss of Mn SOD, it is necessary to ascertain whether tumor cells produce $O_2^{\cdot-}$.

B. $O_2^{\cdot-}$ Production by Tumor Cell Mitochondria

Many studies have been done on the production of superoxide radicals in tumor cells. These have been by many different techniques. The first studies were done by looking at the capacity of cells to produce superoxide. The first such study was done by Dionisi et al.[2] They used the epinephrine assay to study submitochondrial particles. They found that in the presence of antimycin and succinate, both normal and tumor mitochondria produced superoxide radicals. Beef heart submitochondrial particles produced 0.85 nmol adrenochrome per minute per milligram. Ehrlich ascites tumor cell submitochondrial particles produced 0.90 nmol $min^{-1}mg^{-1}$, while Morris hepatoma 3924A produced 3.86 nmol $min^{-1}mg^{-1}$. SOD prevented most of the adenochrome formation. They also found that submitochondrial particles from normal tissue which contain little superoxide dismutase form less hydrogen peroxide then intact mitochondria which contain superoxide dismutase. Intact tumor mitochondria, which have little superoxide dismutase, did not form hydrogen peroxide, but only superoxide radicals. A stoichiometric relationship was found between superoxide radicals formed in normal mitochondrial fragments and the hydrogen peroxide produced in the presence of superoxide dismutase, with a ratio of 2, as expected from the dismutation reaction.

Several other studies soon followed this one which confirmed the idea that tumor cells have the capacity to produce superoxide radicals. Our group at the University of Iowa showed that submitochondrial particles from mouse H6 hepatoma cells had a large capacity to produce superoxide radicals.[30] In the presence of succinate plus antimycin A, they produced 2.88 nmol adrenochrome per minute per milligram, while in the presence of reduced nicotinamide adenine dinucleotide (NADH) plus antimycin A or rotenone, only about one half the amount of adrenochrome was produced per minute per milligram. SOD largely inhibited the formation of adrenochrome, showing the reactions to be due to superoxide radicals.

We have also studied the capacity of rat liver tissue to produce superoxide radicals.[17] Normal liver, slow-growing Morris hepatoma, and fast-growing Morris hepatoma submitochondrial particles all had the capacity to produce superoxide radicals in the presence of succinate plus antimycin A. Fast-growing, undifferentiated Morris hepatomas produced only about a third as much superoxide per minute per milligram as normal liver or slow-growing Morris hepatomas. NADH did not lead to production of $O_2^{\cdot-}$ by either normal liver or hepatoma submitochondrial particles.

One disturbing feature of the above studies is that they require uncouplers or inhibitors of electron transport in order to measure $O_2^{\cdot-}$ production. One study has succeeded without such agents. Intact mitochondria from mouse sarcoma 180 cells have been shown to produce superoxide radicals in the presence of NADH.[15] β-Lapachone stimulated this production in the presence of NADH. Microsomes from this tumor also produced superoxide, with reduced nicotinamide adenine dinucleotide phosphate

(NADPH) being twice as effective as NADH in stimulating this production. β-Lapachone also stimulated microsomes to produce $O_2^{\cdot-}$.

C. $O_2^{\cdot-}$ Production by Tumor Cell Nuclei

Two other studies have been done on the capacity of tumor cells to produce superoxide. These studies have focused on a particularly important organelle of the cell — the nucleus. Bartoli et al. were the first to report superoxide and H_2O_2 production by tumor nuclei.[13] Nuclei isolated from Ehrlich-Lettrê ascites tumor cells catalyzed the cooxidation of epinephrine to adrenochrome in the presence of NADPH. Adrenochrome formation was sensitive to superoxide dismutase (0.8 μM), but not scavengers of hydroxyl radicals (10 mM mannitol or 34 mM ethanol) or singlet oxygen (20 μM diphenylfuran). Addition of NADPH also initiated the production of H_2O_2. The authors also found superoxide dismutase in these tumor cell nuclei, although they felt the amount was not enough to fully protect the nucleus from endogenous superoxide radicals.

Peskin et al. have found a novel type of superoxide generating system in the nuclear membranes from mouse hepatoma 22a ascites cells.[31] Addition of NADPH or NADH to an aerobic suspension of nuclear membranes from these cells initiates $O_2^{\cdot-}$ production, as measured by adrenochrome formation. Adrenochrome formation was inhibited 79% upon addition of 1.7 μg/mℓ of SOD. Boiled SOD was not inhibitory. The oxidation of adrenalin could not be induced by succinate plus antimycin, nor was it affected by rotenone, antimycin, and protamine when initiated by NADH or NADPH; these controls ruled out the possibility of contaminating mitochondria being the source of $O_2^{\cdot-}$ radicals. The $O_2^{\cdot-}$ generating system of hepatoma 22a nuclei was entirely different from that of Ehrlich-Lettrê ascites tumor cell nuclei as well as that of liver microsomes which require NADPH.

Patton et al. have recently shown that purified normal hamster hepatic nuclei also produce superoxide.[32] NADPH is required. They have also determined that these nuclei contain copper-zinc SOD which can be almost entirely removed by extensive washing. They suggested that this nuclear SOD is in equilibrium with cytoplasmic SOD. They also presented evidence which demonstrated that the flavoprotein FAD-monooxygenase is responsible for most of the nuclear $O_2^{\cdot-}$ production.

D. $O_2^{\cdot-}$ Production by Whole Cells (or Nearly So)

All of the above studies involve the use of isolated organelles to measure superoxide. Two studies have been made which attempt to measure $O_2^{\cdot-}$ in preparations closer to the situation that exists in tissues. Wickramsinghe et al.[33] have measured $O_2^{\cdot-}$ produced in a supernatant fraction of human renal clear-cell adenocarcinoma as well as apparently normal tissue from the same patient. The supernatants were obtained by homogenization in sucrose (0.25 M) followed by centrifugation at 600 × g for 1 min. The latter centrifugation removes nuclei, cell debris, and unbroken cells. Using the reduction of nitroblue tetrazolium (NBT) as a test for $O_2^{\cdot-}$, it was found that the renal tumor produced far less $O_2^{\cdot-}$ than normal kidney. Since SOD was not used to ascertain how much of the NBT reduction was due to $O_2^{\cdot-}$, this study must be viewed with caution.

Dinescu-Romalo and Mihai have performed the only study of whole homogenates of tissue.[16] The lactate dehydrogenase catalyzed oxidation of NADH by superoxide was used to measure superoxide anion production. Rat Guêrin T_8 ascites tumors produced superoxide at a far slower rate than normal rat liver. Whether normal rat liver is an appropriate control for this tissue is not indicated. Isolated tumor mitochondria also produced far less superoxide per minute than did isolated normal liver mitochondria. When a tumor was injected in a normal rat, the rate of superoxide production increased substantially in homogenates from liver and from spleen and thymus lymphocytes. Isolated mitochondria from normal liver also produced superoxide at a higher rate when a tumor was present. They could measure no differences in the SOD activity of the liver after tumor implantation, so differences in SOD activity could not explain the differences in superoxide production.

The results of these studies indicate that tumor cells have the capacity to produce $O_2^{\cdot -}$ at the same rate as normal cells. However, the last two studies seem to indicate that they do not do so under physiological conditions. Future studies with more specific radical indicators are needed to elucidate this question.

V. OTHER ANTIOXIDANT ENZYMES IN TUMOR CELLS

It should be realized that superoxide dismutase is only part of the cell's defense mechanism against oxidative damage. Catalase and glutathione peroxidase are two other very important enzymes in the total defense mechanism. Activities of both of these enzymes have been found to be diminished in tumor cells. A comprehensive survey would be out of place in this volume, so we will simply discuss some recent representative papers. Ono has measured catalase activities in a wide variety of rat Morris hepatomas.[34] All tumors examined had low catalase activity compared to normal liver except for one very slow growing well-differentiated tumor. Catalase activity was inversely correlated with growth rate. Wickramsinghe et al. have measured catalase activity from a human renal clear cell adenocarcinoma and from supposedly normal tissue from the same patient.[33] Catalase activity was about four times higher in the normal tissue than the tumor tissue. Bozzi et al. have found both catalase and glutathione peroxidase activities to be drastically lowered in various mouse and rat tumors when compared to normal liver.[12] Interestingly, livers from tumor-bearing mice also had low activities of catalase and peroxidase. Glutathione peroxidase activity was also found to be low in both cytosol and mitochondria of transplanted hepatoma 27 cells compared to liver cells.[8] Docampo et al. could find no catalase activity in mouse sarcoma 180 ascites cells, but could demonstrate SOD activity in these cells.[15] Sun and Cederbaum have demonstrated that catalase activity is significantly lower in hepatocellular carcinoma HC-252 than in normal liver.[35] Pinto et al. have measured catalase activity in normal liver and in diethylnitrosamine-induced liver tumors.[36] Catalase activity was found to be three times lower in tumor tissue than normal tissue. Using these facts, Bozzi and colleages have proposed a tentative model of the tumor cell.[37,38] They feel the ratio between SOD and the enzymes decomposing H_2O_2 appears to be crucial. It was observed that tumors have lower specific SOD activity than liver or red blood cells, but their content of catalase is practically zero, and in some tumor types the decrease of glutathione peroxidase relative to normal control cells is greater than that of SOD. In these cells. the metabolic flow of oxygen derivatives is driven toward accumulation of H_2O_2. These conclusions should be taken with some caution, as in at least one tumor, the activity of catalase appears to be normal (the slow-growing Morris hepatoma mentioned above). However, all of the above studies taken together seem to indicate that the activities of all three enzymes are usually low in tumor cells. It is not known whether one of the enzymes is more important than others in the cancer phenotype. The fact that Mn SOD activity is lowered in the slow-growing Morris hepatoma while catalase activity is normal argues for a more central role of SOD. Studies of all three enzymes in various tumors should be undertaken to answer this question.

VI. THEORIES OF CANCER

A. Radical Damage Theory

We have proposed a model of cancer[39] which is based on the fact that tumor cells have diminished Mn SOD coupled with normal superoxide production. It was proposed that transformation occurs as a normal stem cell differentiates. Transformation occurs by a change in nuclear DNA (mutation) or its expression (nonmutational or reprogramming). This change in the expression of DNA prevents the increase in Mn SOD normally seen as the cell matures. Superoxide levels increase as normal cells mature and are not decreased after transformation. This failure to increase Mn SOD along with increased $O_2^{\cdot-}$ levels results in a cell with increased fluxes of $O_2^{\cdot-}$. Superoxide itself, or radicals derived from it, causes damage to key subcellular structures or causes a shift in biochemistry to avoid or compensate for such damage. These changes are responsible for many of the phenotypic properties of the cancer cell — including the inability to differentiate and to lose cell division potential.

At the same time or at some other time, changes occur which affect the sensitivity of the cancer cell to oxygen. The cancer cell either is not susceptible to inhibition of cell proliferation by oxygen or it is not susceptible to oxygen-induced killing. The latter must be at least partly true because the cancer cell has no Mn SOD. According to current theory, if it were a normal cell, it would necessarily die.

We have further proposed a two-step model for cancer which involves both SOD and catalase and peroxidase.[40] In this two-step model, it is argued that two changes are needed to form a cancer cell: (1) inhibition of differentiation and attainment of immortality; and (2) loss of control of cell proliferation. Either of these conditions by itself is not enough to form a tumor cell. An immortal cell is not a tumor cell because it is subject to control of its proliferation. A cell which has lost its control of proliferation is not a tumor cell because it will divide only a certain number of times and then will cease division. We hypothesized that the attainment of immortality involves Mn SOD while the control of cell proliferation involves catalase and peroxidase. This theory is called the radical damage theory, because radical damage prevents cells from fully differentiating. Consistant with this theory is the fact that many radical scavengers cause differentiation of tumor cells. This will be discussed in Section VIII.D of this chapter. Also consistent with this theory are the abnormalities and damage seen with electron microscopy in tumor mitochondria.[17]

B. Radical Differentiation Theory

The previously mentioned theories fit much of the experimental data regarding cancer. However, they are based on the fact that tumor cells produced a substantial amount of superoxide. This seemed a reasonable hypothesis as $O_2^{\cdot-}$ is currently thought to be formed by random leakage of electrons. However, there is recent evidence that shows that tumor cells may produce little $O_2^{\cdot-}$. Sun and Cederbaum have measured oxidative enzymes in normal rat liver, tumor-bearing rat liver, and hepatoma HC-252.[35] Studies of the mitochondrial enzymes indicated that the activities of cytochrome oxidase and succinate dehydrogenase were threefold lower in tumor homogenates than in liver homogenates. Monoamine oxidase activity could not be detected in HC-252; mixing experiments indicated no inhibitor was present in HC-252. Activities of the peroxisomal enzymes, urate oxidase, D-amino acid oxidase, and L-α-hydroxy acid oxidase were either undetected in the tumor or were 12-fold lower than in liver homogenates. The activity of xanthine oxidase, a cytoplasmic enzyme, was five- to sixfold lower in the tumor. Catalase activity in the tumor was also lower than in liver. These enzyme activities in the liver of tumor-bearing rats were in the same range as those of normal rat liver, except that D-amino acid oxidase activity was slightly lower, and catalase activity was markedly lower and varied over a wide range.

The authors go on to suggest that tumor cells have lowered capacity to produce $O_2^{\cdot-}$ and H_2O_2 due to a lack of enzymes which produce these entities. They suggest that lower activities of oxidative enzymes in tumors may generate a lower oxidative environment at the cellular level and may provide a more agreeable condition for rapid cell proliferation of neoplastic cells. They also suggest that higher activities of oxidative enzymes in normal cells may generate a higher oxidative environment at the cellular level and thus may lead to a decrease in the ability of the normal cell to proliferate, which is considered aging at the cellular level.

The above model is consistent with the basic tenets of our proposed model of cancer that cancer is caused by: (1) an inability to differentiate and consequent attainment of immortality; and (2) lack of growth control. In our earlier work, we hypothesized that inability to differentiate and attainment of immortality was caused by cell damage due to excess $O_2^{\cdot-}$. However, if $O_2^{\cdot-}$ is not produced by tumor cells, this could not be true. In this case, we would hypothesize that a tumor cell fails to differentiate because it produces little $O_2^{\cdot-}$ or H_2O_2.[41] The tumor cell is immortal because its low capacity to produce superoxide and H_2O_2 means that it does not have to differentiate irreversibly upon division. Because of lack of differentiation, the tumor cell never loses its capacity for cell division. In this model, normal cell differentiation would be induced by irreversible changes in DNA caused by $O_2^{\cdot-}$ or H_2O_2. As a cell goes through its program of differentiation, different modulators of gene expression would be sequentially removed, exposing the naked DNA to chemical modification by oxygen radicals or their derivatives. Without $O_2^{\cdot-}$, this irreversible change in DNA would not occur and differentiation would be blocked. Thus, it is easy to reconcile either view with the basic postulates of our model. Only more experimental evidence on $O_2^{\cdot-}$ production in tumors will tell us which model is correct.

Consistent with this theory is the fact that many anticancer drugs that produce $O_2^{\cdot-}$ also partially and sometimes fully differentiate tumor cells. Thus, actinomycin D, mitomycin C, and adriamycin have been shown to differentiate human myeloid leukemia cells.[42]

C. Oxygen Toxicity and Susceptibility to Lipid Peroxidation

Any theory of cancer must explain two rather remarkable phenotypic properties of tumor cells. These are resistance to oxygen toxicity and resistance to lipid peroxidation. We shall see that both of these properties are more consistent with the radical differentiation theory than the radical damage theory, but either theory could be reconciled with these properties.

1. Oxygen Toxicity

Two studies have shown that oxygen inhibits growth of populations of normal cells, but not populations of transformed cells.[43,44] Goetz has shown that hamster embryo cells die out in the presence of high concentrations of oxygen, while cells transformed by chemical carcinogens continue to proliferate in O_2.[43] Mitchell et al. have shown that in an oxygen-enriched environment, normal fibroblasts are maintained without proliferation, while Rous sarcoma virus-infected fibroblasts continue to proliferate.[44]

Since tumor cells are low in SOD, catalase, and peroxidase, this resistance to oxygen toxicity is exceedingly difficult to explain. The problem is solved if one assumes cancer cells produce $O_2^{\cdot-}$ in only very small amounts. Remember, studies to date of $O_2^{\cdot-}$ in tumor cells have usually measured only the capacity to produce $O_2^{\cdot-}$ and not the actual in vivo production. Lack of oxygen toxicity in the tumor cell is consistent with the radical damage theory only if one assumes that tumor cells are more resistant to oxygen-free radicals. This brings us to lipid peroxidation.

2. Susceptibility to Lipid Peroxidation

Tumor cells are less susceptible to lipid peroxidation, oxygen-radical induced or otherwise. A large number of tumor cells have been reported to be resistant to lipid peroxidation, although there is at least one exception.[45] Many of these reports have shown that addition of homogenates of peroxidation-resistant cells strongly inhibit lipid peroxidation in biological membranes that ordinarily undergo this reaction. As a recent important example of such studies, Bartoli and Galeotti have examined lipid peroxidation in tumor microsomal membranes and mitochondria.[46] Microsomes and mitochondria from Morris hepatomas 3924A (fast growing) and 44 (slow growing) and Ehrlich ascites tumor cells exhibited a NADPH-dependent peroxidation of endogenous lipids lower than that of corresponding fractions from rat liver. Moreover, the $O_2^{\cdot-}$ and ascorbate-dependent lipid peroxidation was decreased in microsomes from the two Morris hepatomas. The peroxidative activity was inversely related to the growth rate, with the fast-growing tumors least susceptible to peroxidation. It was suggested that tumor cells have a high capacity for cell division because they never generate the peroxides necessary to halt cell division. Bartoli et al.[18] have also studied the lipid composition and fluidity of tumor microsomal membranes. These parameters were changed in relation to the growth rate. This lipid-to-protein ratio and the degree of fatty acid unsaturation decreased gradually from rat liver to hepatoma 44 and 3924A microsomes. The different lipid composition was also reflected by differences in the physical environment of the bilayer, as indicated by the data with the spin labeled fatty acids. Thus, tumor cells appear to have abnormal membranes.

Lack of susceptibility to lipid peroxidation seems more consistent with the radical differentiation than the radical damage theory. Tumor cells are low in the antioxidant enzymes, SOD, catalase, and peroxidase. If $O_2^{\cdot-}$ is produced in high amounts in tumor cells, it is likely that other lipid soluble antioxidants like vitamin E would be depleted, as they would be the only antioxidant agents left. Depletion of all antioxidents would make the membranes more susceptible to lipid peroxidation. If $O_2^{\cdot-}$ production is low in tumor cells, then these antioxidants would be high in the membranes due to lack of use, thus making the membrane resistant to lipid peroxidation. The radical damage theory could still hold if one assumes a basic change in membrane structure to combat the increased radicals. (As shown above, tumor cell membranes are abnormal, but is this cause or effect?) The radical damage theory could also be viable if different radical chemistry occurred in the tumor cell than in the normal cell. For instance, the tumor cell might produce less $\cdot OH$ from $O_2^{\cdot-}$ and thus might be more resistant to oxygen toxicity. However, the radical damage theory requires two assumptions (increase in $O_2^{\cdot-}$ and decreased susceptibility to radicals), while the radical differentiation theory requires only one assumption (decreased $O_2^{\cdot-}$) and thus seems more attractive.

VII. THE ROLE OF SUPEROXIDE RADICAL IN CARCINOGENESIS

A. Introduction

In the preceding section, we have been examining models which explain the cancer phenotype. These models do not concern themselves with the causative agents of cancer such as radiation, viruses, and chemicals; nor do these theories deal with the mechanism whereby a normal cell becomes a tumor cell. These questions are rightfully involved in the study of carcinogenesis.

One area of cancer research which has received little attention is the role of oxygen radicals in carcinogenesis. Carcinogens have repeatedly been shown to produce free radicals,[47] and antioxidants lower dramatically the incidence of tumors after exposure to carcinogens.[48] The mechanism of these observations has remained obscure.

B. Tumor Initiation

Recent work has begun to shed some light on how tumor cells are formed. One of the big advances has been the theory of initiation-promotion. Carcinogenesis has been found to occur in the two steps — the initiation step and the promotion step. Compounds which cause initiation are usually thought to bind to DNA and to cause irreversible changes. Promoters, on the other hand, are thought to bind to cell membranes and to be reversible. The order of applying the compounds is important with it being necessary to apply the initiator first. Promoters are thought to enhance the activity of initiators. Some agents act as both initiators and promoters (e.g., both chemicals and radiation). Many think the two-step mechanism occurs with all carcinogenic agents, but chemical carcinogens most clearly follow this example.

Confusingly, both initiators and promoters are found to involve radicals. Floyd et al. have recently demonstrated the presence of oxygen-free radicals ($\cdot OH$) after incubation of rat liver nuclei and microsomes with an initiator, a nitrosamine.[49] Several studies have recently been completed with the use of fluorescent light as a carcinogen. Parshad et al. have shown that exposure of mouse cell cultures to fluorescent light produced chromatid breaks and exchanges and DNA cross-links.[50-53] H_2O_2 was inferred to be the causative agent because the light-induced chromosome damage could be minimized by lowering the oxygen concentration or by adding glutathione, ascorbic acid, or catalase. Susceptibility to light-induced chromosome damage increased with time in culture and seemed to be associated with or requisite for the spontaneous malignant transformation of mouse cell lines.[54] All three cell lines followed during long-term culture that either became tumorigenic or showed cytological evidence of neoplastic transformation developed a concomitant increase in susceptibility. In three other cell lines, susceptibility to light-induced chromatid damage was significantly increased in the spontaneously transformed malignant cells as compared with their nonneoplastic precursors. It was shown that the increased susceptibility observed was not simply the result of long-term culture, since three other nonneoplastic cell lines after prolonged culture were significantly less susceptible than their malignant counterparts. The authors speculated that increased susceptibility to light-induced chromatid damage could result from impaired DNA repair or from loss of defense mechanisms for destroying H_2O_2 or scavenging free radicals. In another study, light and oxygen were shown to increase the frequency of malignant transformation in mouse cell lines.[55]

Parshad et al. have also studied fluorescent light-induced chromosome damage in human IMR-90 fibroblasts.[56] These cells are not known to undergo spontaneous neoplastic transformation and are exceedingly difficult to transform by carcinogenic agents. Exposure of these cells to fluorescent light also induced chromatid breaks and exchanges. The chromatid damage was found to be caused largely by the production of H_2O_2, since it could be prevented almost completely by the addition of catalase, and since exogenous H_2O_2 also induced chromatid breaks. The authors concluded that H_2O_2 formed in the cells was the damaging agent because cells illuminated in saline showed the same extent of damage as cells irradiated in culture medium. Addition of selenite to the cultures during light exposure significantly decreased the chromatid damage in a dose-related manner, and this action was thought to be due to elevation of glutathione peroxidase. The free hydroxyl radical, $\cdot OH$, appeared to be partially responsible for the light-induced chromatid damage, since of the free radical scavengers tested (mannitol, vitamin E, and DMSO), only mannitol significantly decreased the light-induced chromatid damage. In a previous study SOD was without effect in a mouse system.[52] Thus, both $\cdot OH$ and H_2O_2 (but not $O_2^{\cdot-}$) formed within the cell were thought to be the agents that directly or indirectly caused chromatid damage. Parshad et al. have recently studied why mouse cells show increased susceptibility to fluorescent light-induced chromatid breaks after malignant transformation.[57] They hypothesized that increased susceptibility could be due to loss or inactivation of catalase or from impaired capacities to repair DNA damage. No consistent changes in catalase activity after neoplastic transformation could be noted. However, impaired capacities to repair

DNA after transformation were observed. Thus, the increased susceptibility of the malignant cell to fluorescent light-induced DNA damage was shown to be due to lack of repair of DNA damage.

One interesting point about the above study is that it provides evidence about our hypothesis of the nature of cancer.[40] We have hypothesized that a cancer cell must have two properties: (1) an attainment of immortality; and (2) a loss of control of cell division. A cell line will have the first property because it is by definition "immortal". The first property was hypothesized to be associated with SOD, while the second property was felt to be associated with catalase and peroxidase. In the present study, a cell line was used which should have low SOD and be immortal. However, according to our theory, catalase should drop precipitously upon transformation. However, catalase did not drop upon transformation, probably because it had already dropped upon being put into culture. Catalase activities were about 35-fold less in the tissue culture cells than observed in a fresh homogenate of C3H embryo tissue. Catalase activity was felt by Parshad et al. to drop precipitously on explantation of embryo tissue to culture since decreased activity was already apparent in a 9-day-old primary culture. Catalase activity may drop before transformation because of some effect of culturing. In any case, low SOD and catalase are associated with malignancy, a fact consistent with our theory. This system may be an unusual one also because catalase does not change with growth cycle as predicted by our theory. Thus, the specific activity of catalase in a suspension culture of a hamster cell line increased to maximum 24 hr after serial transfer to a fresh medium,[58] while the cells in the present study did not show this behavior.

C. Tumor Promotion

In the previous studies, it is unclear if fluorescent light and oxygen are acting as initiators, or promoters, or both.There have been a number of recent studies on the role of oxygen free radicals in tumor promotion. Tumor promoters are agents which do not themselves cause tumors, but enhance the production of tumors by initiators. It has been known for some time that phorbal myristate acetate (PMA) stimulated the production of $O_2^{\cdot-}$ in white cells.[59] The relationship of this stimulation to tumor promotion went unnoticed until the paper by Archer.[60] In this paper a theory for the carcinogenicity of fibers and films was presented. Fibers and films are known to induce tumors by physical rather than chemical means. In other words, chemicals are not released by these materials to cause tumors. Archer proposed that the carcinogenicity of fibers and films can be explained by frustrated phagocytosis. Neutrophils, in trying to engulf and kill the foreign body, release $O_2^{\cdot-}$. It was postulated that these radicals induce somatic mutations in a manner similar to ionizing radiation. This process thus includes both initiation and promotion components.

Witz et al. have examined the role of superoxide radicals in classical tumor promotion.[61] They have hypothesized that tumor promotion by phorbol myristate acetate is mediated "at least in part by reactive oxygen species derived from phagocytic cells which infiltrate mouse skin as part of the inflammatory response induced by tumor promoters." This hypothesis was based on studies which showed that protease inhibitors capable of inhibiting tumor promotion also inhibit the stimulation of phagocytic cells, including macrophages and polymorphonuclear leukocytes (PMN). In addition, the degree of stimulation of PMN by phorbol esters had been previously shown to correlate with their promoting activity. Since promotion by phorbol esters in the mouse skin system is also inhibited by retinoids, the effect of retinoids on the stimulation of PMN by the tumor promoters PMA, mezerein, and teleocidin B was examined. It was found that all-trans retinol, retinyl acetate, and retinoic acid were effective inhibitors of $O_2^{\cdot-}$ production in PMN stimulated with PMA. Retinol similarly inhibited cells stimulated with mezerein or teleocidin B. No-effect concentrations of the protease inhibitor antipain potentiated the inhibitory effect of low levels of retinol. Higher concentrations of retinol and antipain resulted in an additive or less than additive inhibitory effect. Other researchers have also shown that dexamethasone, another known inhibitor of mouse skin tumor promotion, inhibited $O_2^{\cdot-}$ production by phagocytic cells.[62]

Kensler and Trush have also observed inhibition of phobol ester stimulation by retinoic acid (and 5,6-epoxyretinoic acid).[63] Instead of measuring $O_2^{\cdot-}$ production, they measured chemiluminescence (CL), an index of any reactions mediated by $O_2^{\cdot-}$ and $^1O_2$. The tumor promoter 12-*O*-tetracanoylphorbol-13-acetate (TPA) was found to be a potent stimulator of CL by human polymorphonuclear leukocytes; treatment with TPA (100 ng/m*ℓ*) provoked a CL response that peaked within 5 min and persisted for over 30 min. The respose was proportional to concentration over the range of 1 to 100 ng/m*ℓ*. The ability of different phorbol diesters to stimulate both CL and $O_2^{\cdot-}$ production correlated with their relative activities as tumor promoters in vivo. Nonphorbol diester tumor promoters such as iodoacetic acid, anthralin, and Tween 60® were inactive in this system. The TPA-mediated stimulation of CL was inhibited by retinoids; cells preincubated for 15 min with 100 μM retinoic acid showed only a marginal CL response to TPA. Addition of retinoic acid to resting polymorphonuclear leukocytes resulted in a transient burst of CL without concomitant $O_2^{\cdot-}$ release. A similar CL response was seen when retinoic acid was incubated with potassium superoxide in a cell-free system. 5,6-Epoxyretinoic acid, an even more effective inhibitor of TPA-stimulated CL than retinoic acid when added simultaneously with TPA, did not undergo these two CL reactions. It was concluded that retinoic acid may undergo oxidative activation to a species that can exert enhanced antipromoter ractivities.

The latest report on $O_2^{\cdot-}$ and tumor promotion was done by Goldstein et al.[64] Comparison was made of the ability of the potent tumor promoter phorbol myristate acetate (PMA), as well as less active PMA analogs and nonphorbol ester tumor promoters, to stimulate superoxide anion radical ($O_2^{\cdot-}$) production by human polymorphonuclear leukocytes (PMN). The rate of $O_2^{\cdot-}$ production was found to correlate with the tumor-promoting activity of the phorbol esters and not to correlate with their inflammatory activity. Two analogs of phorbol ester inactive in tumor promotion were also inactive in stimulating $O_2^{\cdot-}$ production in PMN.

However, nonphorbol ester promoters did not correlate well. Acetic acid, a weak promoter in mouse skin, did not stimulate $O_2^{\cdot-}$ production. Mezerein, a nonphorbol ester tumor promoter, stimulated $O_2^{\cdot-}$ production slightly more than PMA even though it is less effective as a tumor promoter. Teleocidin B produced $O_2^{\cdot-}$ at the same rate as PMA and also had equal potency as a tumor promoter.

Thus, both initiation and promotion seem to involve the production of oygen-free radicals. Why this should be so remains unclear and a topic for future work. Equally unclear is a fundamental question; since both initiators and promoters produce oxygen free radicals, and since Mn SOD and possibly Cu-Zn SOD are inducible enzymes (induced by $O_2^{\cdot-}$), why are tumor cells low in SOD? One would expect tumor cells to be high in SOD since the enzyme should be induced. Another unanswered question is whether lowering Mn SOD makes a cell likely to develop cancer. As pointed out in an earlier paper, there does seem to be some relationship between low Mn SOD and the development of tumors in humans.[5]

VIII. USE OF SUPEROXIDE DISMUTASE IN THE TREATMENT OF CANCER

A. Introduction

In the previous sections we have discussed the loss of SOD in malignancy, and possible implications and significance. It should be realized that the role of this loss is still uncertain and remains speculation. However, even if the loss of Mn SOD is not central to the formation of cancer or expression of the cancer phenotype, I feel it will still remain important in cancer because of its significance in cancer therapy. I feel knowledge of superoxide dismutase and other antioxidants are essential to rational treatment of cancer. In the next few sections I will explain why this is true.

Three different ways have been proposed to use the difference in superoxide dismutase levels to advantage in treating tumor cells:

1. Increasing the flux of $O_2^{\cdot-}$ to the tumor cell
2. Inhibiting SOD
3. Adding back the lost SOD activity[30]

Each of these methods will be discussed in detail.

B. Increased Flux of $O_2^{\cdot-}$ in Tumor Cells

1. Introduction

Increased production of $O_2^{\cdot-}$ should preferentially kill tumor cells over normal cells because of lowered SOD activities in tumor cells. Many presently used antitumor agents produce $O_2^{\cdot-}$ and apparently already operate on this principle. However, it should be cautioned that there are little data to show that tumor cells are more susceptible to these agents than normal cells. Common sense seems to dictate that this must be so if an agent is to work, but more data are needed in this area. I will now examine in detail the various agents shown to produce oxygen radicals.

2. Bleomycin and Its Analogs

Bleomycin is a glycopeptidic antibiotic used for the treatment of cancer. In vitro, a major effect of bleomycin is the introduction of strand breaks into DNA. Efficient breakage of DNA by bleomycin requires the presence of reducing agents, such as 2-mercaptoethanol, dithiothreitol, ascorbate, or hydrogen peroxide.[65,66] Although reducing agents alone can be injurious to DNA,[67,68] the combination of bleomycin and a reducing agent is far more effective in degrading DNA than is either bleomycin or reducing agent alone.[69,70] Limited damage to DNA may be observed with high concentrations of bleomycin alone.[70,71]

The role of the reducing agents have been examined in great detail. Bleomycin has been shown to bind to DNA in the absence of reducing agents.[72] Superoxide radical can act as a reducing agent to stimulate DNA degradation by bleomycin.[73,74] It has also been shown that Fe(II) can substitute for the reducing agents.[75] Bleomycin and Fe(II) together were far more efficient in cleaving DNA than either species alone. This result has been verified.[76] This latter article was the first to suggest that hydroxyl radical may be responsible for bleomycin-induced damage to DNA; however, no convincing evidence was given.

Fe(III) cannot replace Fe(II) in the degradation of DNA by bleomycin in the absence of reducing agent, but in the presence of reducing agent, either Fe(II) or Fe(III) greatly stimulate DNA degradation by bleomycin.[75] Conversely, metal chelators and other metal ions inhibit DNA degradation.[74] These observations have led to a proposed model for the action of bleomycin.[74] In this model, bleomycin is first proposed to bind to DNA, an event which occurs without the necessity of metal ion or reducing agent. Fe(II) is then proposed to attach to the bleomycin and form a ternary complex. This ternary complex can then form some species which degrades DNA. Reducing agents enhance the reaction by regenerating Fe(II) from Fe(III) and thus continuing the reactions; in particular, $O_2^{\cdot-}$ has been proposed to form Fe(II) from Fe(III).[77,78]

These observations led two groups almost simultaneously to identify the toxic species produced.[79,80] Sugiura and Kikuchi were chronologically the first to identify the toxic species as either superoxide or hydroxyl radical.[79] Using the technique of spin trapping (described in Chapter 5 of this volume), bleomycin-A_2 was shown to form the $O_2^{\cdot-}$ and $\cdot OH$ spin adducts. With *N-tert*-butyl-α-phenylnitrone (BPN) as the spin trap, at low concentrations of bleomycin-Fe(II) complex, $O_2^{\cdot-}$ spin adduct was detected, while at high concentrations, $\cdot OH$ was detected. This observation is consistent with the observation that DNA degradation by Fe(II)-bleomycin is inhibited by SOD at low concentrations of bleomycin, but not at high concentrations.[81] Co(II) complex, but not Cu(II) and Zn(II) complexes with bleomycin, also produced $\cdot OH$ adduct.

At nearly the same time, a second paper appeared which used spin trapping to identify the presumably active species from Fe(II)-bleomycin.[80] Buettner and Oberley observed $\cdot OH$ adduct from bleomycin-Fe. Ethylenediaminetetraacetate (EDTA) abolished the signal, while H_2O_2 enhanced it by a factor of 10. However, in contrast to the earlier work, even at low concentrations of bleomycin-Fe(II), only $\cdot OH$ was observed and not $O_2^{\cdot-}$. The discrepancy is at present unresolved, but one difference in the two studies was that a different spin trap was used in the second study (5,5-dimethyl-1-pyrroline-1-oxide [DMPO]).

Sugiura has used spin trapping to further study the bleomycin-Fe(II) complex.[82] He has found that the $\cdot OH$ spin adduct produced at high concentrations of bleomycin was inhibited by catalase, showing H_2O_2 to be involved. He proposed that $O_2^{\cdot-}$ is first formed from bleomycin-oxygen and then $\cdot OH$ is formed by a Haber-Weiss reaction. However, this model does not take into account the fact that $O_2^{\cdot-}$ is apparently only produced at low concentrations of bleomycin and not at high concentrations. Sugiura went on to show that hydroxyl radicals produced from high concentrations of bleomycin-Fe(II) oxidize 2,6-di-*tert*-butyl-*p*-cresol and α-tocopherol to form the corresponding phenoxy and tocopherol radicals, respectively.

Compounds similar to bleomycin have been studied by spin trapping.[83] Depruvamide bleomycin has significantly lower biological activity than bleomycin.[84] Sugiura has shown it produces only about 1/100 the hydroxyl radical signal as bleomycin.[83]

Tallysomycin, a third generation bleomycin analog, has been found to be a potent antitumor agent, in some cases possessing greater antitumor activity than bleomycin.[85] The mechanism of action of tallysomycin appears to be similar to that of bleomycin in that it produces strand breaks in purified isolated DNA and intracellular DNA similar to bleomycin.[86] However, there has been difficulty in correlating antitumor activity of tallysomycin with the breakage of DNA in vitro.[86] Bleomycin seems to be able to break DNA at a higher rate than tallysomycin even though its antitumor activity may not be as great in a particular system. Buettner and Oberley were the first to study free radical production by tallysomycin.[87] Tallysomycin produced $\cdot OH$ spin adduct when Cu(II) was present, but not when Fe(II) was present. EDTA abolished the signal, but superoxide dismutase or catalase did not. Apparently, $\cdot OH$ can be formed without $O_2^{\cdot-}$ or H_2O_2 as intermediates.

Sugiura has obtained quite different results in the same system.[88] Bleomycin or tallysomycin-Cu(I) but not Cu(II) systems produced hydroxyl radicals. Bleomycin or tallysomycin-Fe(II) systems also produced $\cdot OH$. Sulfhydryl compounds inhibited the signals from Cu(I), but not from Fe(II). Sugiura concluded that tallysomycin and bleomycin have similar modes of action.

Lown's group have found similar results to Sugiura.[89] Looking first at bleomycin, they used the ethidium fluorescence method of measuring strand breakage in conjunction with PM2-covalently closed circular DNA (CCC-DNA).[76] It was shown that bleomycin damages by reactions involving $O_2^{\cdot-}$, H_2O_2 and $\cdot OH$. This was deduced from the facts that SOD, catalase, and free radical scavengers (e.g., isopropyl alcohol) inhibited DNA cleavage. Fe(II) and oxygen were required for this breakage. Similar results were found with tallysomycin.[89] Fe(II), reducing agents, and oxygen were all required. Other metal ions such as Cu(II) inhibited the DNA cleavage. SOD, catalase, and hydroxyl radical scavengers also inhibited the DNA damage, although the inhibition was not complete. Spin trapping showed the production of $\cdot OH$, with this radical almost completely suppressed by SOD and catalase.

Why the latter two studies differ from the original in the requirement for metal ion is still unknown. Perhaps, a different preparation of tallysomycin or the presence of an impurity caused the different results. Although there is still some discrepancy in metal ion requirements, all researchers agree that $\cdot OH$ is formed by this antibiotic.

3. Quinone Anticancer Agents

a. Introduction

A large variety of quinone compounds are effective antitumor agents. Many of these are discussed individually below. In the case where one study discussed more than one compound, all of the compounds used in that study will be discussed together, even if not in the appropriate section.

b. Adriamycin (Doxorubicin)

Adriamycin is an anthracycline antibiotic. Lown's group was the first to study the free radical nature of adriamycin toxicity.[90] Using the ethidium bromide fluorescence assay, it was found that adriamycin and daunorubicin bound to covalently closed circular DNA nick the latter when reduced by sodium borohydride. The degradation is dependent on oxygen and strongly inhibited by SOD, catalase, and sodium benzoate, indicating the intermediacy in the cleavage of $O_2^{\cdot-}$, H_2O_2, and $\cdot OH$, respectively.

Thayer examined adriamycin in a system containing submitrochondrial particles.[91] Adriamycin (doxorubicin) stimulated the formation of $O_2^{\cdot-}$ in submitochondrial particles isolated from bovine heart. $O_2^{\cdot-}$ was detected by oxygen uptake, the reduction of acetylated cytochrome *c*, and cooxidation of epinephrine to adrenochrome; all of the processes were sensitive to SOD. Addition of rotenone plus NADH to submitochondrial particles caused the formation of 4 nmol of superoxide per minute per milligram of protein. Adriamycin at a concentration of 400 μM stimulated the rate of $O_2^{\cdot-}$ formation sixfold to 25 nmol $min^{-1}mg^{-1}$. H_2O_2 accumulated as a final reaction product. Measurements of the relative catalase activity of blood-free tissues of rabbits and rats indicated that heart contained only 2 to 4% of the catalase activity of liver or kidney. It was suggested that an enhanced production of $O_2^{\cdot-}$ and H_2O_2 and the relatively low catalase content of heart tissue may be factors in the cardiotoxicity induced by adriamycin chemotherapy.

Goodman and Hochstein have also observed generation of free radicals by adriamycin and daunomycin.[92] NADPH and purified P450 reductase caused oxygen consumption, partially inhibitable by catalase, from the antitumor drugs daunomycin and adriamycin in excess of the amount of the drug present. A reduction-autoxidation cycle of quinone groups was postulated. During this cycle, the cooxidation of sulfite may be initiated. This latter oxidation is partially inhibited by SOD, suggesting that $O_2^{\cdot-}$ is formed. Catalase also inhibited, suggesting that H_2O_2 is likewise formed. Rat liver microsomes also catalyzed this redox cycling, which is accompanied by the peroxidation of lipids. Those authors also suggested that the formation of oxygen radicals followed by lipid peroxidation may be the basis of the cardiotoxic effects of these drugs.

Henderson et al. have studied the effects of adriamycin and daunomycin on erythrocytes.[93] Addition of the compounds to human erythrocyte suspensions resulted in the stimulation of hexose monophosphate activity that was markedly impaired in the absence of oxyhemoglobin. Evidence for generation of H_2O_2 by these compounds was provided by oxidation of reduced glutathione, by ^{14}C-formate oxidation, and by the catalase-aminotriazole trapping technique. It was hypothesized that reactive oxygen radicals were the precursors of the H_2O_2 formed and that a similar interaction with oxymyoglobin may occur in the heart to produce oxygen radicals that injure cardiac myocytes.

Someya and Tanaka have studied the DNA strand scission induced by adriamycin and aclacinomycin A.[94] The binding of these drugs to PM2 DNA and the consequent cleavage of DNA was demonstrated by agarose gel electrophoresis, using an ethidium bromide assay. Adriamycin was shown to induce single strand scission of DNA in the presence of a reducing agent, but aclacinomycin A caused a much smaller amount of DNA breakage. The DNA cleavage was enhanced by Cu(II) and especially by Fe(II), but not significantly by Ni(II), Zn(II), Mg(II), or Ca(II). The DNA degradation was dependent upon the concentration of the anthracyclines and $CuCl_2$. The degree of DNA cleavage at 0.04m*M* adriamycin was similar to that of 0.4 m*M* aclacinomycin A in the presence of 1 m*M* NADPH and 0.4 m*M* $CuCl_2$. The anthracycline-induced DNA cleavage was stimulated by H_2O_2, but partially inhibited by nitrogen gas atmosphere, SOD, catalase, and potassium iodide ($\cdot OH$ scavenger). The authors suggested that both the free radical of anthracycline antibiotics and $\cdot OH$ were responsible for DNA degradation by these compounds.

Kalyanaraman et al. have used electron spin resonance to study the free radicals produced by microsomal incubations containing adriamycin, daunorubicin, and mitomycin C.[95] When oxygen concentrations were high, $O_2^{\cdot -}$ spin adduct was observed; this was inhibited by SOD. When oxygen concentrations were low, semiquinone radicals were observed. The authors suggested that the time sequence of the appearance of the spin-trapped superoxide and the semiquinone radical metabolites indicated that air oxidation of the semiquinone was responsible for the $O_2^{\cdot -}$ formation.

c. Daunorubicin (Daunomycin)

Daunorubicin is an anthracycline antibiotic similar to adriamycin. Some of its radical chemistry has been discussed in the preceding section. Lown et al. have studied $O_2^{\cdot -}$ production of daunorubicin in relation to its cardiotoxicity.[96] 5-Iminodaunorubicin, when reductively activated, produced single-strand scission in PM2 CCC-DNA by a mechanism which proceeds via production of $O_2^{\cdot -}$ and $\cdot OH$. Under comparable conditions, 5-iminodaunorubin produced much less nicking than daunorubicin. The reason for this was investigated. Using *N*-cyclohexyl-5-iminoquinizarine as a model, assignment of polarographic waves to the quinone moiety of 5-iminodaunorubicin was possible. The electrochemical results indicated that 5-iminodaunorubicin is more difficult to reduce than daunorubicin and that reoxidation of the reduced form, 5,11-dihydro-5-iminodaunorubicin, is much more difficult than reoxidation of reduced daunorubicin. The results suggest a correlation between the diminished generation of $O_2^{\cdot -}$ by 5-iminodaunorubicin and its observed suppressed cardiotoxicity relative to other anthracyclines.

d. Streptonigrin

Streptonigrin is a metabolite of *Streptomyces flocculus* which possesses the aminoquinone moiety. Cone et al. first examined the degradation of DNA by streptonigrin.[97] The production of single-strand breakage in PM2 CCC-DNA by NADH-reduced streptonigrin was demonstrated using the ethidium bromide fluorescence assay. Fe(II) was found to be the most efficient cofactor in cleaving DNA, with Cu(II) less so, and Co(II) having no effect. In the absence of added metal (realizing some metal is probably present as contaminant), streptonigrin DNA cleavage was inhibited by SOD and catalase and H_2O_2 enhanced the cleavage. KI also inhibited DNA damage. In the presence of Cu(II), SOD no longer inhibited, although catalase did but required higher levels than in the absence of Cu(II). The authors speculated that in the latter case, $O_2^{\cdot -}$ may be produced in pairs at the site of the streptonigrin-Cu(II) complex giving rise to H_2O_2 immediately which can then react to form $\cdot OH$.

Lown and Sim have also studied a group of substituted 5,8-quinolinequinones which exhibit antineoplastic activity and which are structurally related to the antitumor antibiotic streptonigrin.[98] These compounds induce single-strand cleavage of PM2 CCC-DNA when reductively activated. The cleavage detected by the ethidium fluorescence assay was specifically enhanced by Fe(II) and Cu(II) and was inhibited by SOD, catalase, and by free radical scavengers (including $\cdot OH$ scavenger). Independent generation of $O_2^{\cdot -}$ by xanthine-xanthine oxidase also cleaved PM2 CCC-DNA. A correlation between the rate of DNA cleavage and inhibition of Walker carcinosarcoma 256 growth was observed with these compounds.

e. Mitomycin

Lown et al. have studied DNA cleavage by mitomycin C.[99] Mitomycin C, reduced chemically, was shown to induce single-strand cleavages as well as monoalkylation and covalent cross-linking in PM2 covalently closed circular DNA. The inhibition of this cleavage by SOD and catalase and by free radical scavengers (including $\cdot OH$) suggested that the degradation of DNA is largely due to $O_2^{\cdot -}$. The following reaction scheme was suggested:

$$\text{Mitomycin C} + NaBH_4 \longrightarrow \text{Mitomycin } CH_2 \qquad (1)$$

$$\text{Mitomycin } CH_2 + O_2 \longrightarrow \text{Mitomycin } \dot{C}H + HO_2\cdot \qquad (2)$$

$$HO_2\cdot \rightleftharpoons H^+ + O_2^{\cdot -} \qquad (3)$$

$$2HO_2\cdot \longrightarrow H_2O_2 + O_2 \qquad (4)$$

$$O_2^{\cdot -} + H_2O_2 \longrightarrow HO\cdot + HO^- + O_2 \qquad (5)$$

In contrast to the behavior of streptonigrin, mitomycin C was not found to inactivate the protective enzymes SOD or catalase.

Lown and Weir have investigated mitomycin B also.[100] Mitomycin B, reduced *in situ* with sodium borohydride ($NaBH_4$), nicked PM2 CCC-DNA rapidly but less efficiently than mitomycin C. This reaction was suppressed by SOD, catalase, and free radical traps (including $\cdot OH$), showing the intermediacy of $O_2^{\cdot -}$, H_2O_2, and $\cdot OH$. DNA was cleaved by mitomycin B, to which it is covalently attached, as well as by the free antibiotic.

Lown et al. have studied mitomycin B, mitomycin C, and streptonigrin using electron spin resonance to identify more rigorously the radicals produced.[101] Reduced mitomycin C in aqueous solution exposed to air gave a 36-line ESR spectrum of the semiquinone identified by computer simulation. Incubation of this radical with the spin trap *N-tert*-butyl-α-phenylnitrone (PBN) gave the $\cdot OH$ spin adduct identified by independent generation of $\cdot OH$. This adduct was also formed from similar treatment of DNA to which mitomycin C had been covalently attached. Incubation of the semiquinone from mitomycin C, mitomycin B, or streptonigrin with catalase or with SOD inhibited the generation of $\cdot OH$, implying the intermediacy of H_2O_2 and $O_2^{\cdot -}$ in its formation. The formation of $\cdot OH$ was similarly inhibited by EDTA, suggesting trace metal involvement. The authors concluded that these results are consistent with the generation by the semiquinone antibiotics in vivo of $\cdot OH$, already implicated in the degradation of DNA.

f. Carminic Acid

Carminic acid is the principal component of the food dye cochineal obtained from the insect *Nopalea coccinellifera.* Lown et al.[102] have found that in contrast to adriamycin and daunorubicin, carminic acid does not bind to DNA, but nicks it slowly, more rapidly when reduced *in situ,* and still more rapidly when prereduced at the quinone moiety. The nicking process requires oxygen, and is selectively inhibited by SOD, catalase, and hydroxyl radical scavengers. The intermediacy of ·OH is supported by observation of ·OH spin adduct with BPN. The single-strand scission of DNA by carminic acid requires two adjacent hydroquinone moieties in the chromophore, since reduced methyl tetra-*O*-methylcarminate is without effect, although it binds weakly to DNA. The authors attribute the fact that daunorubicin and adriamycin produce more extensive DNA strand scission than carminic acid under comparable conditions of prereduction and on a molar basis to the assistance of intercalative binding in the case of anthracyclines.

g. β-Lapachone

β-Lapachone (3,4-dihydro-2,2-dimethyl-2H-naphthol[1,2-b]-pyran-5,6-dione) is an antimicrobial and antitrypanosomal *o*-naphthoquinone. This compound also possesses significant antitumor activity. Addition of β-lapachone to sarcoma 180 mouse tumor cells induced the formation of semiquinone radical.[15] β-Lapachone stimulated $O_2^{\cdot-}$ and H_2O_2 produced in the mitochondrial fraction in the presence of NADH. This drug also increased $O_2^{\cdot-}$ and H_2O_2 produced by the microsomal fraction in the presence of NADPH. Incubation of sarcoma 180 cells with β-lapachone stimulated lipid peroxidation and resulted in a decrease in viability of these cells. α-Tocopherol reduced this toxicity. The authors postulated the basic mechanism of this compound in sarcoma cells to be (1) reduction of the drug at mitochondrial and microsomal membranes to a semiquinone form; (2) autoxidation of the semiquinone free radical with primary production of $O_2^{\cdot-}$ and (3) production of H_2O_2 via SOD and the production of ·OH from $O_2^{\cdot-}$ and H_2O_2. The subsequent lipid peroxidation was postulated to cause cell death.

h. General Theory of Action of Quinone Anticancer Agents

Bachur et al. have proposed a general mechanism for activation of quinone anticancer drugs to free radicals.[103] They examined adriamycin, daunorubicin, carminomycin, rubidazone, noglamycin, aclacinomycin A, and steffimycin (benzanthraquinones); mitomycin C and streptonigrin (*N*-heterocyclic quinones); and lapachol (naphthoquinone). These quinone drugs augmented the flow of electrons from NADPH to molecular oxygen as measured by enhanced NADPH oxidation and oxygen consumption. This reaction was catalyzed by microsomal protein and produced a free radical intermediate form of the drugs as determined by ESR. Microsomes from mouse and rat liver, heart, lung, and spleen and mouse L1210 and P388 tumors all catalyzed and augmented oxygen consumption. Several nonquinone anticancer drugs were tested and found inactive. Since quinone anticancer drugs are associated with the chromosomal damage that appears to be dependent on metabolic activation of these drugs, they proposed that intracellular activation of these drugs to a free radical state may be primary to their cytotoxic activity. "As free radicals, these drugs, because of their high affinity and selective binding to nucleic acids, have the potential to be 'site-specific free radicals' that bind to DNA or RNA and either react directly or generate oxygen-dependent free radicals such as superoxide radical or hydroxyl radical to cause the damage associated with their cytotoxic actions." Bachur's group has also shown that rat liver NADPH cytochrome P450 reductase can also catalyze the single-electron reduction of these quinone antibiotics to a semiquinone free radical.[104] This radical can then transfer a single electron to oxygen to form $O_2^{\cdot-}$. In Bachur's theory two characteristics are necessary for cytoxicity. First, the drug must be capable of biochemical activation to a free radical state. Second, the drug must have an affinity for an essential macromolecule, such as DNA, RNA, or other molecules. Coenzyme Q would satisfy the first requirement, but not the second, and so is not a good antitumor agent.

Sinha and Chignell have provided evidence for this model.[105] Chemical reduction of adriamycin and daunorubicin formed the same partially reduced free radicals previously reported by microsomal activation. In vitro incubation of the chemically activated free radical intermediates with DNA resulted in covalent binding of these drugs to DNA. This, the radical form of the drug can bind DNA, as postulated by Bachur.

The problem with Bachur's theory is that if the drugs are activated in the microsomes, how will a reactive-free radical then reach the crucial site, presumably DNA? Bachur has realized this limitation, and has recently shown that nuclei can themselves activate these drugs.[106] His theory now seems consistent.

i. Hypoxic Vs. Oxygenated Cells

As briefly described above, many of these drugs will produce different radicals depending on the amount of oxygen present. This can be an advantage or disadvantage in treating the tumor depending upon the situation. Kennedy et al. have shown that mitomycin C can kill in vitro tumor cells, but the rate of killing is much higher for cells grown in chronic hypoxia than for normally aerated cells.[107] This may be very useful because most tumors contain a hypoxic fraction which is extremely resistant to treatment. This resistance is thought to be for several reasons. Because hypoxic tumor cells exist at relatively great distances from blood vessels, many drugs may not reach the tumor. In addition, agents which are rapidly metabolized may not reach hypoxic cells in sufficient concentrations to produce biological effects. Furthermore, many hypoxic cells may not be susceptible to those anticancer drugs which are active only against proliferating cells; this is because many hypoxic tumor cells are probably quiescent or are cycling with prolonged cell cycle times. Last, because of low oxygen, it maybe difficult to achieve high oxygen free production in these cells.

This concept has been carried further by Sartorelli's group at Yale.[108] They have classified a large number of drugs based on their cytotoxicity toward EMT6 mouse mammary tumor cells under aerobic and hypoxic conditions. Drugs were placed into three categories. Drugs that were preferentially toxic under oxygenated conditions included bleomycin, procarbazine, streptonigrin, actinomycin D, and vincristine. Drugs preferentially toxic to cells under hypoxic conditions include mitomycin C and adriamycin. On the basis of other reports, the authors feel that the glucose analogs, 5-thio-D-glucose and 2-deoxy-D-glucose, and the radiosensitizers, misonidazole and metronidazole, can be placed in this category. Several antineoplastic agents showed no preferential toxicity to cells under oxygenation or hypoxia. These were: 1,3-bis(2-chloroethyl)-7-nitrosourea, 1-(2-chloroethyl)-3-cyclohexyl-2-nitrosurea, *cis*-diaminedichoroplatinum(II), 5-fluorouracil, and methotrexate. The authors present the concept that future chemotherapeutic regimens for the treatment of solid tumors should include agents that kill both hypoxic and aerobic cells, as well as proliferating and nonproliferating cells. An example of such a regimen would be (1) a bioreductive alkylating agent such as mitomycin C to kill hypoxic cells; (2) an agent such as bleomycin or nitrosourea to kill any nonproliferating oxygenated cells; and (3) X-irradiation and/or an agent to kill actively proliferating oxygenated cells.

4. Other Agents

a. Camptothecin

Camptothecin is a cytotoxic alkaloid derived from the bark and stem wood of *Camptotheca acuminata*. Lown and Chen have studied the scission of DNA by photoactivated camptothecin.[109] Aqueous solutions of camptothecin, its sodium salt, or a number of derivatives produced single-strand breaks in PM2 CCC-DNA when irradiated with 360-nm light. The overall DNA breakage showed an inverse dependence on oxygen, and there is evidence for at least two reaction mechanisms. Photosensitization of the alkaloid may generate radicals which attack DNA or, in the presence of oxygen, generate hydroperoxy radicals. In the anaerobic pathway, photodecarbonylation of the alkaloid may generate a diradical which can collapse to the racemized camptothecin hemicetal or abstract hydrogen atoms from DNA leading to strand scission. In the presence of oxygen, the alternative aerobic pathway occurred, leading to the generation of $HO_2\cdot$, H_2O_2, and $\cdot OH$, which can then attack DNA. The intermediacy of these three species was shown by inhibition of scission with SOD and catalase, and spin trapping. Heat was also found to activate the alkaloid to produce similar effects.

b. Actinomycin D

Actinomycin D is a cyclic pentapeptide antibiotic. When incubated with microsomal proteins, actinomycin D has been shown to generate a free radical intermediate, which stimulates oxygen uptake with the formation of $O_2^{\cdot -}$.[110] Sinha and Cox have prepared N^2-(4-[2,2,6,6-tetramethyl-1-piperidinyl oxyl]) actinomycin D and the related 1,3-diaminopropane analogs.[111] These compounds were found to be more active than actinomycin D in vivo against P-388 leukemia cells in spite of poor DNA binding.[112] The analogs were found to be more active in stimulating O_2 uptake and the formation of $O_2^{\cdot -}$ than the parent compound. It was concluded that there was a good correlation between antitumor activity and $O_2^{\cdot -}$ production in these compounds.

c. Hematoporphyrin

Hematoporphyrin derivative activated by red light has been used to cure experimental tumor systems by Dougherty and co-workers.[113] Recently various human tumors have been shown to respond to such photochemotherapy.[114] Singlet oxygen has been invoked as the cytotoxic agent.[115]

However, two recent reports have suggested that hydroxyl radicals may be the cytotoxic agent. Hariharan et al. used production of thymine ring saturation products of the 5,6-dihydroxydihydrothymine type or thymine glycols in the DNA of cells as a measure of generation of $\cdot OH$.[116] $\cdot OH$ had previously been shown to be mainly responsible for production of thymine glycols when cells are exposed to ionizing radiation. Both bacterial cells *(Micrococcus radiodurans)* and mammalian cells (Chinese hamster V-79) were used in this study. The results clearly demonstrated significant production of thymine glycol only when cells were exposed to both hematoporphyrin and red light. Singlet oxygen did not produce thymine dimers in a previous study.[117] Dimethylsulphoxide, a $\cdot OH$ scavenger, reduced the yield of thymine glycols. The authors felt the results showed the production of $\cdot OH$ by hematoporphyrin plus light.

Buettner and Oberley have used spin trapping to investigate the species produced by hematoporphyrin plus light.[118] The illumination of hematoporphyrin in the presence of 50 m*M* DMPO resulted in $\cdot OH$ adduct generation. When SOD was incorporated into the reaction mixture at 200 unit/mℓ, the $\cdot OH$ signal was suppressed. Heat-inactivated SOD and catalase (2500 units/mℓ) were without effect. When 30% by volume of the water in the solution was replaced with ethyl alcohol, the hydroxyethyl spin adduct was seen, a known product of reaction of ethyl alcohol with $\cdot OH$. In the reaction mixture, 1 *M* azide resulted in the formation of azide radical spin adduct, which has also been shown to be formed from $\cdot OH$. 1,4-Diazobicyclo(2.2.2) octane (DABCO) at 1 *M* concentration resulted in $\sim$ 50% reduction in signal intensity. This seems to suggest a singlet oxygen mechanism; however, the authors point out that DABCO also is an effective radical quencher. Diethylenetriaminepentaacetic acid at 1 m*M* concentration had no effect on the signals observed. These observations in total were thought to be consistent with the hypothesis that hemotoporphyrin in the presence of light produced $O_2^{\cdot-}$, which in turn formed $\cdot OH$ by a non-Haber-Weiss process.

d. Protoporphyrin

Protoporphyrin in the presence of light will produce cytotoxic products.[119] Because β-carotene ameliorates many of the effects caused by protoporphyrin, singlet oxygen has been proposed as the causative agent.[120] However, other evidence has suggested a radical mechanism.[121,122] Buettner and Oberley have used spin trapping to show that protoporphyrin plus light produced $HO_2\cdot$ adduct; this signal was suppressed by SOD.[123] The $HO_2\cdot$ signal eventually turned into the $\cdot OH$ adduct, although the presence of the latter radical was not proven conclusively by looking at the reaction of $\cdot OH$ with various compounds.

Very recently, it has been shown that photoactivated protoporphyrin destroys erythroleukemia, myelocytic leukemia, and Burkitt's lymphoma cells.[124] Porphyrin-treated cells exposed to light showed complete inhibition of RNA and DNA synthesis. Exposure of cells to porphyrin in the dark also resulted in some inhibition of DNA and RNA synthesis, although the effect was much greater in the light. The authors suggest that even though the cytotoxic effect of protoporphyrin in the dark was lower than that of photoactivated porphyrin, protoporphyrin may have therapeutic potential in vivo even without illumination.

e. Ozone

Although the mechanism of toxicity of ozone is still uncertain, most agree that this compound causes production of oxygen free radicals.[125] It has recently been shown that ozone selectively inhibits the growth of human cancer cells.[126] The growth of human cancer cells from lung, breast, and uterine tumors was selectively inhibited in a dose-dependent manner by ozone in ambient air during 8 days of culture. The presence of ozone at 0.3 to 0.5 part per million inhibited cancer cell growth 40 and 60%, respectively, while normal human diploid fibroblasts were unaffected at these levels. Exposure to ozone at 0.8 ppm inhibited cancer cell growth more than 90% and control cell growth less than 50%. The authors concluded that the mechanisms for defense against ozone damage are impaired in human cancer cells and suggest that ozone may be useful in therapy against lung cancer.

f. Ionizing Radiation

Ionizing radiation is one of the oldest cancer treatment modalities. Radiation has been known for some time to have greater cytotoxic effects in the presence of oxygen (oxygen effect) and attempts have been made (such as hyperberic oxygenation) to exploit this fact in cancer therapy. It has long been known that ionizing radiation produced $O_2^{\cdot-}$, but the significance of this was unclear. Oxygen radicals have been suspected as a causative agent in this oxygen effect, but compelling evidence was not obtained until recently. The main evidence that ionizing radiation produced $O_2^{\cdot-}$ and that this radical is involved in toxicity is that SOD has been found to be a radioprotector. Petkau and Chelack were the first to show radioprotection by SOD in whole organisms.[127] Oberley et al. showed that SOD lowered significantly the oxygen enhancement ratio following irradiation of *E. coli* B.[128] Misra and Fridovich provided evidence that $\cdot OH$ generated from $O_2^{\cdot-}$ and H_2O_2 was the cause of oxygen-induced radiation toxicity.[129] This conclusion was reached because SOD, catalase, mannitol, and histidine all diminished the oxygen enhancement ratio. In this study it was found that under conditions of high cell density, the oxygen enhancement ratio was greatly lowered and SOD did not protect. This was hypothesized to occur because the track of ionizing radiation interacted directly with the *E. coli* and not by diffusion of $O_2^{\cdot-}$ and $\cdot OH$ from the ionization track. McLennan et al. have examined the effect of X-radiation on viability and function of alveolar macrophages.[130] In this system, SOD, catalase, and diethylenetriamepentaacetic acid gave significant protection, whereas ethylenediametetraacetic acid and mannitol provided little or no protection. Hydroxyl radicals generated from $O_2^{\cdot-}$, H_2O_2, and iron were felt to be the agent of oxygen-induced cell damage by ionizing radiation. SOD has also been shown to protect whole mice against ionizing radiation.[131] Protection was much greater if given after irradiation than if given before. The nature of this postirradiation protection is still under investigation.

Kong and Davison have done an elegant study on the interactions between O_2, H_2O, $\cdot OH$, e^-, and $O_2^{\cdot-}$ in free radical damage to biological systems.[132] They measured the permeability of resealed erythrocyte ghosts and on the activity of membrane-bound glyceraldehyde-3-phosphate dehydrogenase. Protection against ionizing radiation was afforded by formate, nitrogen, catalase, and SOD. They proposed a dual mechanism for the oxygen effect in irradiated biological systems. Admission of oxygen to an anaerobic system was proposed to cause the conversion of e^- and $\cdot H$ to $O_2^{\cdot-}$. Because of this, interaction between $O_2^{\cdot-}$ and H_2O_2 lead to the formation of $\cdot OH$. In addition, they propose that the presence of oxygen accelerates the ensuing degration of membranes by causing greatly increased branching of radical chain reactions subsequent to initial damage.

A comprehensive review of this subject is out of place in the chapter. I would like to conclude by saying that this area is far from settled. Many qualified researchers have shown that SOD does not radioprotect in some situations and thus feel that $O_2^{\cdot-}$ does not cause the oxygen effect.[133]

5. *Role of Antioxidants*

a. Introduction

In the preceding sections, we have seen how many antitumor tumor agents are thought to produce $O_2^{\cdot-}$ and its reaction products and that these products are responsible for much of the toxicity. In this section, we will discuss how antioxidant levels in tissues affect the efficacy of treatment by these agents.

b. Normal Cell Damage

First, all of the previously discussed antitumor agents also produce $O_2^{\cdot-}$ in normal cells. In fact, one of the major limitations in using these drugs is killing of normal cells. Thayer was one of the first to suggest that the cardiac toxicity seen with adriamycin was due to the low level of catalase in heart tissue.[91] In vivo studies have shown that doxorubicin (adriamycin) administration leads to lipid peroxidation in cardiac tissue and that this peroxidation process can be blocked by alpha tocopherol.[134] Free radical scavengers have also proved effective in lessening the acute pathologic changes observed after adriamycin in mouse,[134] rat,[135] and rabbit heart.[136]

One apparent paradox presented in the early work was that although the conversion of adriamycin to a free radical metabolite is greater in the liver compared to the heart microsomal system, only cardiac tissue developed clinically significant pathologic changes. For this reason, Doroshow et al. have examined the enzymatic defenses of the mouse heart against oxygen radicals.[137] Cardiac muscle was found to contain 10 times less catalase and nearly four times less SOD than liver. Glutathione peroxidase activities were similar in the two tissues. Assay of glutathione peroxidase in the heart after 6 weeks of selenium depletion revealed a >80% drop in enzyme activity. This selenium-deficient state led to enhanced doxorubicin toxicity at a dose of 15 mg/kg i.p. Doxorubicin administration in selenium-sufficient animals resulted in a dose-dependent decrease in cardiac glutathione peroxidase activity; the decrease in enzyme activity lasted 72 hr after 15 mg/kg i.p. In contrast, cardiac SOD and hepatic SOD and glutathione peroxidase were unaffected by this dose of doxorubicin. The authors concluded that the major pathway in cardiac tissue for detoxification of reactive oxygen metabolites is via the concerted action of SOD and selenium-dependent glutathione peroxidase. The latter enzyme may be depleted either by a selenium-deficient diet or by doxorubicin treatment, leaving the heart with limited mechanisms for disposing of H_2O_2 or lipid peroxides.

Consistent with the role of glutathione peroxidase is the work of Olson et al.[138] They examined the role of glutathione (GSH) and soluble sulfhydryl groups in the toxicity of adriamycin. A single injection of adriamycin (15 mg/kg i.p.) caused a statistically significant fall in liver ($p<0.05$), heart ($p<0.02$), and erythrocyte ($p<0.01$) GSH levels. Treatment with cysteamine protected against the fall in soluble SH groups in heart tissue. Cysteamine (50 mg/kg i.p., every 8 hr for 6 days) or *N*-acetylcysteine (100 mg/kg, i.p., 1 hr before and 7 hr after adriamycin) protected against lethality in Swiss ICR-HA mice and decreased the appearance of microscopic myocardial lesions. When endogenous levels of GSH were depleted by diethyl maleate (300 mg/kg i.p., every 8 hr for 4 days), adriamycin lethality was markedly potentiated. Diethyl maleate alone did not cause death. The authors suggest that the GSH system may be involved in adriamycin toxicity because: (1) adriamycin significantly lowers GSH levels; (2) the lowering of cardiac SH groups by adriamycin can be prevented by cysteamine; and (3) adriamycin toxicity can be potentiated by diethylmaleate, a GSH depletor, and reduced by cysteamine or *N*-acetyl cysteine, SH donors.

Besides producing oxygen radicals in normal cells, some anticancer agents may also inhibit SOD in normal cells. Auclair et al.[139] have measured SOD levels in red blood cells (RBC) of patients treated with nitrogen mustard. Intravascular hemolysis was common in patients treated with nitrogen mustard. Patients having intravascular hemolysis had much lower RBC SOD than normal subjects or in treated patients without hemolysis.

c. *Protection of Normal Tissue by Added Antioxidants*

If oxygen radicals are the cause of normal cell damage and if low antioxidant levels in target tissue are potentiating factors, then these facts suggest that adding back antioxidants should prevent normal tissue damage. Indeed, as noted above, vitamin E has already been shown to protect heart muscle against adriamycin. McGuinness et al. have studied the effect of superoxide dismutase on the kidney damage caused by *cis*-platinum.[140] The rationale behind this study was that platinum (IV) has been shown to stimulate the production of $O_2^{\overline{\cdot}}$ by phagocytes or xanthine oxidase.[141] In rats, daily administration of a subcutaneous dose of 2.5 mg/kg of Cu-Zn SOD lowered the nephrotoxicity of a 5.0-mg/kg dose of the antitumor agent *cis*-diamine-dichloroplatinum II (*cis*-platinum). The toxicity of *cis*-platinum was studied by comparative measurement of blood urea nitrogen, body weight, and comparative renal histology.

One of the criticisms raised against adding antioxidants to prevent normal cell damage is that they may also prevent tumor cell killing, since both normal and tumor cell killing may be caused by similar radical mechanisms. However, Gutteridge and Fu have recently proposed that normal and tumor cell killing by bleomycin may be due to different targets.[142] They postulated that tumor cell killing is due to DNA damage caused by oxygen radicals, while normal cell killing is due to membrane lipid peroxidation initiated by oxygen radicals. They have, moreoever, found that antioxidant phenols with reducing properties inhibited lipid peroxidation by bleomycin but greatly enhanced bleomycin-iron dependent damage to DNA. They suggest that clinical administration of these phenolic compounds with reducing properties could potentiate the tumor damaging properties of bleomycin while at the same time protecting normal lipid membranes from deleterious-free radical damage.

SOD has also been used to protect against normal tissue damage by ionizing radiation. In a double-blind, placebo-controlled study on 42 patients, it was demonstrated that Cu-Zn SOD injected 15 to 30 min after completion of a daily dose of fractionated radiation ameliorated the side effect of high energy radiation therapy of bladder tumors.[143] It was observed that the SOD did not interfere with the killing of the tumor. It was felt that tumor cell damage occurred immediately during irradiation so that SOD would not interfere, whereas the normal cell damage was due to later inflammation, which SOD could ameliorate.

d. Manipulation of Endogenous Antioxidants

One of the other problems with the idea of adding antioxidants to protect normal tissue is the difficulty in getting the drug to the right tissue at the proper time and for the proper duration. One possibility to avoid this problem is to manipulate enzyme concentrations in target tissues before treatment. Lapidus et al. have shown that it is possible to raise antioxidant levels in rat lung tissue and thus protect against subsequent exposure to radiation.[144] Antioxidant enzymes were raised by exposure to endotoxin and/or exposure to 95% O_2 for 48 hr. Exposure to 3000 rads resulted in less damage to lungs (as evidenced by weight change and lung pathology) which had elevated activities of antioxidant enzymes. Thus, it might be possible to raise antioxidant enzyme levels in an affected normal tissue and thus protect this tissue against later damage by an antitumor agent.

This same principle might also work against the clinician in some cases. D'Allessandro et al. have shown that adriamycin can cause an induction of catalase activity in the hearts of mice treated with this drug.[145] If this is also true in tumors, subsequent treatments after catalase has been induced may not be as effective. Knowledge of antioxidant enzyme levels as a function of time in normal and tumor cells may be necessary to accurately plan antitumor therapy with $O_2^{\cdot-}$ producing agents, especially with regards to fractionation.

e. Endogenous Antioxidant Enzyme Activities in Tumors

Besides the considerations discussed above, knowledge of antioxidant enzyme activities in tumor tissue may give a clear indication of which tumors are suceptible to oxygen radical-producing anticancer agents. Well-differentiated tumors apparently have high activities of both Mn SOD and catalase, although as discussed earlier, Mn SOD activity in the well-differentiated Morris hepatoma is still less than that of normal liver. Those tumors with high antioxidant enzyme levels are not as likely to be killed by agents such as bleomycin as tumors with low antioxidant enzyme levels. Knowledge of antioxidant enzyme activities could be used to predict which tumors are susceptible to oxygen radical producers.

6. Nature of Toxic Species

Most of the previous studies have stressed the production of $O_2^{\cdot-}$ and thus $\cdot$OH and the low SOD levels as the causative factor in the toxicity of these antitumor drugs. However, an alternative theory has been presented. Bozzi et al. have theorized that H_2O_2 is the actual important agent in causing toxicity to tumor cells.[37,146] In the first of their studies, oxygen uptake and lactate production of different strains of ascites tumor cells were assayed after exposure to an extracellular photochemical source known to produce oxygen free radicals (flavin mononucleotide [FMN]).[37] The various cells showed differential sensitivity to the treatment, ranging from nearly full inactivation of Ehrlich cells to nearly full resistance of Yoshida cells. Glucose plus succinate added after treatment reestablished basal oxygen uptake capacity, suggesting the cell membrane was the primary site of damage. This was confirmed by dye-permeabilization and protein leakage in sensitive cells. SOD (0.3 μM) or $\cdot$OH radical scavengers such as benzoate, glycerol, or mannitol at concentrations as high as 0.1 M had no effect on the decline of basal oxygen uptake by these cells. Catalase (0.15 μM) protected all types of tumors if added before irradiation and 0.1 to 1 mM H_2O_2 could substitute for FMN plus light. They conclude that H_2O_2 is the only species which caused cell damage. The authors also surveyed antioxidant enzymes in these cells. They conclude that the levels of glutathione peroxidase and its subsidiary enzymes, glucose-6-phosphate dehydrogenase, glutathione reductase, and glutathione synthetase correlate most closely with the degree of damage seen in the various cell types.

The second study by Bozzi et al. dealt with the effect of daunomycin on tumor cells.[146] In this study, 0.5 mM daunomycin caused severe inhibition of respiration in Ehrlich ascites tumor cells, whereas Yoshida ascites cells were almost as resistant as rat hepatocytes. They felt an inverse relationship existed in the two types of tumor cells (which are both catalase deficient) between the extent of cellular damage brought about by $O_2^{\cdot-}$ and the efficiency of the glutathione-mediated H_2O_2-detoxifying system. Thus, H_2O_2 is again thought to be the damaging species.

These studies are extremely interesting but flawed in their interpretation. Inhibition of oxygen consumption is not necessarily equitable with cell death if it is reversible. What they are observing may be a natural metabolic phenomenon unrelated to cell death. Pathways involve glucose and H_2O_2, and inhibition of respiration have been hypothesized to control cell division.[40] Moreover, the tumor cells may be susceptible to H_2O_2 only because SOD is low, resulting in damage from $O_2^{\cdot-}$. More studies in this area are urgently needed.

C. Inhibition of SOD

1. Diethyldithiocarbamate

This idea has mostly been that of Lin and associates. Their concept is that normal cells have both Cu-Zn and Mn SOD activity, whereas tumor cells have mostly only Cu-Zn SOD activity. If this Cu-Zn SOD is inhibited, tumor cells should be much more susceptible to oxygen radicals than normal cells which still have Mn SOD. Lin et al. have studied the use of diethyldithiocarbamate (DDC) to inhibit SOD. DDC has been shown to inhibit Cu-Zn SOD activity both in vivo and in vitro. It also has been shown to increase the radiation sensitivity of erythrocytes.[147] Irradiation of human erythrocytes at 20 to 30 krad did not cause hemolysis up to 2 days after irradiation. When the erythrocytes were treated with 2.4 mM DDC for 90 min at 37°C prior to irradiation, they became sensitive to radiation. DDC treatment 1.5 hr prior to irradiation was found to inhibit SOD by more than 80%.

In another study, they found that the cytotoxic effect on DON cells (Chinese hamster cells) was dependent on DDC concentration and exposure time.[148] After 8 to 10 days of incubation with 10^{-9} M DDC, no change in DON cell survival was noted. However, incubation with 10^{-4} M DDC yielded marked inhibition of cell survival. The inhibitory effect of 2.4×10^{-4} M DDC could not be reversed after 2 hr incubation. When DDC-treated cells were irradiated, they did not survive as well as cells that were treated with radiation or DDC alone. The combined effects of DDC and hyperthermia were dramatic. Cells treated 5 min at 43°C with 10^{-4} M DDC or 15 min at 43° with 10^{-4} M DDC showed a decrease in survival of 40 and 95%, respectively. In a similar study, Lin et al. showed that DDC enhanced oxygen toxicity of Chinese hamster and mouse cells.[149] Lastly, they have proposed and shown that DDC enhanced the toxicity of bleomycin.[150] Using Chinese hamster cells (V79), DDC treatment (0.1 mmol/ℓ, 30 min at 37°C) eliminated the characteristic resistance portion of the bleomycin survival curve. Since the DDC bleomycin survival curve then became more exponential, the cytotoxicity became dose dependent. There was a greater than 20-fold net increase in cytotoxicity to 30 μg/mℓ bleomycin (60 min) if the cells are pretreated with 0.1 mmol/ℓ DDC for 30 min at 37°C. The ability of DDC to radiosensitize has been verified by others. Rigas et al. have shown that DDC enhanced the killing of human lymphocytes by radiation.[151] Westman and Marklund have shown that DDC enhanced the radiation-induced killing of Chinese hamster cells.[152]

There remains a lot of work to be done in this area. For instance, to be effective this drug must be taken up by tumor cells and should show some selectivity for tumor cells over normal cells — either in uptake or sensitivity to killing. Professor A. C. H. Kogen has provided valuable data in this area.[153] He has studied the use of DDC in the treatment of Ehrlich ascites carcinoma in the mouse (EAC) and Jensen's sarcoma in the rat (JC). DDC at 0.25 mg/g and 0.5 mg/g inhibited the growth of EAC by 20 and 35%, respectively. Similar results were obtained with JC. DDC also increased the survival of mice bearing EAC; median survival increased 30%. Hyperbaric oxygen potentiated the effects of DDC on JC at a dose of 0.7 mg/g. These results remove some of the doubts regarding DDC and suggest a potential for therapeutic use.

2. Streptozotocin

One other antitumor agent can inhibit SOD. Streptozotocin, an agent obtained from *Streptomyces achromogenes,* has been used to treat inoperable cancer of pancreatic β-cells. This drug has been shown to decrease SOD activity in rat retinae and erthrocytes by 30 to 40% from control levels 5 days after the induction of diabetes with this compound.[154] The decline is SOD was shown to be unrelated to diabetes because treatment with insulin failed to affect retinal SOD values. Homogenates from lungs, liver, brain aortas, kidneys, and whole eyes and lens of streptozotocin-diabetic rats showed no change in SOD activity. Retinas (rat and bovine) and erythrocytes (human and rat) also showed decreased SOD activity following in vitro incubation with streptozotocin. Whether streptozotocin decreases tumor cell SOD remains to be shown.

D. Addition of Superoxide Dismutase to Tumor Cells

A cancer cell has many phenotypic properties different from that of normal cells. One of these properties is of course the loss of SOD. Our group at Iowa has been interested in the effects of adding back the lost SOD activity. This interest is twofold: (1) by adding back the lost SOD activity we can see which, if any, of the phenotypic properties of the cancer cell will be affected by SOD (such as growth rate, metastatic potential, etc.); (2) adding SOD to tumor cells represents a new, largely unstudied mode of cancer therapy. In this technique, we are trying not to kill tumors, but to cause them to stop dividing and possibly to differentiate.

However interesting this mode of therapy may be, it is technically difficult.[155] Natural Cu-Zn SOD does not penetrate well into cells and has a short circulartory lifetime. For this reason, Cu-Zn SOD was not expected to have much effect on tumors. We found that Cu-Zn SOD had a small, but significant effect on tumor growth and animal survival in one mouse tumor model,[30] but not in another.[156] For this reason, we have concentrated on small molecular weight compounds with SOD-like activity. Cu(II) (3,5-diisopropylsalicylate)$_2$ (Cu Dips) was one of the first we tested.[156] Cu Dips caused an increase in mean survival of 85%, an increase in median survival of 88%, and an increase in maximum survival of greater than 140% in mice bearing transplanted Ehrlich ascites carcinoma.[156] Moreover, Cu Dips caused both a large inhibition of cell division and a large increase in cell differentiation in in vitro mouse neuroblastoma cells.[157]

However, the interpretation of those results are not without difficulty. Cu Dips loses its SOD activity in the presence of excess ligand such as bovine serum albumin (BSA)[158] This might cause it to lose its activity in vivo. Moreoever, Cu chelates cause many other effects.[159] Hopefully, other model compounds with SOD-like activity will be developed in the near future.

The idea of treating tumor cells by differentiation rather than by killing, I feel, is a good one. This type of therapy will avoid the problems of normal cell damage seen with current drugs. I feel the ideal therapy in this case would be to give compounds which increase or mimic SOD activity with compounds which also increase or mimic catalase and peroxidase activities. There is already evidence that such compounds do cause at least partial differentiation of tumor cells. Dimethyl sulfoxide (DMSO, ·OH scavenger) will differentiate several tumor cell lines.[5] Retinoic acid, which has recently been shown to decrease $O_2^{\cdot-}$ production in vivo,[60,63] also differentiates and inhibits the growth of several tumor cell lines. Increased oxygen tension, which may increase $O_2^{\cdot-}$ and thus induce SOD, has been shown to differentiate mouse neuroblastoma cells.[160] Vitamin E will also differentiate these cells.[161] Recently hemin, used in the synthesis of catalase, has been shown to cause partial differentiation of neuroblastoma[162] and erythroleukemia cells.[163]

E. Interferon

Interferon is one of the most widely publicized new anticancer drugs. Recent evidence has linked interferon and SOD. Two studies have linked the genes coding for sensitivity to interferon to Cu-Zn SOD in mouse and man[164] and have mapped these genes to mouse chromosome 16[164,165] and human chromosome 21.[166] The significance of this genetic link remains to be established.

IX. SUMMARY

Mn SOD activity seems to be lowered in all tumor cells. Cu-Zn SOD activity is usually, but not always, diminished. The rate of production of $O_2^{\cdot-}$ by tumor cells remains unclear. However, the role of oxygen radicals in carcinogenesis and in cancer therapy establishes the importance of these entities. Knowledge of $O_2^{\cdot-}$ and SOD levels may help us to prevent and treat cancer.

ACKNOWLEDGMENTS

The author would like to acknowledge that many of the ideas and work presented in this chapter are not his alone. In particular, Dr. Garry R. Buettner was responsible for working out much of the free radical chemistry involved. Dr. Terry D. Oberley deserves credit for the two-step theory of cancer as well as a good deal of the cell biology used. Dr. Susan Leuthauser is responsible for development of the idea that tumors produce a substance which inhibits Mn SOD (toxohormone). Much of the work was accomplished by a score of capable people including Drs. Isabel Bize, Shailendra Sahu, and Dean Loven. Mr. Loren Schutt has provided invaluable technical assistance. Last, I would like to thank Susan Barnett and Carol Kieffer for typing this manuscript.

REFERENCES

1. **Yamanaka, N. Y. and Deamer, D.,** Superoxide dismutase activity in WI-38 cell cultures. Effects of age, trypsinization, and SV-40 transformation, *Physiol. Chem. Phys.*, 6, 95, 1974.
2. **Dionisi, D., Galeotti, T., Terranove, T., and Azzi, A.,** Superoxide radicals and hydrogen peroxide formation in mitochondria from normal and neoplastic tissues, *Biochem. Biophys. Acta*, 403, 292, 1975.
3. **Sahu, S. K., Oberley, L. W., Stevens, R. H., and Riley, E. F.,** Superoxide dismutase activity of Ehrlich ascites tumor cells, *J. Natl. Cancer Inst.*, 8, 1125, 1977.

4. **Oberley, L. W., Bize, I. B., Sahu, S. K., Leuthauser, S. W. H. C., and Gruber, H. E.**, Superoxide dismutase activity of normal murine liver, regenerating liver, and H6 hepatoma, *J. Natl. Cancer Inst.*, 61, 375, 1978.
5. **Oberley, L. W. and Buettner, G. R.**, Role of superoxide dismutase in cancer: a review, *Cancer Res.*, 39, 1141, 1979.
6. **Yamanaka, N., Ota, K., and Utsumi, K.**, Changes in superoxide dismutase activities during development, aging, and transformation, in *Biochemical and Medical Aspects of Active Oxygen*, Hayaishi, O. and Asada, K., Eds., University Park Press, Baltimore, 1978, 183.
7. **Oberley, L. W., Sahu, S. K.,and Cheng, F. H. F.**, Superoxide dismutase activity in normal mouse brain, prenatal brain, and neuroblastoma cells, *Proc. Am. Assoc. Cancer Res.*, 19, 147, 1978.
8. **Peskin, A. V., Koen, Y. M., and Zbarsky, I. B.**, Superoxide dismutase and glutathion peroxidase activities in tumors, *F.E.B.S. Lett.*, 78, 41, 1977.
9. **Peskin, A. V., Zbarsky, I. B., and Konstantinov, A. A.**, An examination of the superoxide dismutase activity in tumor tissue, *Dokl. Akad. Nauk. S.S.S.R.*, 229, 751, 1976.
10. **Tyler, D. D.**, Polarographic assay and intracellular distribution of superoxide dismutase in rat liver, *Biochem. J.*, 147, 493, 1975.
11. **Lankin, U. Z. and Gurevich, S. M.**, Inhibition of the peroxidation of lipids and detoxification of lipoperoxides by protective enzymes (superoxide dismutase, glutathione peroxidase, and glutathione reductase) in experimental malignant growth, *Dokl. Akad. Nauk S.S.S.R.*, 226, 705, 1976.
12. **Bozzi, A., Mavelli, I., Finazzi Agro, A., Strom, R., Wolf, A. M., Modovi, B., and Rotilio, T.**, Enzyme defense against reactive oxygen derivatives. II. Erythrocytes and tumor cells, *Mol. Cell. Biochem.*, 10, 11, 1976.
13. **Bartoli, G. M., Galeotti, T., and Azzi, A.**, Production of superoxide anions and hydrogen peroxide in Ehrlich ascites tumor cell nuclei, *Biochim. Biophys. Acta*, 497, 622, 1977.
14. **VanBalgooy, J. N. A. and Roberts, E.**, Superoxide dismutase in normal and malignant tissues in different species, *Comp. Biochem. Physiol.*, 62B, 263, 1979.
15. **Docampo, R., Cruz, F. S., Boveris, A., Muniz, R. P. A., and Esquivel, D. M. S.**, β-Lapachone enhancement of lipid peroxidation and superoxide anion and hydrogen peroxide formation by sarcoma 180 ascites tumor cells, *Biochem. Pharm.*, 28, 723, 1979.
16. **Dinescu-Romalo, G. and Mihai, C.**, Superoxide production and superoxide dismutase in several tissues from normal and Guérin T_8 ascites tumor-bearing rats, *Cell Mol. Biol.*, 25, 101, 1979.
17. **Bize, I. B., Oberley, L. W., and Morris, H. P.**, Superoxide dismutase and superoxide radical in Morris hepatomas, *Cancer Res.*, 40, 3686, 1980.
18. **Bartoli, G. M., Bartoli, S., Galeotti, T., and Bartoli, E.**, Superoxide dismutase content and microsomal lipid composition of tumors with different growth rates, *Biochim. Biophys. Acta*, 620, 205, 1980.
19. **Tarakhovskii, A. M., Glinskii, G. V., and Shliakhovenko, V. A.**, Superoxide dismutase activity in tumor and liver tissues of tumor-bearing animals, *Ukr. Biochem. J.*, 52, 628, 1980.
20. **Petkau, A., Monasterski, L. G., Kelly, K., and Friesen, H. G.**, Modification of superoxide dismutase in rat mammary carcinoma, *Res. Comm. Chem. Pathol. Pharmacol.*, 17, 125, 1977.
21. **Loven, D. P., Oberley, L. W., Rousseau, F. M., and Stevens, R. H.**, Superoxide dismutase activity in 1,2-dimethylhydrazine-induced rat colon adenocarcinoma, *J. Natl. Cancer Inst.*, 6, 377, 1980.
22. **Yamanaka, N., Nishida, K., and Ota, K.**, Increase of superoxide dismutase activity in various human leukemia cells, *Physiol. Chem. Phys.*, 11, 253, 1979.
23. **Sykes, J. A., McCormack, F. X., Jr., and O'Brien, T. J.**, A preliminary study of the superoxide dismutase content of some human tumors, *Cancer Res.*, 38, 2759, 1978.
24. **Galeotti, T., Barrello, S., Seccia, A., Farallo, E., Bartoli, G. M., and Serri, F.**, Superoxide dismutase content in human epidermis and squamous cell epithelioma, *Arch. Dermatol. Res.*, 267, 83, 1980.
25. **Baur, G. and Wendel, A.**, The activity of the peroxide-metabolizing system in human colon carcinoma, *J. Cancer Res. Clin. Oncol.*, 97, 267, 1980.
26. **Marklund, S. and Westman, G.**, CuZn Superoxide Dismutase and Mn Superoxide Dismutase in Human Tissues and Human Malignant Tumors, presented at Conf. Oxygen Oxy-Radicals Chem. Biol., Austin, Tex., May 25 to 29, 1980.
27. **Leuthauser, S. W. H. C., Oberley, L. W., and Oberley, T. D.**, Lowered superoxide dismutase activity in distant organs of tumor-bearing mice, *J. Natl. Cancer Inst.*, submitted.
28. **Busch, H.**, Toxohormone, in *The Biochemistry of the Cancer Cell*, Academic Press, New York, 1962, 241.
29. **Fridovich, I.**, Superoxide dismutases, *Annu. Rev. Biochem.*, 44, 147, 1975.
30. **Oberley, L. W., Leuthauser, S. W. H. C., Buettner, G. R., Sorenson, J. R. J., Oberley, T. D., and Bize, I. B.**, The use of superoxide dismutase in the treatment of cancer, in *Active Oxygen and Medicine*, Autor, A. P., Ed., Academic Press, New York, in press.
31. **Peskin, A. V., Zbarsky, I. B., and Konstantinov, A. A.**, A novel type of superoxide generating system in nuclear membranes from hepatoma 22a ascites cells, *F.E.B.S Lett.*, 117, 44, 1980.

32. **Patton, S. E., Rosen, G. M., and Rauckman, E. J.,** Superoxide production by purified hamster hepatic nuclei, *Mol. Pharm.*, 18, 588, 1980.
33. **Wickramsinghe, R. H., Reddy, P. K., Klein, L., and Villee, C. A.,** Superoxide anions and other components of human renal adenocarcinoma, *Clin. Biochem.*, 9, 24, 197.
34. **Ono, T.,** Enzyme patterns and malignancy of experimental hepatomas, in *Biological and Biochemical Evaluation in Experimental Hepatomas,* Yoshida, T., Ed., Japanese Cancer Association, Tokyo, 1966, 189.
35. **Sun, A. S. and Cederbaum, I.,** Oxidoreductase activities in normal rat liver, tumor-bearing rat liver, and hepatoma HC-252, *Cancer Res.*, 40, 4677, 1980.
36. **Pinto, R. E., Duarte, M. F., and Roda-Santos, M. L.,** Aerobic glutathione oxidation and catalase activity in hepatomous liver of rats treated with diethylnitrosamine, *I.R.C.S. Med. Sci.*, 8, 97, 1980.
37. **Bozzi, A., Mavelli, I., Modovi, B., Strom, R., and Rotilio, G.,** Differential sensitivity of tumor cells to externally generated hydrogen peroxide. Role of glutathione and related enzymes, *Cancer Biochem. Biophys.*, 3, 135, 1979.
38. **Rotilio, G., Bozzi, A., Mavelli, I., Mondovi, B., and Strom, R.,** Tumor cell as models for oxygen-dependent cytotoxicity, in *Biological and Clinical Aspects of Superoxide and Superoxide Dismutase,* Bannister, W. H. and Bannister, J. V., Eds., Elsevier/North-Holland, New York, 1980, 118.
39. **Oberley, L. W., Oberley, T. D., and Buettner, G. R.,** Cell differentiation, aging, and cancer: the possible roles of superoxide and superoxide dismutases, *Med. Hyp.*, 6, 249, 1980.
40. **Oberley, L. W., Oberley, T. D., and Buettner, G. R.,** Cell division in normal and transformed cells: the possible role of superoxide and hydrogen peroxide, *Med. Hyp.*, 7, 21, 1981.
41. **Oberley, L. W. and Oberley, T. D.,** The role of superoxide radical in carcinogenesis, *Med. Hyp.*, in preparation.
42. **Lotem, J. and Sachs, L.,** Potential pre-screening for therapeutic agents that induce differentiation in human myeloid leukemia cells, *Int. J. Cancer,* 25, 61, 1980.
43. **Goetz, I.,** Oxygen toxicity in normal and neoplastic cells in culture, *In Vitro,* 11, 382, 1975.
44. **Mitchell, R. S., Elgas, R. J., and Balk, S. D.,** Proliferation of Rous sarcoma virus-infected, but not of normal chicken fibroblasts in oxygen-enriched environment: preliminary report, *Proc. Natl. Acad. Sci. U.S.A.*, 73, 1265, 1976.
45. **Arneson, R. M., Aloyo, V. J., Germain, G. S., and Chenevey, J. E.,** Antioxidants in neoplastic cells. I. Changes in the antioxidative capacity of mouse neuroblastoma cells measured by a single-phase assay, *Lipids,* 13, 383, 1978.
46. **Bartoli, G. M. and Galeoti, T.,** Growth-related lipid peroxidation in tumor microsomal membranes and mitochondria, *Biochim. Biophys. Acta,* 574, 537, 1979.
47. **Ts'o, P. O., Caspary, W. J., and Lorentzen, R. J.,** The involvement of free radicals in carcinogenesis, in *Free Radicals in Biology,* Vol. 3, Pryor, W. A., Ed., Academic Press, New York, 1977, 21.
48. **Ts'o, P. O., Caspary, W. J., and Lorentzen, R. J.,** The involvement of free radicals in carcinogenesis, in *Free Radicals in Biology,* Vol. 3, Pryor, W. A., Ed., Academic Press, New York, 1977, 294.
49. **Floyd, R. A., Soong, L. M., Stuart, M. A., and Reigh, D. L.,** Spin trapping of free radicals produced from nitrosamine carcinogens, *Photochem. Photobiol.*, 28, 857, 1978.
50. **Gantt, R., Parshad, R., Ewig, R. A. G., Sanford, K. K., Jones, G. M., Tarone, R. E., and Kohn, K. W.,** Fluorescent light-induced DNA cross-linkage and chromatid breaks in mouse cells in culture, *Proc. Natl. Acad. Sci. U.S.A.*, 7, 3809, 1978.
51. **Parshad, R., Sanford, K. K., Jones, G. M., Price, F. M., and Taylor, W. G.,** Oxygen and light effects on chromosomal aberrations in mouse cells in vitro, *Exp. Cell Res.*, 104, 199, 1977.
52. **Parshad, R., Sanford, K. K., Jones, G. M., and Tarone, R. E.,** Fluorescent light-induced chromosome damage and its prevention in mouse cells in culture, *Proc. Natl. Acad. Sci. U.S.A.*, 75, 1830, 1978.
53. **Sanford, K. K., Parshad, R., and Gantt, R.,** Light and oxygen effects on chromosomes, DNA and neoplastic transformation of cells in culture, in *Nutritional Requirements of Cells in Culture,* Katsuta, H., Ed., Japan Scientific Societies Press, Tokyo, 1978, 117.
54. **Parshad, R., Sanford, K. K., Tarone, R. E., Jones, G. M., and Baeck, A. E.,** Increased susceptibility of mouse cells to fluorescent light-induced chromosome damage after long-term culture and malignant transformation, *Cancer Res.*, 39, 929, 1979.
55. **Sanford, K. K., Parshad, R., Jones, G., Handleman, S., Garrison, C., and Price, F.,** Role of photosensitization and oxygen in chromosome stability and "spontaneous" malignant transformation in culture, *J. Natl. Cancer Inst.*, 63, 1245, 1979.
56. **Parshad, R., Taylor, W. G., Sanford, K. K., Camalier, R. F., Gantt, R., and Tarone, R. E.,** Fluorescent light-induced chromosome damage in human IMR-90 fibroblasts. Role of hydrogen peroxide and related free radicals, *Mutation Res.*, 73, 115, 1980.
57. **Parshad, R., Sanford, K. K., Jones, G. M., Tarone, R. E., Hoffman, H. A., and Grier, A. H.,** Susceptibility to fluorescent light-induced chromatid breaks associated with DNA repair deficiency and malignant transformation in culture, *Cancer Res.*, 40, 4415, 1980.

58. **DeLuca, C.**, Studies on the modulation of catalase activity in mammalian cell culture, *Exp. Cell Res.*, 44, 403, 1966.
59. **Johnston, R. B., Jr. and Lehmeyer, J. E.**, The involvement of oxygen metabolites from phagocytic cells in bactericidal activity and inflammation, *Superoxide and Superoxide Dismutases,* Michelson, A. M., McCord, J. M., and Fridovich, I., Eds., Academic Press, London, 1977, 291.
60. **Archer, V. E.**, Carcinogenicity of fibers and films: a theory, *Med. Hyp.,* , 127, 1979.
61. **Witz, G., Goldstein, B. D., Amoruso, M., Stone, D. S., and Troll, W.**, Retinoid inhibition of superoxide anion radical production by human polymorphonuclear leukocytes stimulated with tumor promoters, *Biochem. Biophys. Res. Commun.*, 97, 883, 1980.
62. **Goldstein, I. M., Perez, H. D., and Chernoff, D. M.**, Effects of adrenal corticosteroids on polymorphonuclear leukocytes, in *Advances in Inflammation Research,* Weissman, G., Samuelsson, B., and Paoletti, R., Eds., Raven Press, New York, 1979, 515.
63. **Kensler, T. and Trush, M. A.**, Inhibition of phorbol ester-stimulated chemiluminescence in human polymorphonuclear leukocytes by retinoic acid and 5,6-epoxyretinoic acid, *Cancer Res.*, 41, 216, 1981.
64. **Goldstein, B. D., Witz, G., Amoruso, M., Stone, D. S., and Troll, W.**, Stimulation of human polymorphonuclear leukocyte superoxide anion production by tumor promoters, *Cancer Lett.*, 11, 257, 1981.
65. **Takeshita, M., Horwitz, S. B., and Grollman, A.P.**, Bleomycin, an inhibitor of vaccinic virus replication, *Virology,* 60, 455, 1974.
66. **Onishi, T., Iwata, H., and Takagi, Y.**, Effects of reducing and oxidizing agents on the action of bleomycin, *J. Biochem. (Tokyo),* 77, 745, 1975.
67. **Bode, V. C.**, Single-strand scissions in circular and linear λ DNA by the presence of dithiothreitol and other reducing agents, *J. Mol. Biol.*, 26, 125, 1967.
68. **Rhaese, H. and Freese, E.**, Chemical analysis of DNA alterations. I. Base liberation and backbone breakage of DNA and oligdeoxyadenylic acid induced by hydrogen peroxide and hydroxylamine, *Biochim. Biophys. Acta,* 155, 476, 1967.
69. **Shirakawa, I., Azegami, M., Ishii, S. I., and Umezawa, H.**, Reaction of bleomycin with DNA. Strand scission of DNA in the absence of sulfhydryl or peroxide compounds, *J. Antibiot.*, 24, 761, 1971.
70. **Umezawa, H., Akasura, H., Oda, K., and Hori, S.**, The effect of bleomycin on SV40 DNA: characteristics of bleomycin action which produces a single scission in a superhelical form of SV40 DNA, *J. Antibiot.*, 26, 521, 1973.
71. **Haidle, C.W.**, Fragmentation of deoxyribonucleic acid by bleomycin, *Mol. Pharmacol.*, 7, 645, 1971.
72. **Chien, M., Grollman, A. P., and Horwitz, S. B.**, Bleomycin-DNA interactions: fluorescence and proton magnetic resonance studies, *Biochemistry,* 16, 3641, 1977.
73. **Ishida, R. and Takahashi, T.**, Increased DNA chain breakage by combined action of bleomycin and superoxide radical, *Biochem. Biophys. Res. Comm.*, 66, 1432, 1975.
74. **Sausville, E. A., Peisach, J., and Horwitz, S. B.**, Effect of chelating agents and metal ions on the degradation of DNA by bleomycin, *Biochemistry,* 17, 2740, 1978.
75. **Sausville, E. A., Peisach, J., and Horwitz, S. B.**, A role for ferrous ion and oxygen in the degradation of DNA by bleomycin, *Biochem. Biophys. Res. Commun.*, 73, 814, 1976.
76. **Lown, J. W. and Sim, S. K.**, The mechanism of the bleomycin-induced cleavage of DNA, *Biochem. Biophys. Res. Commun.*, 77, 1150, 1977.
77. **Fong, K. L., McCay, P. B., Poyer, J. L., Keele, B. B., and Misra, H.**, Evidence that peroxidation of lysosomal membranes is initiated by hydroxyl free radicals produced during flavin enzyme activity, *J. Biol. Chem.*, 248, 7792, 1973.
78. **Buettner, G. R., Oberley, L. W., and Leuthauser, S. W. H. C.**, The effect of iron on the distribution of superoxide and hydroxyl radicals as seen by spin trapping and on the superoxide dismutase assay, *Photochem. Photobiol.*, 28, 693, 1978.
79. **Sugiura, Y. and Kikuchi, T.**, Formation of superoxide and hydroxyl radicals in iron(II)-bleomycin-oxygen system: electron spin resonance detection by spin trapping, *J. Antibiot.*, 31, 1310, 1978.
80. **Oberley, L. W. and Buettner, G. R.**, The production of hydroxyl radical by bleomycin and iron(II), *F.E.B.S. Lett.*, 97, 47, 1979.
81. **Sausville, E. A., Stein, R. W., Peisach, J., and Horwitz, S. B.**, Properties and products of the degradation of DNA by bleomycin and iron(II), *Biochemistry,* 17, 2746, 1978.
82. **Sugiura, Y.**, Production of free radicals from phenol and tocopherol by bleomycin-iron(II) complex, *Biochem. Biophys. Res. Commun.*, 87, 649, 1979.
83. **Sugiura, Y.**, Comparison of metal complexes between depyruvamide bleomycin and bleomycin: an important effect of axial donor on metal coordination and oxygen activation, *Biochem. Biophys. Res. Commun.*, 88, 913, 1979.
84. **Takahashi, K., Ekimoto, H., Aoyagi, S., Koyu, A., Kuramochi, H., Yoshioka, O., Matsuda, A., Fujii, A., and Umezawa, H.**, Biological studies on the degradation products of 3-[(S)-1′-phenylethylamino] propylaminobleomycin: a novel analog (pepleomycin), *J. Antibiot.*, 32, 36, 1979.

85. **Bradner, W. T.,** BV-2231, a third-generation bleomycin: preclinical studies, in *The Bleomycins-Current Status and New Developments,* Carter, S. K., Crooke, S. T., and Unezawa, H., Eds., Academic Press, New York, 1978.
86. **Strong, J. E. and Crooke, S. T.,** DNA breakage by tallysomycin, *Cancer Res.,* 38, 3322, 1978.
87. **Buettner, G. R. and Oberley, L. W.,** The production of hydroxyl radical by tallysomycin and copper(I), *F.E.B.S. Lett.,* 101, 333, 1979.
88. **Sugiura, Y.,** The production of hydroxyl radical from copper(I) complex systems of bleomycin and tallysomycin: comparison with copper(II) and iron(II) systems, *Biochem. Biophys. Res. Commun.,* 90, 375, 1979.
89. **Lown, J. W. and Joshua, A. V.,** Interactions of the glycopeptide antitumor antibiotics bleomycin and tallysomycin with deoxyribonucleic acid in vitro, *Biochem. Pharm.,* 29, 521, 1980.
90. **Lown, J. W., Sim, S., Majumdar, K. C., and Chang, R. Y.,** Strand scission of DNA by bound adriamycin and daunorubicin in the presence of reducing agents, *Biochem. Biophys. Res. Commun.,* 76, 705, 1977.
91. **Thayer, W. S.,** Adriamycin stimulated superoxide formation in submitochondrial particles, *Chem. Biol. Interactions,* 19, 265, 1977.
92. **Goodman, J. and Hochstein, P.,** Generation of free radicals and lipid peroxidation by redox cycling of adriamycin and daunomycin, *Biochem. Biophys. Res. Commun.,* 77, 797, 1977.
93. **Henderson, C. A., Metz, E. N., Balcerzak, S. P., and Sagone, A. L., Jr.,** Adriamycin and daunomycin generate reactive oxygen compounds in erythrocytes, *Blood,* 52, 878, 1978.
94. **Someya, A. and Tanaka, N.,** DNA strand scission induced by adriamycin and aclacinomycin A, *J. Antibiot.,* 32, 839, 1979.
95. **Kalyanaraman, B., Perez-Reyes, E., and Mason, R. P.,** Spin-trapping and direct electron spin resonance investigations of the redox metabolism of quinone anticancer drugs, *Biochim. Biophys. Acta,* 630, 119, 1980.
96. **Lown, J. W., Chen, H., and Plambeck, J. A.,** Diminished superoxide anion generation by reduced 5-iminodaunorubicin relative to duanorubicin and the relationship to cardiotoxicity of the anthracycline antitumor agents, *Biochem. Pharm.,* 28, 2563, 1979.
97. **Cone, R., Hasan, S. K., Lown, J. W., and Morgan, A. R.,** The mechanism of the degradation of DNA by streptonigrin, *Can. J. Biochem.,* 54, 219, 1976.
98. **Lown, J. W. and Sim, S.,** Studies related to antitumor-antibiotics. VIII. Synthesis of streptonigrin analogues and their single strand scission of DNA, *Can. J. Biochem.,* 54, 446, 1976.
99. **Lown, J. W., Begleiter, A., Johnson, P., and Morgan, A. R.,** Studies related to antitumor antibiotics. V. Reactions of mitomycin C with DNA examined by ethidium fluorescence assay, *Can. J. Biochem.,* 54, 110, 1976.
100. **Lown, J. W. and Weir, G.,** Studies related to antitumor antibiotics. XIV. Reaction of mitomycin B with DNA, *Can. J. Biochem.,* 56, 296, 1978.
101. **Lown, J. W., Sim, S., and Chen, H.,** Hydroxyl radical production by free and DNA-bound aminoquinone antibiotics and its role in DNA degradation. Electron spin resonance detection of hydroxyl radicals by spin trapping, *Can. J. Biochem.,* 56, 1042, 1978.
102. **Lown, J. W., Chen, H., Sim, S., and Plambeck, J. A.,** Reactions of the antitumor agent carminic acid and derivatives with DNA, *Bioorg. Chem.,* 8, 17, 1979.
103. **Bachur, N. R., Gordon, S. L., and Gee, M. V.,** A general mechanism for microsomal activation of quinone anticancer agents to free radicals, *Cancer Res.,* 38, 1745, 1978.
104. **Bachur, N. R., Gordon, S. L., Gee, M. V., and Kon, H.,** NADPH cytochrome P-450 reductase activation of quinone anticancer agents to free radicals, *Proc. Natl. Acad. Sci.,* 76, 954, 1979.
105. **Sinha, B. K. and Chignell, C. F.,** Binding mode of chemically activated semiquinone free radicals from quinone anticancer agents to DNA, *Chem. Biol. Interactions,* 28, 301, 1979.
106. **Bachur, N. R., Friedman, R. D., and Gee, M. V.,** Nuclear catalyzed antibiotic free radical formation, *Proc. Am. Assoc. Cancer Res.,* 20, 128, 1979.
107. **Kennedy, K. A., Rockwell, S., and Sartorelli, A. C.,** Preferential activation of mitomycin C to cytotoxic metabolites by hypoxic tumor cells, *Cancer Res.,* 40, 2356, 1980.
108. **Tercher, B. A., Lazo, J. S., and Sartorelli, A. C.,** Classification of antineoplastic agents by their selective toxicities toward oxygenated and hypoxic tumor cells, *Cancer Res.,* 41, 73, 1981.
109. **Lown, J. W. and Chen, H.,** Studies on the effects of the antitumor agent camptothecin and derivatives on deoxyribonucleic acid. Mechanism of the scission of deoxyribonucleic acid by photoactivated camptothecin, *Biochem. Pharm.,* 29, 905, 1980.
110. **Bachur, N. R., Gee, M. V., and Gordon, S. L.,** Enzymatic activation of activation of actinomycin D (ACT-D) to free radical state, *Proc. Am. Assoc. Cancer Res.,* 19, 75, 1978.
111. **Sinha, B. K. and Cox, M. G.,** Stimulation of superoxide formation by actinomycin D and its N^2-substituted spin-labeled derivatives, *Mol. Pharm.,* 17, 432, 1979.
112. **Sinha, B. K., Cox, M. G., Chignell, C. F., and Cysyk, R. L.,** Synthesis and biological properties of N^2-substituted spin-labeled analogues of actinomycin D, *J. Med. Chem.,* 250, 5622, 1975.

113. **Dougherty, T. J., Grindey, G. B., Fiel, R., Weishaupt, K. R., and Boyle, D. G.,** Photoradiation therapy. II. Cure of animal tumors with hematoporphyrin and light, *J. Natl. Cancer Inst.,* 55, 115, 1975.
114. **Dougherty, T. J., Kaufman, J. E., Goldfarb, A., Weishaupt, K. R., Boyle, D., and Mittleman, A.,** Photoradiation therapy for the treatment of malignant tumors, *Cancer Res.,* 38, 2628, 1978.
115. **Weishaupt, K. R., Comer, C. J., and Dougherty, T. J.,** Identification of singlet oxygen as the cytotoxic agent in photo-inactivation of a murine tumor, *Cancer Res.,* 36, 2326, 1976.
116. **Hariharan, P. V., Courtney, J., and Eleczko, S.,** Producation of hydroxyl radicals in cell systems exposed to haematoporphyrin and red light, *Int. J. Radiat. Biol.,* 37, 691, 1980.
117. **Cadet, J. and Teoule, R.,** Comparative study of oxidation of nucleic acid components by hydroxyl radicals, singlet oxygen and superoxide anion radicals, *Photochem. Photobiol.,* 28, 661, 1978.
118. **Buettner, G. R. and Oberley, L. W.,** The apparent production of superoxide and hydroxyl radicals by hematoporphyrin and light as seen by spin-trapping, *F.E.B.S. Lett.,* 121, 161, 1980.
119. **Spikes, J. D.,** Porphyrins and related compounds as photodynamic sensitizers, *Ann. N. Y. Acad. Sci.,* 244, 496, 1975.
120. **Hsu, J., Goldstein, B. D., and Harber, L. C.,** Photoreactions associated with in vitro hemolysis in erythropoietic protoporphyria, *Photochem. Photobiol.,* 13, 67, 1971.
121. **Schothorst, A. A., Van Stevenick, J., Went, L. N., and Suurmond, D.,** Photodynamic damage of the erythrocyte membrane caused by protoporphyrin in protoporphyria and in normal red blood cells, *Clin. Chim. Acta,* 39, 161, 1972.
122. **Girotti, A. W.,** Photodynamic action of protoporphyrin IX on human erythrocytes: cross linking of membrane proteins, *Biochem. Biophys. Res. Comm.,* 72, 1367, 1976.
123. **Buettner, G. R. and Oberley, L. W.,** Superoxide formation by protoporphyrin as seen by spin trapping, *F.E.B.S. Lett.,* 98, 18, 1979.
124. **Malik, Z. and Djaldetti, M.,** Destruction of erythroleukemia, myelocytic leukemia, and Burkitt's lymphoma cells by photoactivated protoporphyrin, *Int. J. Cancer,* 26, 495, 1980.
125. **Menzel, D. B.,** The role of free radicals in the toxicity of air pollutants (nitrogen oxides and ozone), in *Free Radicals in Biology,* Vol. 2., Pryor, W. A., Eds., Academic Press, New York, 1976, 189.
126. **Sweet, F., Kao, M., Lee, S. D., Hager, W. L., and Sweet, W. E.,** Ozone selectively inhibits growth of human cancer cells, *Science,* 209, 931, 1980.
127. **Petkau, A. and Chelack, W. S.,** Protection of *Acholeplasma laidlawii* B by superoxide dismutase, *Int. J. Radiat. Biol.,* 26, 421, 1974.
128. **Oberley, L. W., Lindgren, A. L., Baker, S. A., and Stevens, R. H.,** Superoxide ion as the cause of the oxygen effect, *Radiat. Res.,* 68, 320, 1976.
129. **Misra, H. P. and Fridovich, I.,** Superoxide dismutase and the oxygen enhancement of radiation lethality, *Arch. Biochem. Biophys.,* 176, 577, 1976.
130. **McLennan, G., Oberley, L. W., and Autor, A. P.,** The role of oxygen-derived free radicals in radiation-induced damage and death of nondividing eucaryotic cells, *Radiat. Res.,* 84, 122, 1980.
131. **Petkau, A., Chelack, W. S., and Plesbach, S. D.,** Protection of postirradiated mice by superoxide dismutase, *Int. J. Radiat. Biol.,* 29, 297, 1976.
132. **Kong, S. and Davison, A. J.,** The role of interactions between O_2, H_2O_2, $\cdot OH$, e^-, and $O_2^{\cdot -}$ in free radical damage to biological systems, *Arch. Biochem. Biophys.,* 704, 18, 1980.
133. **Samuni, A. and Czapski, G.,** Radiation-induced damage in *Escherichia coli* B: the effect of superoxide radicals and molecular oxygen, *Radiat. Res.,* 76, 624, 1978.
134. **Myers, C. E., McGuire, W. P., Liss, R. H., Ifrim, I., Grotzinger, K., and Young, R. C.,** Adriamycin: the role of lipid peroxidation in cardiac toxicity and tumor response, *Science,* 197, 165, 1977.
135. **Sonnevald, P.,** Effect of α-tocopherol on the cardiotoxicity of adriamycin in the rat, *Cancer Treat. Rep.,* 62, 1033, 1978.
136. **Van Fleet, J. F., Greenwood, L., Ferrans, V. J., and Rebar, A. A.,** Effect of selenium-vitamin E on adriamycin-induced cardiomyopathy in rabbits, *Am. J. Vet. Res.,* 39, 997, 1978.
137. **Doroshow, J. H., Locker, G. Y., and Myers, C. E.,** Enzymatic defenses of the mouse heart against reactive oxygen metabolites. Alterations produced by doxorubicin, *J. Clin. Invest.,* 65, 128, 1980.
138. **Olson, R. D., MacDonald, J. S., VanBoxtel, C. J., Boerth, R. C., Harbison, R. D., Slonim, A. E., Freeman, R. W., and Oates, J. A.,** Regulatory role of glutathione and soluble sulfhydryl groups in the toxicity of adriamycin, *J. Pharm. Exp. Ther.,* 21, 450, 1980.
139. **Auclair, C., Dhermy, D., and Boivin, P.,** Superoxide dismutase deficiency in patients with nitrogen mustard therapy-induced intravascular hemolysis, *Cancer Chemother. Pharmacol.,* 4, 281, 1980.
140. **McGinness, J. E., Proctor, P. H., Demopoulos, H. B., Hokanson, J. A., and Kirkpatrick, D. S.,** Amelioration of cis-platinum nephrotoxicity by orgotein (superoxide dismutase), *Physiol. Chem. Phys.,* 10, 267, 1978.
141. **Oyanagui, T.,** Stimulatory effect of platinum (IV) ion on the production of superoxide radical from xanthine oxidase and macrophages, *Biochem. Pharmacol.,* 26, 473, 1977.

142. **Gutteridge, J. M. and Fu, X.,** Enhancement of bleomycin-iron free radical damage to DNA by antioxidants and their inhibition of lipid peroxidation, *F.E.B.S. Lett.,* 123, 71, 1981.
143. **Edsmyr, F., Huber, W., and Meander, K. B.,** Orgotein efficacy in ameliorating side effects due to radiation therapy. I. Double-blind, placebo controlled trial in patients with bladder tumors, *Curr. Ther. Res.,* 19, 198, 1976.
144. **Lapidus, R., Ahmad, K., and Frank, L.,** Antioxidant enzymes and protection against lung irradiation, *Clin. Res.,* 28, 894, 1980.
145. **D'Allessando, N., Dusonchet, L., Crosta, L., Crescimamo, M., and Rausa, L.,** Does catalase play a role in adriamycin induced cardiotoxicity?, *Pharmacol. Res. Commun.,* 12, 441, 1980.
146. **Bozzi, A., Mavelli, I., Mondovi, B., Strom, R., and Rotilio, G.,** Differential cytotoxicity of daunomycin in tumor cells is related to glutathione-dependent hydrogen peroxide metabolism, *Biochem. J.,* 194, 369, 1981.
147. **Stone, D., Lin, P.-S., and Kwock, L.,** Radiosensitization of human erythrocytes by diethyldithiocarbamate, *Int. J. Radiat. Biol.,* 33, 393, 1978.
148. **Lin, P.-S., Kwock, L., and Butterfield, C. E.,** Diethyldithiocarbamate enhancement of radiation and hyperthermic effects on Chinese hamster cells in vitro, *Radiat. Res.,* 77, 501, 1979.
149. **Lin, P.-S., Kwock, L., Lui, P., and Hefter, K.,** Enhancement of oxygen toxicity by diethyldithiocarbamate on Chinese hamster and mouse cells, *Int. J. Radiat. Oncol. Biol. Phys.,* 5, 1699, 1979.
150. **Lin, P.-S., Kwock, L., and Goodchild, N. T.,** Copper, superoxide radical, diethyldithiocarbamate, and bleomycin toxicity, *Lancet,* 7, 777, 1979.
151. **Rigas, D. A., Eginites-Rigas, C., Bigky, R. H., Stankova, L., and Head, C.,** Biphasic radiosensitization of human lymphocytes by diethyldithiocarbamate: possible involvement of superoxide dismutase, *Int. J. Radiat. Biol.,* 38, 217, 1980.
152. **Westman, G. and Marklund, S.,** Diethyldithiocarbamate, a superoxide dismutase inhibitor, decreases the radioresistance of Chinese hamster cells, *Radiat. Res.,* 83, 303, 1980.
153. **Kogan, A. C. H.,** Experimental research of a new concept in cancer chemotherapy whereby tumor cells are converted into anaerobic-like cells which do not tolerate an oxygen environment, in *Current Problems in Experimental Chemotherapy of Tumors,* report to The Academy of Sciences of the U.S.S.R., 1980, 113.
154. **Crouch, R., Kimsey, G., Priest, D. G., Sarda, A., and Buse, M. G.,** Effect of streptozotocin on erythrocyte and retinal superoxide dismutase, *Diabetologica,* 15, 53, 1978.
155. **Petkau, A., Chelack, W. S., Kelly, K., Barefoot, C., and Monasterski, L.,** Tissue distribution of bovine ^{125}I-superoxide dismutase in mice, *Res. Commun. Chem. Path. Pharm.,* 17, 125, 1977.
156. **Leuthauser, S. W. H. C., Oberley, L. W., Oberley, T. D., and Sorenson, J. R. J.,** Antitumor effect of a copper coordination compound with superoxide dismutase activity, *J. Natl. Cancer Inst.,* 66, 1077, 1981.
157. **Oberley, L. W., Sahu, S. K., and Sorenson, J. R. J.,** Copper coordination compounds cause morphological differentiation of in vitro neuroblastoma, *Biochem. Biophys. Res. Commun.,* submitted.
158. **Leuthauser, S. W. H. C.,** Antitumor Activities of Superoxide Dismutase and Copper Coordination Compounds, Ph.D. thesis, University of Iowa, Iowa City, December 1979.
159. **Sorenson, J. R. J.,** The anti-inflammatory activities of copper complexes, in *Model Ions in Biological Systems,* Vol. 14, Siegel, H., Ed., in press.
160. **Erkell, L. J.,** Dfferentiation of mouse neuroblastoma cells under increased oxygen tension, *Exp. Cell. Res.,* 48, 374, 1980.
161. **Prasad, K. N., Ramanujam, S., and Gaudreau, D.,** Vitamin E induces morphological differentiation and increases the effect of ionizing radiation on neuroblastoma cells in culture, *Proc. Soc. Exp. Biol. Med.,* 161, 570, 1979.
162. **Ishii, D. N. and Maniatis, G. M.,** Haemin promotes rapid neurite outgrowth in cultured mouse neuroblastoma cells, *Nature,* 274, 372, 1978.
163. **Malik, Z., Halbrecht, I., and Djaldetti, M.,** Regulation of hemoglobin synthesis, iron metabolism, and maturation of Friend leukemia cells by 5-aminolevulinic acid and hemin, *Differentiation,* 13, 71, 1979.
164. **Cox, D. R., Epstein, L. B., and Epstein, C. J.,** Genes coding for sensitivity to interferon (IfRec) and soluble superoxide dismutase (SOD-1) are linked in mouse and man and mop to mouse chromosome 16, *Proc. Natl. Acad. Sci. U.S.A.,* 77, 2168, 1980.
165. **Lin, P.-S., Slate, D. L., Lawyer, F. C., and Ruddle, F. H.,** Assignment of the murine interferon sensitivity and cytoplasmic superoxide dismutase genes to chromosome 16, *Science,* 209, 285, 1980.
166. **McKusick, V. A. and Ruddle, F. H.,** The statusof the gene map of the human chromosome, *Science,* 196, 390, 1977.

APPENDIX

ABBREVIATIONS

ADP: adenosine 5′-diphosphate
ATP: adenosine 5′-triphosphate
BPN: *N-tert*-butyl-α-phenylnitrone
BSA bovine serum albumin
CCP: cytochrome *c* peroxidase
CD: circular dichroism
CDAA: *N,N*-diallyl, 2-choroacetamide
CDMPO: 2-carboxy-5,5-dimethyl-1-pyrroline-*N*-oxide
CL: chemiluminescence
CMU: monuron
CNA: 9-cyanoanthracene
CuDIPS Cu(II)(3,5-diisopropylsalicylate)
Cu-Zn SOD: copper and zinc-containing superoxide dismutase
cyclic AMP: 3′,5′-cyclic adenylic acid
cyt *c:* cytochrome *c*
DABCO: 1,4-diazobicyclo; [2.2.2]octane
DCA: 9,10-dicyanoanthracene
DCIP: dichlorophenol-indophenol
DCMU: dichlorophenyldimethylurea or diuron
DDC: diethyldithiocarbamate
DETAPAC: diethylenetriaminepentaacetic acid
DMF: dimethylformamide
DMH: 1,2-dimethylhydrazine
DMPO: 5,5′-dimethyl-1-pyrroline 1-oxide
DMSO: dimethylsulfoxide
DPBF: 1,3-diphenylisobenzofuran
E.D.: electron donor
EDTA: ethylenediamine tetraacetate, or tetraacetic acid
EAC: Ehrlich ascites carcinoma
EPR: electron paramagnetic resonance
or
ESR: electron spin resonance
FAD: flavin adenine dinucleotide
FeSOD: iron-containing superoxide dismutase
FMN: flavin mononucleotide
GC: gas chromatography
GDH: glyceraldehyde-3-phosphate dehydrogenase
GLC: gas liquid chromatography
GSH: reduced glutathione
GSSG: oxidized glutathione, or glutathione disulfide
HPLC: high pressure liquid chromatography
HRP: horseradish peroxidase
JS: Jensen's sarcoma
KMBA: 4-keto-mercapto-butyric acid
KO_2: potassium superoxide
LDH: lactate dehydrogenase
MDA: malondialdehyde
4-MeP_4BN: 4-(*N*-methyl pyridinion) *tert*-butyl nitrone
MnSOD: manganese-containing superoxide dismutase
$NaBH_4$: sodium borohydride
NAD^+: oxidized nicotinamide adenine dinucleotide
NADH: reduced nicotinamide adenine dinucleotide
$NADP^+$: oxidized nicotinamide adenine dinucleotide phosphate
NADPH: reduced nicotinamide adenine dinucleotide phosphate
NBT: nitroblue tetrazolium
NHE: normal hydrogen electrode
NMR: nuclear magnetic resonance

NtB:	2-nitroso-2-methylpropane
$^1O_2$ or $^1\Delta_g$	singlet oxygen
PBN=BPN:	α-phenyl-*tert*-butyl nitrone
PMA:	phorbol myristate acetate
PM2 CCC-DNA:	PM2 covalently closed circular DNA
PMN:	polymorphonuclear leukocytes
p-NDA:	*p*-nitrosodimethylaniline
4-POBN:	α-4-pyridyl-*N*-oxide *tert*-butyl nitrone
R.B.:	Rose bengal
RBC:	red blood cell
SOD:	superoxide dismutase
SQc:	ubisemiquinone associated with ubiquinol-cytochrome *c* reductase
SQn:	ubisemiquinone associated with NADH-dehydrogenase
SQs:	ubisemiquinone associated with succinate-ubiquinone reductase
SV40:	simian virus 40
THF:	tetrahydrofuran
Tiron:	1,2-dihydroxybenzene-3,5-disulfonate
TMN:	tetranitromethane
TMPO:	2,5,5-trimethyl-1-pyrroline-*N*-oxide
TPA:	12-*O*-tetracanoyl-phorbol-13-acetate
TSY:	trypticase soy yeast extract medium
UQH_2:	ubiquinol or ubiquinone

INDEX

A

B

C

D

H

I

K

L

M

N

O

T

U

V

W

X